Birth Defects in India

Anita Kar

Editor

Birth Defects in India

Epidemiology and Public Health Implications

 Springer

Editor
Anita Kar
Birth Defects and Childhood Disability Research
Centre
Pune, India

ISBN 978-981-16-1553-5   ISBN 978-981-16-1554-2   (eBook)
https://doi.org/10.1007/978-981-16-1554-2

This Springer imprint is published by the registered company Springer Nature Singapore Pte Ltd.
The registered company address is: 152 Beach Road, #21-01/04 Gateway East, Singapore 189721, Singapore

Preface

Across the world, over 8 million children are born with disabilities and chronic medical conditions caused by congenital anomalies (birth defects), genetic disorders and developmental disabilities. Birth defects like spina bifida, congenital heart defects, orofacial clefts and club foot; genetic disorders like muscular dystrophy and sickle cell anemia; and developmental disabilities like Down syndrome, cerebral palsy, intellectual disability and autism are commonly encountered in community and clinical settings. Ninety-five percent children with these conditions reside in low- and middle-income countries (LMICs). Despite the large numbers, birth defects and developmental disabilities are neglected childhood conditions, under-prioritized by public health services, especially in LMICs. Access to care is limited, so that survivors have a poor quality of life, while parents and caregivers encounter considerable financial expenditure and emotional distress in attempting to provide the best possible care.

Birth defects are unseen in global maternal and child health agendas, and not viewed as public health issues requiring immediate attention. Without the impetus of a global plan, birth defects are not perceptible in national health and social welfare programs of LMICs. Developing public health services for these conditions is a challenging task in resource-limited settings, as key public health activities of prevention, surveillance, medical care, physical rehabilitation, social support and community engagement activities have to be developed in the form of a multidisciplinary, integrated service. Industrialized country models cannot be exactly replicated in resource-constrained settings.

Research is crucial for guiding the development of low-cost birth defects service models in LMICs. Models that utilize existing program activities are relevant, as a birth defect service has to be delivered in the background of other, larger public health issues and considerable resource limitations. Data on the magnitude of birth defects and developmental disabilities, their epidemiology and trends, the medical, rehabilitation and social service needs, survival and quality of life of families, are indispensable in informing the health planning process. Health service models for prevention, surveillance and care need to be developed.

The goal of this book is to approach birth defects from a public health/health service perspective, focusing on essential public health functions and touching upon welfare activities that are essential to support the needs of children with disabilities and lifelong medical conditions. India has initiated services for birth defects, and this book examines and collates existing activities, in an attempt to understand the components that need to be put in place in order to develop a full-fledged service. The need for such a service is urgent, as data suggest that over 500,000 pregnancies are affected each year and around 50,000 newborns die from birth defect-related complications.

This book is intended for readers from the disciplines of public/global health, maternal and child health specialists and disability researchers and from related disciplines in the health sciences such as rehabilitation sciences, nursing and anthropology. Although written with India as a case study, the similarity in health system structure and functions should make it relevant to other LMICs. It is hoped that this public health perspective will bring about a discussion on birth defects as neglected maternal and child health issues, and the need for research to address the issue of childhood disability and chronic conditions, especially in LMICs.

Pune, India   Anita Kar

Acknowledgements

This book originates from the research studies of the contributing authors. Most of these studies were conducted in community settings, in the homes of children with birth defects, genetic disorders and developmental disabilities. The isolation of families, the lack of access to medical care and rehabilitation services, the lack of information and guidance, the exorbitant out-of-pocket expenditure and sometimes the questionable treatments were evident to the researchers. At the same time, the warmth of families, and their belief that the research would change their lives, forms the motivation for this book. The Editor and contributors would like to acknowledge all these families who, over two decades, have opened up their hearts and permitted the researchers to identify the essential components of a public health program that could substantially reduce the day-to-day challenges in their lives. The authors remain deeply indebted to all parents and caregivers and especially to the children, for their cooperation.

The Editor gratefully acknowledges parents who permitted the use of photographs and Dr. Ajay Meher, Dr. Md. Javed, Ariba Imtiaz, Dipthi Benoy, Bhagyashree Radhakrishnan, Trushna Girase and Dhammasagar Ujagare for sharing these photographs. The artwork provided by Kalyani Divgi is sincerely acknowledged.

The contributors would like to acknowledge several colleagues from non-governmental organizations, special schools and many clinicians and public health experts, who have collaborated and contributed to shaping these ideas.

The contributors gratefully acknowledge the authors and sources whose work has been cited in the text and regret their inability to cite all the scholarly work available in the field. The Editor is grateful to colleagues at the School of Health Sciences and the administration and support of Savitribai Phule Pune University where the research on birth defects was conducted. The research is now being continued at the Birth Defects Centre in Pune.

The Editor sincerely acknowledges the assistance of Dr. Joeeta Pal. She has carefully read and edited the manuscript and provided many valuable suggestions. Thanks are also due to Satvinder Kaur and colleagues from Springer Nature publishers for their patience and support. The Editor remains responsible for any omissions that may have appeared in the text.

Introduction: Birth Defects and Public Health in Low and Middle-Income Countries

Birth defects (congenital anomalies/malformations), developmental disabilities and genetic disorders are neglected childhood conditions that remain under-prioritized as public health issues in low- and middle-income countries (LMICs). Majority of these disorders are of prenatal origin. They are responsible for causing disability and complex, life-long medical conditions. Congenital disorders are usually detected at birth, in early childhood or sometimes during pregnancy. Although considered to be rare, they include some frequently encountered childhood disorders like Down syndrome, congenital heart defects, spina bifida, autism, cerebral palsy, sickle cell anemia, thalassemia and muscular dystrophy. A limited number of birth defects can be corrected. Most children require lifelong medical care, and several conditions like muscular dystrophy and thalassemia progressively deteriorate, resulting in premature mortality, despite best of treatment. Other birth defects and developmental disabilities cause intellectual, visual, hearing, speech or movement difficulties. Conditions like autism and attention-deficit/hyperactivity disorder impair socio-emotional and behavioral functioning. Physical rehabilitation therapies are required. Majority of children are dependent on others for daily living activities. Due to these characteristics, birth defects, genetic disorders and developmental disabilities have substantial human and healthcare implications [1]. Even with the best of care, quality of life of children and caregivers is affected. Children require multiple therapies, and most have special education and life-skill needs. Parents, as primary caregivers, have the responsibility of organizing lifelong care and managing multiple visits to medical providers, therapists or special educators. The change from normal family routine, the physical act of caregiving, and the need to manage and coordinate different activities for providing care, takes a toll on the physical and emotional health of the parents. Perception of stigma and significant out-of-pocket expenditure for medical and rehabilitation services affect the quality of life of families.

Birth Defects and Health Systems of Low- and Middle-Income Countries

Birth defects affect families across the world, but their consequences are most acutely perceived in low- and middle-income countries (LMICs). In these settings, public health services for congenital disorders are emerging or have not yet been established. The experience of providing services for non-communicable diseases is recent, as health systems are primarily focused on tackling infectious conditions and conditions arising from nutritional deficiencies. As such, services for addressing childhood disability are not a routine consideration of public health programs. Birth defects services are complex, multi-disciplinary and require integration of a number of activities. A child with Down syndrome, for example, will require routine child health services, and additionally, some children may require treatment for seizures or surgery for congenital heart defects. Physical rehabilitation such as physiotherapy and occupational therapy and special education and skills for nurturing functional independence are necessary for all children with Down syndrome. Parents require guidance on activities of daily living, counseling on child care and child rearing, and information on where services are available. Social welfare support such as supplemental income is essential, as providing care for a child with a disability or with a chronic medical condition involves considerable expenditure.

LMIC health systems are not appropriately equipped to handle such a complex service structure. Several birth defects require management by specialists. Pediatric surgery may be needed, and treatment may involve specialized equipments or high-cost therapeutics. Some of the latter may be orphan drugs and limited in availability. In the background of other public health priorities and chronic under-funding, it may be difficult to organize resources for such specialized activities. Physical rehabilitation services are rudimentary in most LMICs. The convergence between medical and social welfare services and policies, clear definition of roles and responsibilities of each service and clearly laid out referral mechanisms to develop a smoothly functioning service are difficult in light of the fragmented functioning of LMIC health services. Health service responsibilities do not end with just putting in place services for medical and physical rehabilitation. Plans for prevention of these conditions, including establishment of genetic services, a robust surveillance system, and developing skills and competencies for health service staff to manage birth defects and developmental disabilities are required. Education for supporting and guiding families on caring for a disabled or sick child, implementing appropriate regulations for protecting children and families, and ensuring the human rights of children with disabilities are other integral functions. Health and social welfare systems of LMICs need to establish priorities, allocate funds and develop plans for identifying the types and amount of services required and the referral mechanisms between medical, rehabilitation, education and social welfare services, in order to provide a continuum of services for children with these conditions [2]. Research is essential to inform an evidence-based birth defects service.

The public health impact of birth defects has largely remained unappreciated in LMICs, even though global data suggest that nearly 94% of the over 8 million children with congenital disorders are born in these regions [3]. The global child health agenda that drives national priorities is focused on mortality reduction. For example, the Sustainable Development Goal target 3.2 aims at ending all preventable deaths of newborns and children below five years of age by 2030. It recommends the reduction of neonatal mortality to at least 12 per 1000 live births and 25 per 1000 live births among children under five years of age [4]. Birth defects cause considerable numbers of neonatal and child deaths in LMICs, but one of the key public health challenges is in providing lifelong services for children with disabilities and chronic medical conditions. This component, that is the need and mechanism to include medical care and rehabilitation for children with disabilities and chronic medical conditions, does not find a visible place in the global child health dialogue.

The low prevalence of individual birth defects and developmental disabilities has also contributed to their invisibility. The prevalence of congenital heart defects, the most common birth defect, is around 1% [5]. This is significantly lower than the prevalence of preterm birth complications that affects 11% of pregnancies globally [6]. The low prevalence, and a multitude of clinical types included under the umbrella terms birth defects and developmental disabilities, deter health service thinking, in the background of larger numbers of potentially addressable child health issues. Another impediment for developing services is that there is no single strategy to prevent these conditions. Specific protection is known for two birth defects. Preconception folate supplementation reduces the prevalence of neural tube defects [7], while rubella immunization is a specific protection for congenital rubella syndrome [8]. However, several primary prevention strategies are well known, and these utilize health promotion approaches to increase awareness about risk and protective factors, and teratogenic exposures for birth defects [9]. Prenatal screening and genetic testing are other approaches for prevention [10]. Such a package of activities can prevent up to 60% of birth defects [11].

The Need for This Book

Medical management and rehabilitation for specific birth defects and developmental disabilities are well-established health service functions in industrialized countries, where health care is assured to all. But in LMICs, where access to health care is not universal, a public health program for birth defects and developmental disabilities has to be carefully constructed. The program has to keep in mind the existing public health priorities and the limited availability of resources. The service has to include key public health functions of prevention, surveillance, medical care and rehabilitation, identify and develop needed skills, competencies and infrastructural requirements, identify and partner with community-based organizations to enhance the outreach of services, ensure widespread community awareness about these conditions, including disability sensitization programs. Several interventions are low cost and within service capabilities of LMICs. Several activities may already be available through different health programs. A situational review of available services and capabilities of LMICs, the magnitude of birth defects and developmental disabilities, service gaps and opportunities is unavailable. The purpose of this book is to conduct such an analysis, using India as a case study.

Birth Defects and the Public Health Context in India

The need for developing a birth defects service in India is urgent, as the country is undergoing epidemiological transition. The rate of transition is uneven across the country, so that health indicators show a very wide range of values [12]. There are limited sources of data on birth defects and developmental disabilities in India, but public data from 2016 reported 300,000 birth defects and 1.9 million children with developmental delays and disabilities in the age-group of 2 to 18 years [13]. In terms of child mortality, birth defects were the fourth leading cause of mortality in 17 states and union territories of India [14] and the second leading cause of mortality in the state of Kerala. In 2017, congenital anomalies were responsible for over 50,000 neonatal deaths [15]. Data from a national survey reported that 2.2% of the 1.3 billion Indian population are disabled and 30% of the disability was reported to have been present since birth [16]. Modeled estimates point to an equally large magnitude of children surviving with disability caused by congenital disorders in India [15]. Other studies indicate that nearly 70% of children with birth defects survive with disabilities [17]. Such data imply that the public health focus to tackle birth defects should not only target congenital anomaly neonatal and child mortality reduction, but also address the medical and rehabilitation needs of survivors, through a well-organized medical and disability service [17].

India has in place a number of services for birth defects and developmental disabilities. Free-of-cost medical services for selected birth defects and developmental disabilities are available through the Rashtriya Bal Swasthya Karyakram (RBSK) [18]. Early intervention centers have been established in several districts. Patients with hemoglobinopathies and bleeding disorders receive free services, and national guidelines for prevention and control of haemoglobinopathies have been published [19]. Manuals and guidelines have been developed for health service staff, and training activities have been conducted. Social rehabilitation services are available, but they are implemented through non-governmental organizations and difficult to access. There is no referral link between medical and social welfare services. Surveillance for birth defects and developmental disabilities, and a prevention program for these disorders is yet to be initiated. Due to the limited availability of (free-of-cost) government services, medical care and rehabilitation services are primarily obtained from unregulated private providers, through personal expenditure.

Content and Organization of the Book

Birth defects in India: Epidemiology and Public Health Implications examines birth defects and developmental disabilities in terms of public health/health service functions. The public health system and existing maternal and child health services of India form the context for examining the services that are available, and the additional activities that need to be put in place in order to develop a holistic birth defects service. The book explores five broad aspects.

Part I A Public Health Approach

The goal of Part I is to understand the components of a birth defect service. The first article (1. "Birth Defects: A Public Health Approach") introduces birth defects and developmental disabilities and presents the public health framework for health service activities to address these conditions. The second article (2. "Some Common Birth Defects") provides an overview of some of the frequently encountered birth defects. The purpose is to illustrate the medical, rehabilitation and social service needs of children with these disorders. The third article (3. "Thalidomide: Understanding the Responsibilities of a Birth Defects Service") uses lessons from the thalidomide episode of the early 1960s to justify the different components of a birth defect service.

Part II Surveillance, Registries and Magnitude

The five articles in Part II are aimed at understanding the magnitude of birth defects and developmental disabilities in India. The first article in this part (4. "Birth Defects Surveillance in India") explains the methods of birth defect surveillance. It points out several methodological issues that arise while establishing birth defect surveillance in India. The second article in this part (5. "Rare Disease Registries: A Case Study of the Haemophilia Registry in India") describes rare disease registries. It uses the example of the hemophilia registry in India, to illustrate how a global-local partnership has led to a systematically organized disease registry, that is a rich source of epidemiological data. The next two articles (6. "Magnitude of Congenital Anomalies in India and 7. "Magnitude of Developmental Disabilities in India) reviews the data on the magnitude of these conditions, the different sources of data and the uncertainties in estimates that arise due to non-standard study designs. The final article in this part (8. "Magnitude and Characteristics of Children with Congenital Disabilities in India) provides estimates on the magnitude of survivors with congenital disabilities in the country.

Part III Prevention

The three articles of Part III examine the opportunities for preventing birth defects in India. The first chapter (9. "Preventing Congenital Anomalies Through Existing Maternal and Child Health Services in India") describes the risk factors and exposures for congenital anomalies, the prevalence of selected risk factors among women in the reproductive age in India, existing services of the maternal and child health programme and the potential of this service to put together activities for prevention of birth defects. The next article (10. "Neural Tube Defects and Folate Status in India") reviews data on this important micronutrient in Indian women and discusses the prevalence and prevention of neural tube defects in the country. The last article in this part (11. "Haemoglobinopathies: Genetic Services in India") describes the emerging genetic services in India, by discussing the national guidelines for prevention and control of hemoglobinopathies in the country.

Part IV Services

The two articles in Part IV discuss the available services for birth defects and developmental disabilities in India. The first article (12. "Medical, Rehabilitation and Social Welfare Services for Children with Birth Defects and Developmental Disabilities in India") describes the medical services and the physical and social rehabilitation services that are currently available in India. The next chapter (13. "Early Childhood Intervention Services in India") describes early intervention and the sub-optimal functioning of these centers in the country.

Part V Quality of Life

Part V discusses the needs and quality of life of children, caregivers and families. The first article (14. "Birth Defects Stigma") alerts health service providers on how the intersection of disability, prenatal and genetic testing and the dilemma of termination of a pregnancy precipitates perceptions of stigma. The next article (15. "Parenting a Child with a Disability: A Review of Caregivers' Needs in India and Service Implications") points to a paucity of studies documenting the needs of caregivers of children with birth defects and developmental disabilities in India. The final article discusses the quality of life and psychosocial needs of caregivers, using a case study of mother's of boys with haemophilia (16. "Quality of Life and Psychosocial Needs of Caregivers of Children with Birth Defects: A Case Study of Haemophilia").

Pune, India   Anita Kar, Ph.D.

References

1. Christianson A, Howson CP, Modell B (2006) March of Dimes Global Report on birth defects: the hidden toll of dying and disabled children. March of Dimes Birth Defects Foundation, White Plains, New York
2. World Health Organization: Sixty-third World Health Assembly WHA63.17 Agenda item 11.7 21 May 2010 Birth defects. https://apps.who.int/gb/ebwha/pdf_files/WHA63/A63_R17-en.pdf?ua=1 (2010). Accessed 26 Mar 2018
3. Darmstadt GL, Howson CP, Walraven G et al (2016) Prevention of congenital disorders and care of affected children: a consensus statement. JAMA Pediatr 170(8):790–793
4. Desa UN (2016) Transforming our world: the 2030 agenda for sustainable development. https://sustainabledevelopment.un.org/content/documents/21252030%20Agenda%20for%20Sustainable%20Development%20web.pdf . Accessed 15 Mar 2018
5. Zimmerman MS, Smith AGC, Sable CA et al (2020) Global, regional, and national burden of congenital heart disease, 1990–2017: a systematic analysis for the Global Burden of Disease Study 2017. The Lancet Child and Adolescent Health. (Published online January 21, 2020). https://doi.org/10.1016/S2352-4642(19)30402-X. Accessed 18 Mar 2020
6. Vogel JP, Chawanpaiboon S, Moller AB et al (2018) The global epidemiology of preterm birth. Best Pract Res Clin Obstetr Gynaecol 52:3–12
7. van Gool JD, Hirche H, Lax H, De Schaepdrijver L (2018) Folic acid and primary prevention of neural tube defects: a review. Reprod Toxicol 80:73–84
8. Grant GB, Reef SE, Patel M, Knapp JK, Dabbagh A (2017) Progress in rubella and congenital rubella syndrome control and elimination—worldwide, 2000–2016. MMWR Morb Mortal Weekly Report 66(45):1256–1260
9. Taruscio D, Mantovani A, Carbone P et al (2015) Primary prevention of congenital anomalies: recommendable, feasible and achievable. Public Health Genomics 18(3):184–191
10. Ballantyne A, Goold I, Pearn A (2006) Medical genetic services in developing countries: the ethical, legal and social implications of genetic testing and screening. https://apps.who.int/iris/bitstream/handle/10665/43288/924159344X_eng.pdf Accessed 15 Mar 2018
11. Czeizel AE, Intody Z, Modell B (1993) What proportion of congenital abnormalities can be prevented? Brit Med J 306(6876):499–503
12. Dandona L, Dandona R, Kumar GA et al (2017) Nations within a nation: variations in epidemiological transition across the states of India, 1990–2016 in the Global Burden of Disease Study. The Lancet 390(10111):2437–2460
13. Government of India Ministry of Health and Family Welfare (2019) Answers data of Rajya Sabha questions for session 240/yearwise physical status Rashtriya Bal Swasthya Karyakram (RBSK) during 2014–15 and 2015–16. https://data.gov.in/node/3978901/download Accessed 18 Nov 2019
14. Dandona R, Kumar GA, Henry NJ et al (2020) Subnational mapping of under-5 and neonatal mortality trends in India: the Global Burden of Disease Study 2000–17. The Lancet. Published online May 12, 2020. 10.1016/S0140-6736(20)30471-2. Accessed 12 May 2020. https://doi.org/10.1016/S0140-6736(20)30471-2 Accessed 12 May 2020
15. Modell B, Darlison MW, Moorthie S, Blencowe H, Petrou M, Lawn J (2016) Epidemiological methods in community genetics and the modell global database of congenital disorders (MGDb). https://discovery.ucl.ac.uk/id/eprint/1532179/17/Epidemiological%20Methods%20in%20Community%20Genetics%20and%20the%20Modell%20Global%20Database%202017-04.pdf Accessed 11 June 2018
16. Ministry of Statistics and Programme Implementation Persons with Disabilities in India. NSS Report No. 583 (76/26/1) July–December 2018. http://www.mospi.gov.in/sites/default/files/publication_reports/Report_583_Final_0.pdf Accessed 1 Nov 2019
17. Bhide P, Gund P, Kar A (2016) Prevalence of congenital anomalies in an Indian maternal cohort: healthcare, prevention, and surveillance implications. PLoS ONE 11(11): e0166408. https://doi.org/10.1371/journal.pone.0166408 Accessed 9 August 2018

18. Government of India Ministry of Health and Family Welfare National Health Mission (2013) Rashtriya Bal Swasthya Karyakram a child health screening and early intervention services under NRHM, Ministry of Health and Family Welfare. http://nhm.gov.in/images/pdf/pro grammes/RBSK/Operational_Guidelines/Operational%20Guidelines_RBSK.pdf. Accessed 6 Aug 2019
19. Ministry of Health and Family Welfare Government of India Guidelines on Haemoglobinopathies in India, 2016. https://nhm.gov.in/images/pdf/programmes/RBSK/ Resource_Documents/Guidelines_on_Hemoglobinopathies_in%20India.pdf Accessed 3 Mar 2019

Contents

Editor and Contributors

About the Editor

Prof. Anita Kar is the former Director of the School of Health Sciences of Savitribai Phule Pune University (University of Pune), India, and the Founder-Director of the Birth Defects and Childhood Disability Research Centre, Pune. This research NGO is working on identifying contextually appropriate policies and services for congenital disorders and childhood disability in India, as well as advocating for including these conditions in the global maternal and child health dialogue. Dr Kar has a background in human genetics, public health and epidemiology. Her work in the field of birth defects has spanned over two decades and focuses specifically on examining how lack of services affect children, caregivers and families, and asking how birth defects services can be introduced in the resource-constrained health systems of low middle-income countries like India. She was responsible for establishing one of the first University Grants Commission approved Master of Public Health programs in the country.

Contributors

Humaira Ansari, MPH Symbiosis International (Deemed) University, Pune, Maharashtra, India

Prajkta Bhide, PhD Birth Defects and Childhood Disability Research Centre, Pune, India

Amruta Chutke, BAMS, MPH Department of Community Medicine, Bharati Vidyapeeth DTU Medical College, Pune, India

Sumedha Dharmarajan, PhD Birth Defects and Childhood Disability Research Centre, Pune, India

Charuta Gokhale, PhD Tata Institute of Social Sciences, Mumbai., Mumbai, India

Pooja Gund, PhD Indian Institute of Public Health, Hyderabad, India

Anita Kar, PhD Birth Defects and Childhood Disability Research Centre, Pune, India

Juhi Nakade, PhD Birth Defects and Childhood Disability Research Centre, Pune, India

Supriya K. Nikam, BHMS, MPH Birth Defects and Childhood Disability Research Centre, Pune, India

Supriya Phadnis, PhD Healthcare Management, Goa Institute of Management, Poreim, Sattari, Goa, India

Dhammasagar Ujagare, MPH ICMR—National AIDS Research Institute, Pune, India

Abbreviations

5-Methyl-THF	5-methyltetrahydrofolate
AAIDD	American Association on Intellectual and Developmental Disabilities
ADHD	Attention-deficit/hyperactivity disorder
ADL	Activities of daily living
AGS	Annual Global Survey
ANC	Antenatal care
ARMS-PCR	Amplification-refractory mutation system polymerase chain reaction
ASD	Atrial septal defect
ASHA	Accredited Social Health Activist
ASO	Allele-specific oligonucleotide hybridization
BCG	Bacille Calmette–Guerin
BDRI	Birth Defects Registry of India
BMI	Body mass index
BMT	Bone marrow transplantation
CAE	Cellulose acetate electrophoresis
CDC	Centres for Disease Control and Prevention
CE HPLC	Cation-exchange high-performance liquid chromatography
CFC	Clotting factor concentrate
CHDs	Congenital heart defects
CHWs	Community health workers
CI	Confidence interval
COA	Coarctation of the aorta
CRPD	Convention on the Rights of Persons with Disabilities
CSHCN	Children with special healthcare needs
CTEV	Congenital talipes equinovarus
DALYs	Disability-adjusted life years
DCIP	Di-chlorophenolindophenol
DDH	Developmental dysplasia of the hip
DEIC	District Early Intervention Centres
DEPwD	Department of Empowerment of Persons with Disabilities

DSM-5	Diagnostic and Statistical Manual of Mental Disorders, 5th Edition
EAR	Estimated average requirement
EC	Ethics Committee
ECCE	Early childhood care and education
ECD	Early child development
ECLAMC	Estudio Colaborativo Latino Americano de Malformaciones Congenitas
ESIS	Employees State Insurance Scheme
ETL	Epidemiological transition level (states)
EUROCAT	European Surveillance of Congenital Anomalies
FFP	Fresh frozen plasma
fIPV	Fractional Inactivated Poliomyelitis Vaccine
GBD	Global Burden of Disease
GDP	Gross domestic product
GNM	General nurse midwife
GRD	Gastroesophageal reflux disease
HbA2	Hemoglobin A2
Hep-B	Hepatitis B
HFI	Hemophilia Federation of India
HICs	High-income countries
HRQoL	Health-related quality of life
HSTC	Hematopoietic stem cell transplant
HTC	Hemophilia treatment center
ICBDSR	International Clearinghouse for Birth Defects Surveillance and Research
ICD-10	International Statistical Classification of Diseases and Related Health Problems 10th Revision
ID	Intellectual disability
IEC	Information, education and communication
IEF	Isoelectric focusing
IMR	Infant mortality rate
IQ	Intelligence quotient
IQR	Inter-quartile range
LMIC	Low- and middle-income countries
MCCD	Medical certification of cause of death
MCEE	Maternal and Child Epidemiology Estimation Group
MCH	Maternal and child health services
MDS	Million Death Study
MGDb	Modell Global Database of Congenital Disorders
MGRS	Multicentre Growth Reference Study
MoHFW	Ministry of Health and Family Welfare
MTP	Medical termination of pregnancy
NABH	National Accreditation Board for Hospitals and Healthcare Providers
NDD	Neuro-developmental disorders

NESTROFT	Naked eye single tube red cell osmotic fragility test
NIPT	Non-invasive prenatal test
NMO	National Member Organization
NMR	Neonatal mortality rate
NSS	National Sample Survey
NTDs	Neural tube defects
OC	Orofacial cleft
OPV	Oral polio vaccine
PDA	Patent ductus arteriosus
PDD	Pervasive developmental disorder
PMSMA	Pradhan Mantri Surakshit Matritva Abhiyan
PUBOs	Pune Urban Birth Outcomes study
RBC	Red blood cell
RBSK	Rashtriya Bal Swasthya Karyakram
RDs	Rare diseases
RMNCH+A	Reproductive, maternal, newborn, child and adolescent health services
ROP	Retinopathy of prematurity
RPwDA	Rights of Persons with Disabilities Act 2016
SDI	Socio-demographic Index
SEAR-NBBD	Southeast Asia Regional Newborn and Birth Defects Database
SEARO	Southeast Asia Regional Office
SRS	Sample Registration System
TGA	Transposition of the great arteries
tHcy	Total homocysteine
TOF	Tetralogy of Fallot
TOPFA	Terminations of pregnancy after detection of fetal anomalies
UI	Uncertainty interval
UT	Union territory
VSD	Ventricular septal defect
WBDR	World Bleeding Disorders Registry
WFH	World Federation of Hemophilia
WHA	World Health Assembly
WHO	World Health Organization
YLD	Years lived with disability

List of Figures

List of Tables

List of Boxes

Part I
A Public Health Approach

Part I approaches birth defects (and developmental disabilities) from a public health perspective. It describes the public health framework, that is the public health activities to address these conditions (1. "Birth defects : A Public Health Approach"). Using the thalidomide episode of the 1960's as a case study, the article underlines the relevance of the different public health functions that make up a birth defect service (3. "Thalidomide: Understanding the Responsibilities of a Birth Defects Service"). The part is supported by an overview of some frequently encountered birth defects, genetic disorders and developmental disabilities (2. "Some Common Birth Defects"). The purpose of the chapter is to provide an overview of the medical, rehabilitation and social welfare needs for common birth defects.

Chapter 1
Birth Defects: A Public Health Approach

Anita Kar

Abstract Birth defects and developmental disabilities consist of several debilitating clinical disorders that cause disabilities and lifelong medical conditions in children. Down syndrome, congenital heart defects, spina bifida, autism cerebral palsy and congenital hearing impairment are some common examples. Birth defects and developmental disabilities are neglected childhood conditions in low- and middle-income countries (LMICs), with limited medical and rehabilitation services. Survivors have multiple healthcare needs that include lifelong medical care and physical rehabilitation. Social welfare measures like supplemental income, access to appropriate education and employment need to be provided. Interventions to foster the independence of children and adults with disabilities and disabling conditions, and implementation of regulations to ensure social inclusion and protection of rights of children with disabilities are other essential activities. Providing care for children with such special healthcare needs is a challenge for health services across the world. For LMICs, this challenge is significantly pronounced as access to health care is not universal, health systems are under-funded, health service needs are large, and disability services and services for chronic paediatric disorders are unavailable or of limited availability. This article approaches birth defects and developmental disabilities from a public health/health service perspective. It begins by placing the humanitarian, demographic and epidemiological arguments that demand the introduction of birth defects services as integral components of maternal and child health services of LMICs. The article then presents the public health framework, discussing key issues of birth defects surveillance, prevention, care and other essential activities. The final section of the article describes the context, that is the health indicators and services in India, which form the background within which a holistic birth defects service needs to be planned, developed and implemented.

Keywords Congenital · Birth defects · Developmental disabilities · Public health · Health service · India · LMIC

A. Kar (✉)
Birth Defects and Childhood Disability Research Centre, Pune 411020, India

© Springer Nature Singapore Pte Ltd. 2021
A. Kar (ed.), *Birth Defects in India*,
https://doi.org/10.1007/978-981-16-1554-2_1

Birth defects, that is congenital anomalies (congenital malformations), genetic disorders (single gene and chromosomal disorders) and developmental disabilities make up a group of diverse clinical disorders that cause disabilities and disablement in children since birth. Congenital anomalies are "… structural or functional abnormalities, including metabolic disorders, that are present from birth" [1, 2]. Congenital heart defects, spina bifida, orofacial clefts and clubfoot (congenital talipes equinovarus) are some common examples. Haemoglobinopathies like sickle cell anaemia and thalassemia, and Down syndrome are some commonly encountered single gene and chromosomal disorders. Developmental disabilities, which cause social, emotional, behavioural, cognitive and motor impairments in children [3], are commonly encountered as autism spectrum disorders, attention deficit hyperactivity disorder, cerebral palsy, learning disorders, and hearing, speech, cognitive and visual impairments. Despite distinct groups of clinical disorders, these conditions are the single most important cause of childhood disability and cause lifelong and frequently life-threatening and life-limiting medical conditions. These conditions are either obvious at birth, detected in the prenatal period or diagnosed in early childhood.

The severity of disabilities and the disablement varies among children, as also by the type of condition. The disability may have a cosmetic effect (e.g. orofacial clefts) and may cause physical impairment (movement difficulties in children with limb deformities, spina bifida or cerebral palsy), cognitive impairment (Down syndrome), sensory deficits (visual, hearing and speech impairments), socio-emotional and behavioural impairments (autism and attention deficit hyperactivity disorder) or combinations of these impairments/complications. Several of these conditions, such as epilepsy, or conditions like sickle cell anaemia and thalassemia, bleeding disorders like haemophilia or Von Willebrand disease and muscular dystrophies cause chronic medical complications. The need for medical care is lifelong.

From a health services perspective, children with birth defects and developmental disabilities form a distinct group, collectively referred to as Children with Special Healthcare Needs (CSHCN) [4]. CSHCN have requirements that are additional to those routinely offered by child health services. Special healthcare needs include paediatric surgery, management by specialists, physical rehabilitation services such as physiotherapy and occupational therapy, social rehabilitation services such as special education, and other social welfare measures. Parents require counselling and instructions on home care and, as necessary, medical management of the child at home. A referral mechanism between medical and social welfare services is required. Such a complex referral system is challenging to set up and sustain, anywhere across the globe. For health services in low- and middle-income countries (LMICs), the need to put up this integrated service, with complex referral pathways, is especially challenging.

The aetiology of congenital anomalies and developmental disabilities are complex and best described individually. Congenital anomalies are caused by preconception and antenatal factors which affect the development of anatomical structures during foetal development (Fig. 1.1 panel A). Preconception factors include genetic factors that are transmitted from parents to children. Around 20% of congenital anomalies may be caused by single gene and chromosomal disorders. The large majority of

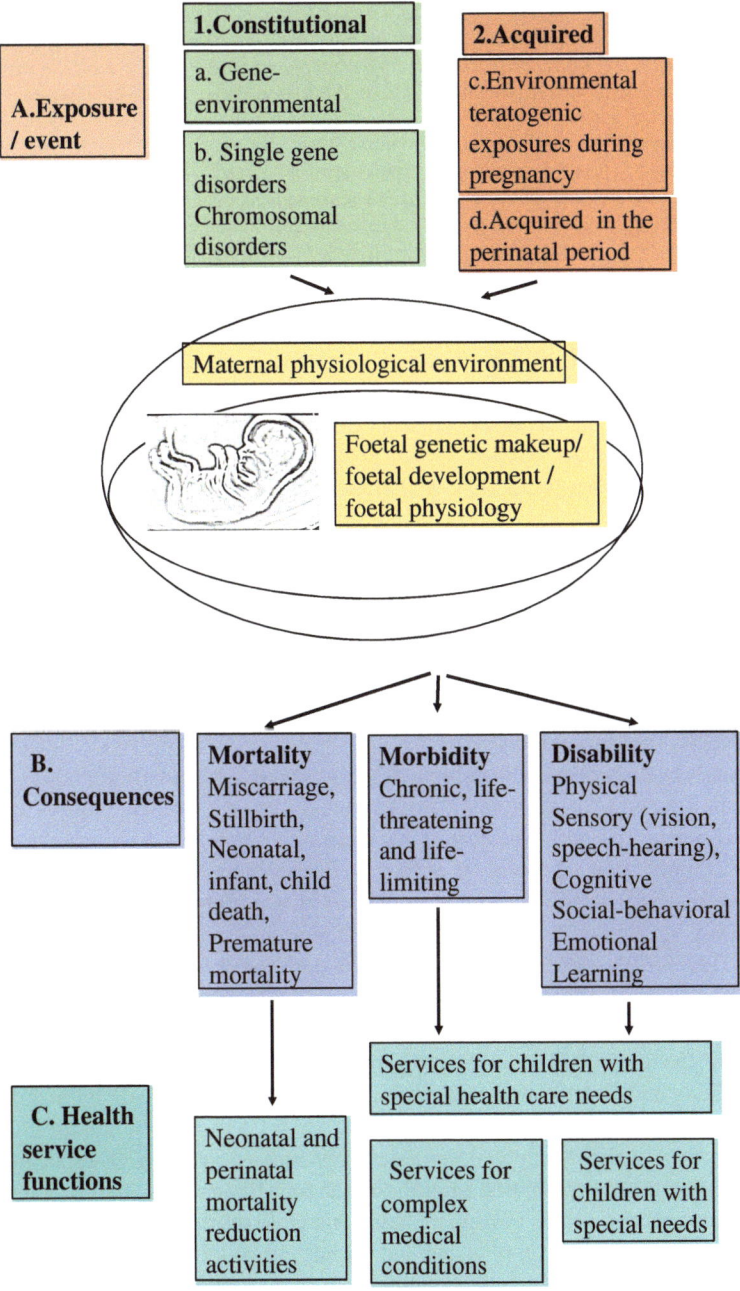

◄**Fig. 1.1 Birth defects aetiology, consequences and health service functions. A** Birth defects
and developmental disabilities arise from preconception, prenatal or perinatal events or exposures.
Majority of birth defects have a genetic component. While most congenital anomalies are caused by
gene–environmental interactions (**a**), some congenital anomalies may be of wholly genetic origin
(**b**). Several congenital anomalies are acquired through maternal infections or teratogenic exposures
during pregnancy (**c**). Some developmental disabilities (such as cerebral palsy) may be acquired by
perinatal insults, such as hypoxia or trauma (**d**). **B** The consequences of birth defects are spontaneous
abortions, stillbirths, neonatal, infant and child mortality and premature adult mortality. The major
public health service requirement is to provide lifelong medical care and services for children and
adults with disabilities (**C**)

congenital anomalies (and developmental disabilities) are caused by gene–environmental factors. For example, low maternal folate levels (the maternal physiological environment) and certain polymorphisms in the foetal one carbon metabolism
pathway (foetal genetic make-up) predispose the foetus to neural tube defects like
anencephaly and spina bifida. Congenital anomalies arising in the antenatal period
are of environmental origin, caused by infections with rubella virus, cytomegalovirus
and Zika virus or exposure to teratogenic drugs, or specific physical and chemical agents in the environment. The cause of majority of congenital anomalies is,
however, unknown (Chap. 9). Environmental/extrinsic factors provide an opportunity for health service interventions to prevent birth defects. Rubella immunization,
folic acid supplementation, strengthening maternal health services and educating
women about protection against teratogenic exposures are examples of preventive
interventions for birth defects.

Several developmental disabilities are also caused by environmental, genetic,
gene–environmental or as yet unknown factors. Some developmental disabilities are
acquired during the perinatal period, for example, cerebral palsy caused by intra-
partum complications. Several other factors, such as low birth weight, prematurity
and poor maternal health status are associated with a higher risk of developmental
disabilities. Strengthening reproductive health services and providing care for premature and low birth weight babies are important steps for prevention of developmental
disabilities.

Figure 1.1 (panel B) shows the consequences of birth defects, which may be severe
enough to cause foetal loss during pregnancy or cause neonatal and child mortality.
Birth defects and developmental disabilities are major causes of childhood disabilities (e.g. orofacial cleft, club foot or cerebral palsy). Most functional birth defects
(such as congenital heart defects) and developmental disabilities (such as intellectual impairment) are diagnosed in early childhood. Figure 1.1 panel C shows that
health service functions go beyond neonatal and child mortality reduction activities.
In addition to medical care, physical and social rehabilitation services for children
with disabilities and chronic medical conditions are needed.

The Need for Birth Defects Services in LMICs

There are several humanitarian, demographic and epidemiological arguments for implementing birth defects services, especially in LMICs. The term birth defects service is being used here to describe generalized services for congenital anomalies and developmental disabilities.

Individual, Health and Social Welfare System Impact

Due to their chronic and disabling nature, birth defects and developmental disabilities impact the quality of life of children and parents, and have substantial implications on health and social welfare services (Fig. 1.2). During the prenatal period, birth defects cause adverse foetal outcomes, that is, miscarriages and stillbirths, which affect health indicators and cause individual distress [5]. Some birth defects can be detected in the prenatal period, using ultrasonography or genetic tests, confronting parents with the difficult choice of pregnancy termination (where permissible by law). At birth, health systems incur the cost of neonatal intensive care for babies with birth defects, many of whom may be preterm born or have low birth weight in addition to medical complications. Paediatric surgical care is necessary to address life-threatening malformations. When such services are not in place, birth defects cause substantial numbers of deaths among newborns and during infancy. In case of severe birth defects, parents are confronted with the traumatic experience of perinatal palliative care.

Studies have reported alterations in the quality of life of parents of children with disabilities or disabling conditions [6] (Chap. 16). The child's disability or chronic illness provokes feelings of loss, grief, shame and guilt, which are reinforced by insensitive social responses. These lead to families experiencing stigma and social isolation (Chap. 14) [7]. Several studies have reported the impaired quality of life (QoL)/health-related quality of life (HRQoL) of children with disabilities and medical conditions. Majority of these studies have been conducted in high-income countries, where healthcare costs are supported to a large extent by the public health system or health insurance. The studies have covered a broad range of clinical conditions, but conclude that parents of children with complex medical conditions and disabilities undergo similar physical and emotional experiences, related to the irreversible nature of these conditions [8]. Breakdown of marital relationships and sibling neglect are reported to be higher in families with children with disabilities or lifelong medical conditions [9]. High levels of physical exhaustion, stress, anxiety and depression result in poorer quality of life and poorer parenting [10].

Children and adults with birth defects and developmental disabilities experience a lower QoL. The prevalence of depression and anxiety problems among children with intellectual disability in the USA was reported to be as high as 35% [11]. A study of adults with congenital heart disease (CHD) reported a higher prevalence of

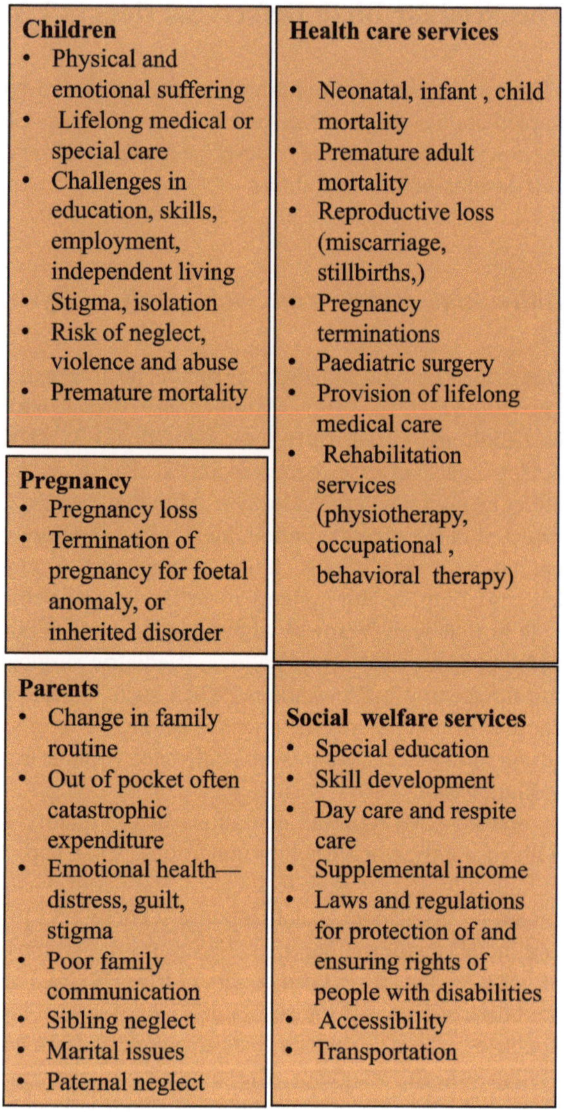

Fig. 1.2 Individual, health
and social welfare system
impact

Children
- Physical and emotional suffering
- Lifelong medical or special care
- Challenges in education, skills, employment, independent living
- Stigma, isolation
- Risk of neglect, violence and abuse
- Premature mortality

Pregnancy
- Pregnancy loss
- Termination of pregnancy for foetal anomaly, or inherited disorder

Parents
- Change in family routine
- Out of pocket often catastrophic expenditure
- Emotional health—distress, guilt, stigma
- Poor family communication
- Sibling neglect
- Marital issues
- Paternal neglect

Health care services

- Neonatal, infant, child mortality
- Premature adult mortality
- Reproductive loss (miscarriage, stillbirths,)
- Pregnancy terminations
- Paediatric surgery
- Provision of lifelong medical care
- Rehabilitation services (physiotherapy, occupational, behavioral therapy)

Social welfare services
- Special education
- Skill development
- Day care and respite care
- Supplemental income
- Laws and regulations for protection of and ensuring rights of people with disabilities
- Accessibility
- Transportation

activity limitation and poorer physical health-related quality of life [12]. A review of the quality of life of children with orofacial clefts identified that severity of speech problems, age of the child, number of surgeries, facial appearance and unmet need for more surgeries were associated with poor QoL [13]. QoL studies for other condition-specific birth defects and developmental disabilities indicated similar impact on the quality of survival of children. Children with disabilities have a higher risk of physical violence and abuse. A systematic review estimated that nearly 27% of children with disabilities experienced some form of violence, with 20% reporting physical violence and 14% reporting sexual violence [14]. The risk of experiencing violence increased 3.68-fold for children with disabilities, as compared to children without disabilities, with the risk of physical violence and sexual violence being increased 3.56 and 2.88 times, respectively. One of the worst forms of abuse, mutilation and killing of children and adults with oculocutaneous albinism, an inherited condition where individuals lack melanin pigment in the skin, hair and eyes, has been reported from some sub-Saharan African countries [15].

Social welfare service functions are essential for ensuring access to mainstream or special education, vocational skills and employment for children and adults with disabilities or disabling conditions. One of the major consequences for families is the financial burden of caregiving, especially as children with special needs and those with complex medical conditions require multiple therapies. Across the world, even in industrialized settings, not all costs of medical care and therapies are reimbursed, so that families encounter out-of-pocket expenditure. In LMICs, nearly all care is funded through personal finances, often resulting in catastrophic financial consequences [16]. Such costs are one of the major reasons for sub-optimal treatment [17]. The loss of the caregivers' employment in order to tend to the child adds to financial hardship. Legal guardianship, to ensure that the child/adult is well protected after the demise of parents/caregivers, is another social service responsibility.

The US National Survey on Children with Special Healthcare Needs (NS-CSHCN) 2009–10 illustrates the situations encountered by families with a child with a special healthcare need [18]. This survey reported that among nearly a quarter of families where the child had a severe disability, parents spent over 10 h in a day caring for and coordinating activities for the care of the child. This included making appointments and taking children for visits to healthcare providers and therapists, and ensuring that care providers exchanged information on the needs of the child. The burden of caregiving was more on families from the lower-income strata. A quarter of the respondents in the survey reported giving up employment or reducing the hours worked in order to care for the child. The likelihood of having to give up their employment in order to stay at home with the child was higher among lower-income families and those with more severely affected children. Over 21% of families reported that they had financial problems, with the problems being higher among low-income families, and those with more severely affected children. Annually, 21% of families reported spending on healthcare costs that were not covered by insurance plans. Low-income families incurred less expenditure, possibly as they had other sources of insurance, unpaid bills or due to delayed care or forgoing care because of limited finances. More than one-third of families with children who were severely impaired reported higher out-of-pocket expenditures. There is a paucity of similar studies that document the impact of birth defects and developmental disabilities on children, adults and parents in LMICs. There can be no doubt that these conditions would severely affect the lives of people in these settings.

Magnitude of Congenital Anomalies

The second argument for implementing birth defects services in LMICs is the magnitude of these conditions. Surveillance systems have been established in industrialized countries that collect data on the magnitude of congenital anomalies and developmental disabilities. Such data are collected through birth defect registries, which were established after the thalidomide incident of the late 1950–early 1960s, further described in Chap. 3. Thalidomide, a teratogenic drug, was used as an anti-emetic by women in the late 1950s, resulting in the birth of large numbers of severely disabled children. The drug was withdrawn from the market in 1962, but the large numbers of children surviving with disabilities and chronic health needs alerted public health agencies on the need for routine monitoring of teratogenic exposures during pregnancy. Birth defects surveillance was established in industrialized countries in the late 1960s in order to provide early warning against new teratogens. Surveillance systems are also in place for monitoring the prevalence of common developmental disabilities. Data on single gene disorders are collected through disease registries, where patient information on specific conditions is stored. Disease registries are often linked to biobanks.

There are several birth defects surveillance systems. The National Birth Defects Prevention Network (NBDPN) in the USA [19] and EUROCAT, a network of population-based registries for the epidemiological surveillance of congenital anomalies in Europe [20], routinely monitor the numbers of affected births. The Latin American Collaborative Study of Congenital Malformations (ECLAMC) is a research programme that investigates risk factors for congenital anomalies [21]. The Southeast Asia Regional Office (SEARO) of the World Health Organization (WHO) has established the Newborn and Birth Defects Database (NBDD) with online reporting from 220 hospitals from 9 countries [22]. The International Clearinghouse for Birth Defects Surveillance and Research (ICBDSR) promotes birth defects surveillance by creating a global network of institutions conducting birth defects surveillance [23].

Congenital anomalies affect 2–3% of births [24], but the prevalence may be as high as 6% when single gene disorders are included in these measures [25]. Estimates suggest that globally, "… 7.9 million children are born with a genetic or a partially genetic condition each year … (and) … several hundred thousands" are born with congenital disorders due to infectious or other environmental exposures in utero [26]. Among 2.7 million deaths occurring globally, one in ten neonatal deaths was caused by a congenital disorder. Congenital disorders caused an estimated 848,000 deaths among children below five years of age. Several severe birth defects end in a stillbirth, with statistics suggesting that as many as 192,000 annual stillbirths may be caused by an underlying congenital condition. These numbers are likely to be underestimates as majority of LMICs are yet to establish birth defects surveillance [26]. Registry data from the USA suggest that 1 in 33 births is affected with a major birth defect, while an Indian study estimated that 1 in 44 pregnancies was affected with a major congenital disorder [27]. Unlike industrialized nations, LMIC data on birth defects and developmental disabilities are primarily available from surveys and ad hoc studies.

Epidemiological Transition

One of the key arguments for putting in place a birth defects service is the increasing numbers of neonatal and child deaths caused by congenital anomalies, especially in South Asian countries. Typically, the proportion of congenital anomalies (and developmental disabilities) remain constant (e.g. congenital anomalies affect 2–3% of births). As other causes of child mortality reduce, birth defects emerge as leading causes of neonatal and child deaths [28]. This relationship can be illustrated with Indian data on mortality rates of prevalent neonatal causes of mortality, between 1990 and 2017. Figure 1.3 shows that while birth defect mortality has remained constant, other common causes of neonatal mortality (preterm birth complications, neonatal

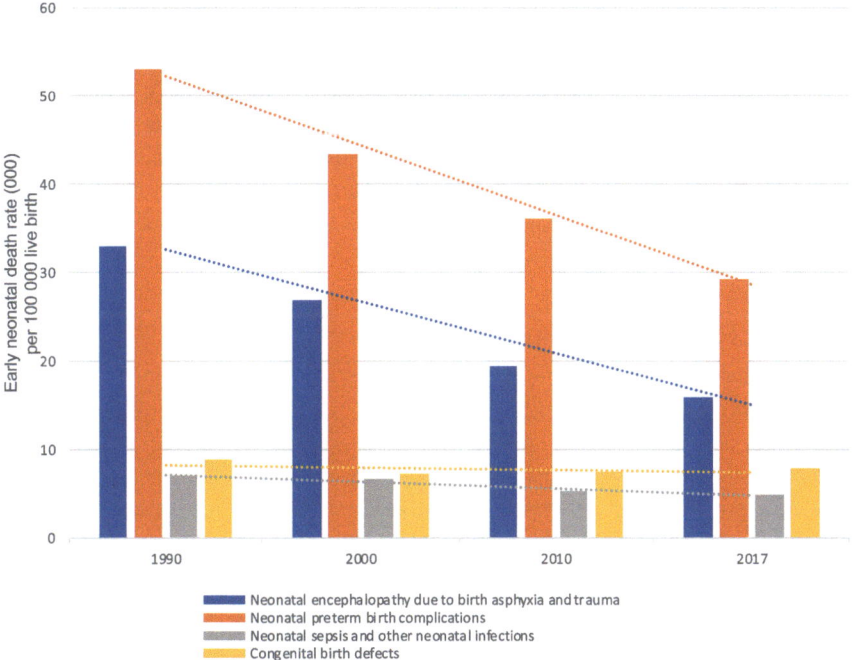

Fig. 1.3 Trends in neonatal mortality by cause in India (1990–2017). The figure shows that between 1990 and 2017, neonatal preterm birth complications, neonatal encephalopathy due to birth asphyxia and trauma, and neonatal sepsis have decreased in India, while the number of deaths caused by congenital anomalies has remained constant. Congenital anomalies now cause higher numbers of mortality than neonatal sepsis [30]

encephalopathy due to trauma and asphyxia, neonatal sepsis and other neonatal infections) have declined, driven by the impetus to achieve the child health targets of the Millennium Development Goal and the Sustainable Development Goals. Continuing public health activities to reduce the prevailing causes of neonatal and child mortality will increase the visibility of deaths caused by congenital disorders [29]. Such trends are already occurring across states of India [30].

Global data on cause-specific child mortality also show such evidences of epidemiological transition. For example, diarrhoea was the third leading cause of mortality in 2000. By 2015, aggressive public health interventions resulted in reduction in diarrhoeal deaths, so that intrapartum complications emerged as the third largest cause of mortality. Current data show that reduction in common causes of neonatal deaths in South Asia has led to substantial increase in numbers of deaths caused by congenital anomalies and preterm birth complications between 2000 and 2015. In China, for example, a 70% reduction in child mortality has altered the cause composition of mortality, with congenital anomalies (19%) and injuries (14%) emerging as the leading causes of child mortality. Liu and colleagues noted a 61% decrease in mortality between 2000 and 2015, while Hug and colleagues from the UN Interagency Group for Child Mortality Estimation reported a 51% decrease in neonatal mortality due to the leading causes between 1990 and 2017 [31].

Table 1.1 and Fig. 1.4 show the estimated causes of mortality among 5.9 million children below five years of age occurring globally in 2015 [29]. About 45% (2.7 million of these 5.9 million deaths) were in the neonatal period. These deaths occurred primarily due to preterm birth complications, pneumonia and intrapartum-related complications. Congenital anomalies were the fourth largest cause of neonatal mortality globally [29]. This transition in the causes of child mortality underlines the need for a public health policy, as well as services to address birth defects.

A Public Health Framework

Public health, the art and science of preventing disease and promoting health and well-being, functions through an organized service, the primary goal of which is to ensure that the population remains healthy and disease free. Public health activities include (a) prevention of diseases or health conditions, and controlling associated risk factors and exposures, (b) health promotion activities, (c) surveillance to inform the magnitude of the conditions/risk factors, (d) provision of medical and other forms of services for treatment/management of the condition, (e) raising community awareness, (f) investing in required infrastructure and supplies, (g) developing knowledge and skills of the health workforce and (h) developing policy and regulations. Public health services are packaged into health programmes with defined goals and objectives, which are routinely monitored, and the interventions are evaluated to ensure that they achieve the predetermined goals and objectives of the programme. Figure 1.5 shows some of the essential public health activities for birth defects.

Table 1.1 Causes of mortality in neonates[a] and children[b] below five years of age globally (in millions). (data from [29])

Cause	0–27 days[a]	1–59 months[b]
Preterm birth complications	6.761 (5.959–7.632)	0.793 (0.436–1.202)
Intrapartum-related events	4.562 (3.937–5.177)	0.388 (0.203–0.657)
Sepsis/meningitis	2.874 (2.005–3.739)	
Congenital anomalies	2.169 (1.861–2.735)	1.496 (1.178–1.851)
Pneumonia	1.134 (0.793–1.713)	5.455 (4.661–6.752)
Tetanus	0.243 (0.129–0.600)	
Diarrhoea	0.121 (0.072–0.504)	3.645 (2.872–4.730)
Injuries		2.341 (1.944–2.938)
Malaria		2.193 (1.613–3.237)
Meningitis		0.825 (0.652–1.157)
AIDS		0.614 (0.541–0.722)
Measles		0.529 (0.274–1.920)
Pertussis		0.387 (0.377–0.427)
Others (other causes originating during the perinatal period, cancers, malnutrition and other specified causes)	1.327 (1.017–1.719)	4.683 (3.835–5.752)

[a]The major causes of neonatal mortality are prematurity, neonatal sepsis, intrapartum birth complications (birth asphyxia and trauma) and congenital anomalies
[b]In the age group of 1–59 months, pneumonia, diarrhoea, malaria and injuries are the prevalent causes of mortality globally

Birth Defect Surveillance

Surveillance is the continuous, systematic collection of data together with its analysis and interpretation in order to provide information for planning and resource allocation. Surveillance data is used to determine the magnitude and trends of the conditions being monitored and the effectiveness of proposed interventions. Birth defects surveillance fulfils these functions by informing public health planners on the numbers and types of prevalent birth defects [31]. By describing the impact of birth defects on routine indicators such as neonatal, infant and child mortality, low birth weight and preterm births, stillbirths and numbers of children with disabilities and with special healthcare needs, birth defects surveillance helps in increasing the visibility of these conditions, and the healthcare and social service requirements. The central function of birth defects surveillance is its role in detecting teratogenic exposures. Birth defects surveillance is further discussed in Chap. 4.

Data on genetic disorders are compiled into registries (Chap. 5). Registry data are collected from specialized centres. For example, centres treating patients

(a)

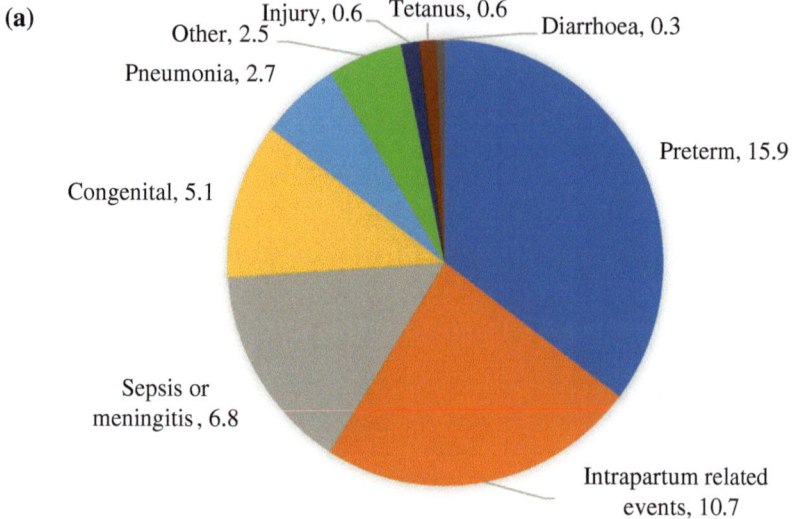

(b)

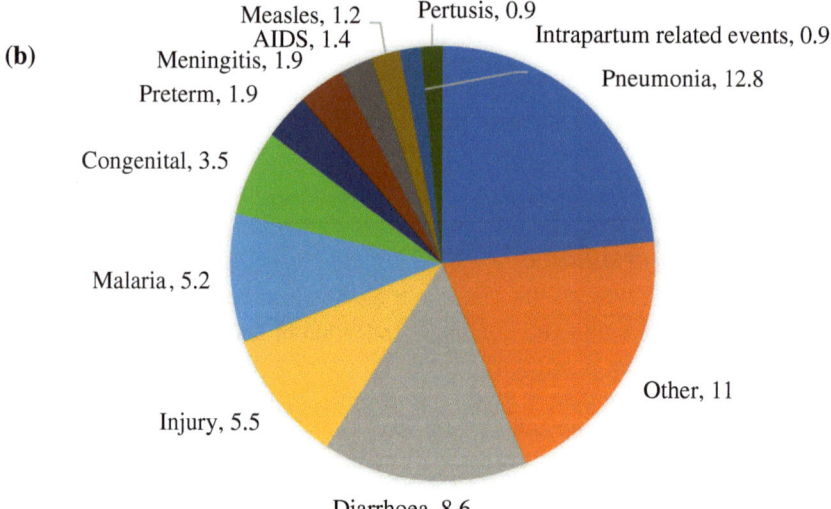

Fig. 1.4 Proportion of global child mortality by cause. (**a**) 0–27 days and (**b**) 1–59 months

with haemophilia routinely report data to a central haemophilia registry. Public health expertise is required for establishing birth defects surveillance and registries, including the selection of appropriate variables that can yield meaningful epidemiological data.

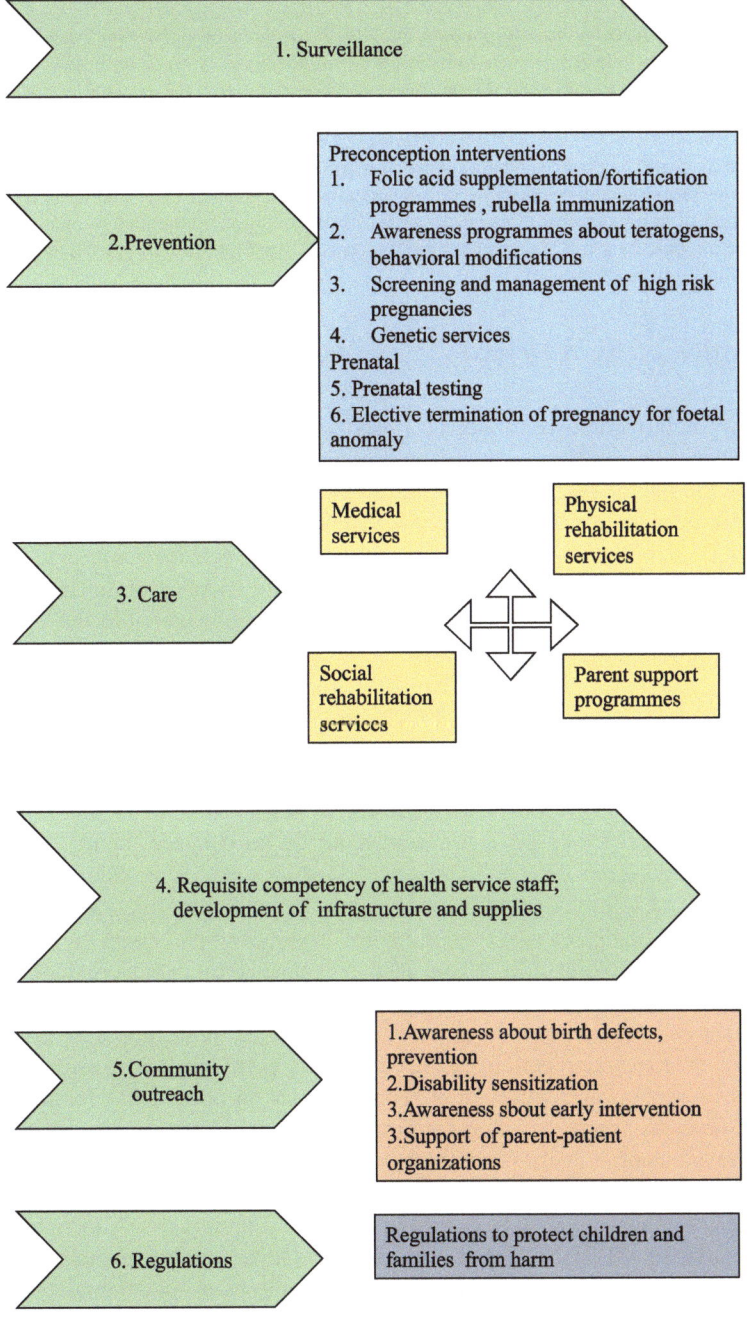

◄Fig. 1.5 Birth defects service components. Public health responsibilities of a birth defects service. These activities include surveillance to measure the numbers and types of birth defects, prevention services (which extend from the preconception period through the prenatal period), provision of care (in the form of an integrated service of medical, physical and social rehabilitation services, parent empowerment and support programmes), development of skills and competency of the public health workforce and developing appropriate infrastructure and supplies, community outreach services (including disability sensitization, awareness about early intervention services and support of non-government organizations for championing the programme), and developing appropriate regulations to protect children from physical harm (e.g. from inappropriate treatment) and families from financial harm

Prevention of Birth Defects

Majority of birth defects cannot be linked to any specific cause, but there are several environmental, genetic and gene–environmental factors that are known to increase the risk of birth defects (Chap. 9). Several modifiable risk factors can be addressed through preconception, and prenatal preventive measures. These include addressing micronutrient (folate) and elemental (iodine) deficiencies prior to conception, increasing awareness about lifestyle factors (avoidance of self-medication, abstaining from tobacco and alcohol intake), improving maternal health status (by managing maternal obesity, pregestational diabetes, hypothyroidism, epilepsy) and enhancing awareness about protection from teratogenic exposures (TORCH, Zika virus infections, teratogenic medications, chemical and physical exposures during pregnancy).

Wholly genetic conditions arise from preconception events (e.g. single gene mutations, pathogenic gene variants, heritable chromosomal anomalies) (although some chromosomal and single gene disorders occur de novo in the foetus). Report of family history, carrier detection and genetic testing services provide the opportunity for making informed reproductive choices. Prenatal ultrasonography can identify congenital malformations, providing opportunities to parents to prepare themselves for the impending birth. Where legal, pregnancy termination provides another alternative, but is associated with ethical and psychological issues.

Prevention of birth defects requires a number of strategies. The first is to target the preconception period, by increasing awareness about pregnancy planning through promotion of contraception. This provides an opportunity for women to address preconception risk factors for birth defects, including correcting folate deficiency/insufficiency. The second strategy aims at increasing awareness about the prevention of teratogenic risk exposures in the prenatal period. A third component is provision of prenatal ultrasonography and genetic testing services. Although genetic testing may be resource intensive, genetic disorders such as the haemoglobinopathies, have been controlled in several Mediterranean countries using community genetic programmes (Chap. 11). A genetic service is required to provide a choice to parents. Where abortion is legal, the prenatal diagnosis of a malformation or a genetic disorder provides parents with a choice of pregnancy termination. Nearly 60% of birth defects are preventable through various primary, secondary and tertiary prevention activities [32, 33]. It is noteworthy that although birth defects prevention activities have not been established in a majority of countries, global data show a small reduction in congenital anomalies. Such reductions may have been achieved through implementation of health interventions for other maternal and child health conditions.

Integrated medical and rehabilitation services

Medical services for birth defects and developmental disabilities extend from the prenatal period through the lifespan. Sick neonates require intensive care. Several congenital malformations require surgical correction. Most medical services are disorder-specific and long term and require careful management by specialists. Providing medical services requires speciality skills, infrastructure and resource investments which may not be available across all LMICs. Children with developmental disabilities require early intervention which includes physical rehabilitation services. Rehabilitation forms an important component of a birth defects service. Disability progression needs to be addressed by providing physiotherapy, speech/auditory services, and behavioural and occupational therapies.

Social welfare services are equally important in ensuring functional independence. Referral between medical and rehabilitation services for children with locomotor, visual, intellectual, hearing or speech impairment is required. Policies and services to ensure education and employment opportunities for children with special needs are essential for inclusion of birth defect survivors. Social welfare services such as supplemental income, providing day care and respite care are important support services for caregivers. An essential component of a birth defects service is to ensure that caregivers receive appropriate information for caring for the child, including training in activities of daily living and guidance on home management of medical conditions. Psychosocial support and counselling of caregivers and referring caregivers to parent support groups form an important component to ensure coping and mental well-being.

One of the most important characteristics of a birth defects service is to ensure that an integrated and comprehensive platform of services is available for families, with an interdisciplinary care team coordinating medical care, rehabilitation therapies and education. Services for care of children need to be done in consultation with caregivers. Ensuring that parents understand the recommended therapies is important to ensure that parents comprehend that children require multiple therapies. A major health service responsibility is to develop a strong referral system to ensure that parents can reach the needed services. It is essential that each component of the service be clearly defined in terms of functions and responsibilities. Most importantly, referral pathways to other services need to be clearly delineated, in order to converge disparate services.

Other Public Health Activities

Sensitizing communities about birth defects and developmental disabilities, and community education on early intervention, and opportunities for physical and social rehabilitation are important responsibilities of a birth defects service. Building capacity and infrastructure for addressing birth defects, developing training modules, developing guidelines and identifying and addressing infrastructural needs and medical and rehabilitation care require public health expertise. Networking with non-governmental/parent–patient organizations ensures further outreach of services. Finally, public health and health service research is integral for developing efficient, low-cost services that are aligned to existing maternal and child health programmes and utilize existing services. Public stewardship therefore is essential in ensuring the development and implementation of a birth defects service.

The World Health Assembly Resolution on Birth Defects

In 2010, the World Health Assembly adopted a resolution on birth defects, identifying these disorders as important causes of child mortality and disability. It provided a set of recommendations which form a comprehensive and feasible plan of action to address birth defects, within the frameworks of existing maternal and child health and other health and welfare programmes [34]. The resolution called for raising awareness about birth defects as significant causes of childhood disability and deaths, and the need for governments to plan and design services and dedicate funds to prevent these conditions and provide care for children. The resolution identified some feasible interventions that could be immediately implemented to prevent birth defects. These include environmental regulations to prevent exposure to teratogenic exposures and increasing maternal health promotion messages on the benefits of rubella vaccination, folic acid supplementation, and tobacco and alcohol avoidance during pregnancy. The resolution recommended the establishment of community genetic and newborn screening programmes. Concomitantly, it emphasized the importance of birth defects surveillance, developing the expertise of health service staff to address these conditions and strengthening birth defects research. It underlined the need to protect human rights of children with disabilities, foster parent–patient organizations and provide assistance and support to families in child care and child-raising efforts.

As a first step towards implementing these recommendations, it was suggested that data collection on mortality and morbidity due to birth defects should be initiated. Other activities were the development of national plans for prevention and management of birth defects within existing maternal and child health programmes, development of ethical and legal guidelines, evolving community genetic services within the primary healthcare system and promotion of research to ameliorate the impact of these conditions [34]. Overall, activities to mobilize this agenda have been slow, but training LMICs in birth defects surveillance has been an ongoing activity.

India as an Example of a LMIC

This section introduces the Indian socio-demographic and health indicators. It provides an overview of the health system structure and function, in order to illustrate the public health context in which a birth defects service has to function.

Socio-demographic and Health Indicators

The challenge of establishing a birth defects service in India is reflected in the geographic and demographic indicators of the country. India is the seventh largest nation in the world, covering 3,287,263 km^2. In 2019, India had an estimated population of 1.3 billion, which is nearly one-fifth of the global population [35]. The country is organized into 28 states and 8 union territories (UTs) (Fig. 1.6). For administrative purposes, each state is further subdivided into districts, sub-districts, towns and municipalities and villages. There are 718 districts in India, 5564 sub-districts (tehsils), 7935 towns/municipalities and 640,867 villages in the country. The population density is 464 persons per square kilometre, varying in different regions. The population of some districts may be as large as the population of a small European nation. The geographic expanse of the country and the population density underline the magnitude of the required service.

A quarter of the Indian population is below the poverty line. At the same time, the lack of economic equity in India is illustrated by the fact that the country ranks fourth in terms of number of billionaires, after the USA, China and Germany. There is a progressively expanding middle class. Among the impoverished populations, eighty percent reside in rural areas, with poverty in villages being 27% as compared to 6% in big cities. Secondary school completion is lower among the poor, and poverty is associated with higher levels of illiteracy (45%). The main source of income of the rural poor is from farm labour, and income is mainly in the form of daily wages.

In 2015–16, the labour force participation was 75% for men and 24% for women. The social composition of the country is varied. While Hindus make up the major religious group (80%), other groups include Muslims (14%), Christians (2%), Sikhs, Buddhists and Jains. There are 23 different official languages in India. The diversity of the country indicates the complexity of public health activities. A birth defects service will have to include public communication that is in the appropriate language and is culturally appropriate.

The life expectancy at birth is 68 years, being higher in urban than rural areas (Table 1.2). The total fertility rate per woman is 2.3, with a higher rate of 2.5 in rural as compared to 1.8 in urban areas. A major issue of concern in India remains the sex ratio, reported at 973 female per 1000 male babies. This skewed sex ratio is caused by the practice of sex-selective abortion. This gender bias is also evident in infant mortality data, with infant mortality rates being higher (38 per 1000 live births) in females than males (35 per 1000 live births). Other selected socio-economic and

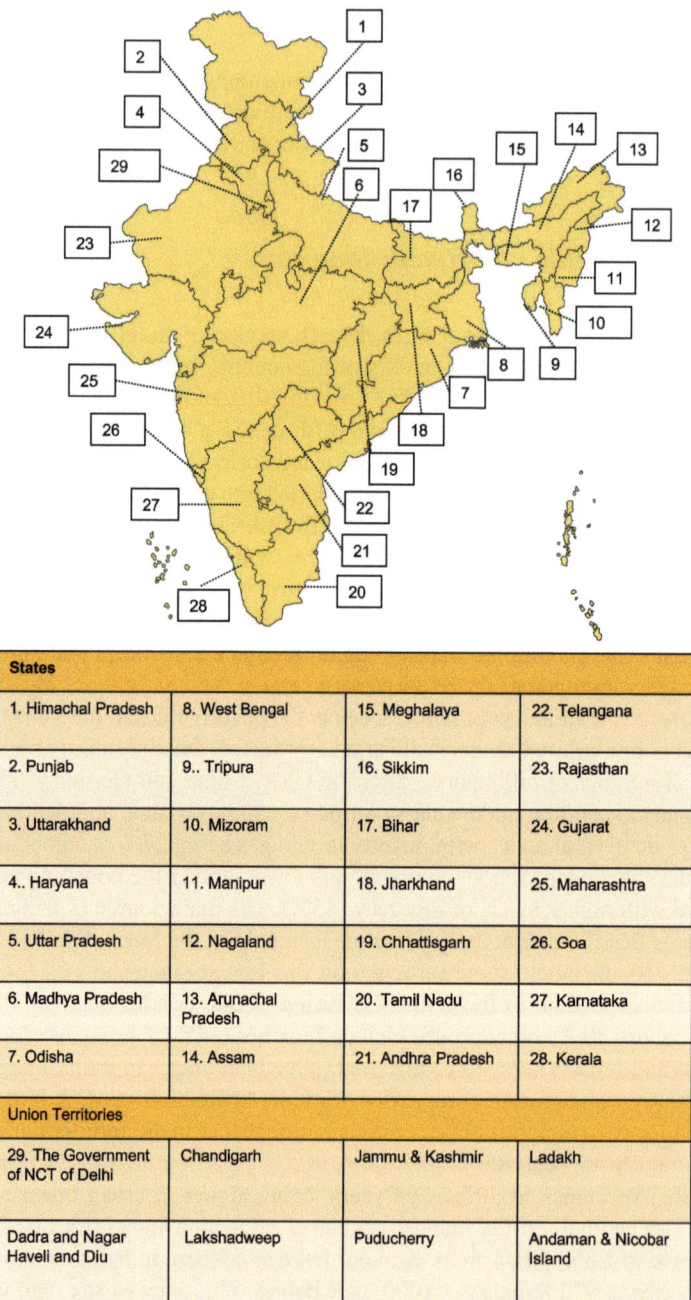

Fig. 1.6 India states and union territories. India is made up of 28 states and 8 union territories. The map is not updated for the recent division of the state of Jammu and Kashmir into 3 union territories (Jammu and Kashmir, Ladakh and Kargil)

Table 1.2 Selected demographic and health indicators of India[a]

	Indicator	
A	*Socio-economic and demographic*	
	Population (2018) [35]	1,311,051,000
	Percent population below poverty line (2011–12) [35]	21.9 (13.7 urban, 21.9 rural)
	Literacy rate in population above 15 years [36]	74.37 (82.37 male, 65.79 female)
	Median age of the population [36]	27.9 (27.2 male, 28.6 female)
	Birth rate 2016[b] [36]	20.4 (17.0 urban, 22.1 rural)
	Death rate[c] [36]	6.4 (5.4 urban, 6.9 rural)
	Expected life at birth 2011–2015 [36]	68.3 (71.9 urban, 67.1 rural)
	Mean age at marriage (2016) [36]	22.2
	Total fertility rate (2016) [36]	2.3 (1.8 urban, 2.5 rural)
	Sex ratio at birth[d] [36]	940
B	*Infrastructure* [37]	
	Households with electricity (%)	88.2 (97.5 urban, 83.2 rural)
	Households with an improved drinking water source	89.9 (91.1 urban, 89.3 rural)
	Households using improved sanitation facility	48.4 (70.3 urban, 36.7 rural)
C	*Government health facilities* [38]	
	Sub-centres	156,231
	Primary health centres	25,650
	Community health centres	5624
	Number of hospitals	23,582
	Number of beds	710,761
	Number of hospitals (rural)	19,810
	Number of beds (rural)	279,588
	Number of hospitals (urban)	3772
	Number of beds (urban)	431,173
	Number of medical colleges	476
	Undergraduate admission capacity per year [36]	52,646 (for 642 colleges)
	Institutions offering General Nurse Midwife (GNM) degree [36]	3215
	GNMs graduating annually [36]	129,926
	Private hospitals accredited by the National Accreditation Board for Hospitals and Healthcare Providers (NABH) [39]	662
D	*Healthcare cost: protection and investment*	
	Households with any usual member covered by a health scheme or health insurance [36]	28.7 (28.2 urban, 28.9 rural)

(continued)

Table 1.2 (continued)

	Indicator	
	Public expenditure on health (as % of GDP) 2015–16 [36]	1.02
E	*Antenatal care* [37]	
	Mothers who had antenatal check-up in the first trimester (%)	58.6 (69.1 urban, 54.2 rural)
	Mothers who had at least 4 antenatal care visits (%)	51.2 (66.4 urban, 44.8 rural)
	Mothers who had full antenatal care (%)	21.0 (31.1 urban, 16.7 rural)
	Institutional births (%)	78.9 (88.7 urban, 75.1 rural)
	Institutional births in public facility (%)	52.1 (54.4 urban, 46.2 rural)
	Annual numbers of births[d] (2018)	24,164,357,000
F	*Child health indicators*	
	Infant mortality rate 2016 [36]	34.0 (35.0 male, 38.0 female) (23.0 urban, 38.0 rural)
	Neonatal mortality rate [36]	24.0 (14.0 urban, 27.0 rural)
	Post-neonatal mortality rate [36]	11.0 (9.0 urban, 11.0 rural)
	Perinatal mortality rate [36]	23.0 (14.0 urban, 26.0 rural)
	Stillbirth rate [36]	4.0 (3.0 urban, 5.0 rural)
	Under five mortality rate [36]	39.4
	Children aged 12–23 months who received most of the vaccinations at public health facilities [37]	82%
	Children age 12–23 months fully immunized (BCG, measles and 3 doses each of polio and DPT) (%) [37]	62.0 (63.9 urban, 61.3 rural)
	Children under 5 years who are stunted (height for age) (%) [37]	38.4 (31.0 urban, 41.2 rural)
	Children under 5 years who are wasted (weight for height) (%) [37]	21.0 (20.0 urban, 21.5 rural)
	Children under 5 years who are underweight (weight for age) (%) [37]	35.8 (29.1 urban, 38.3 rural)

[a]Numbers in square brackets indicate references from which the data have been cited
[b]Rate per 1000 mid-year population
[c]Death rate per 1000 mid-year population (2013)
[d]Number of females per 1000 males

demographic indicators are shown in Table 1.2A, B. The low literacy rates, rural residence of a large proportion of the population and the possibility of environmental teratogenic exposures from an agricultural lifestyle, infectious agents, poor nutrition, micronutrient deficiencies and undetected chronic illness are important considerations for the birth defects service.

Socio-demographic Index and Epidemiological Transition

Indian states are at various levels of development. Table 1.3 shows the socio-demographic index (SDI) of states and union territories of India [40]. A belt of eight low SDI states, stretching across the country, make up nearly half the population of the country. These are also the states with poor health indicators.

This variation in development across states of India is one of the factors contributing to the difference in rates of epidemiological transition across the country. India is a very heterogeneous country in terms of its SDI and health indicators, so that the country has been described as "nations within a nation" [41]. The states/UTs have been categorized into four groups (epidemiological transition level, ETL group states) based on the rate of epidemiological transition, measured by the ratio of disability-adjusted life years (DALYs) of communicable, maternal, neonatal and nutritional disease burden to those from non-communicable diseases and injuries (Table 1.4). A comparison of Tables 1.3 and 1.4 shows the close correlation between SDI levels and epidemiological transition. This heterogeneity in the rate of epidemiological transition provides an opportunity for a birth defects service to be introduced in a phased manner, initiating with high and high-middle ETL states. As discussed later however, the sheer numbers of births in low SDI states imply that in absolute numbers, the affected pregnancies and births may be equal to, if not higher as compared to the higher ETL groups of states.

Table 1.3 States of India categorized by socio-demographic index[a]

High SDI states (>0.6)	Middle SDI states (0.54–0.6)	Low SDI states (≤0.53)
Total population 318 million	Total population 387 million	Total population 675 million
Uttarakhand (0.61)	Andhra Pradesh (0.54)	Bihar (0.43)
Punjab (0.62)	West Bengal (0.54)	Jharkhand (0.49)
Tamil Nadu (0.62)	Tripura (0.54)	Uttar Pradesh (0.49)
Maharashtra (0.62)	Arunachal Pradesh (0.56)	Rajasthan (0.49)
Mizoram (0.62)	Meghalaya (0.56)	Madhya Pradesh (0.49)
Sikkim (0.63)	Karnataka (0.57)	Chhattisgarh (0.51)
Nagaland (0.63)	Gujarat (0.58)	Odisha (0.52)
Himachal Pradesh (0.63)	Telangana (0.58)	Assam (0.53)
Kerala (0.66)	Manipur (0.59)	
Delhi (0.72)	Jammu and Kashmir (0.59)	
Goa (0.74)	Haryana (0.6)	

[a]Numbers in parenthesis indicate SDI scores [40]

Table 1.4 States of India categorized by epidemiological transition level group [41]

High	Higher-middle	Lower-middle	Low
Himachal Pradesh	Haryana	Arunachal Pradesh	Bihar
Punjab	Delhi	Mizoram	Jharkhand
Tamil Nadu	Telangana	Nagaland	Uttar Pradesh
Goa	Andhra Pradesh	Uttarakhand	Rajasthan
Kerala	Jammu and Kashmir	Gujarat	Meghalaya
	Karnataka	Tripura	Assam
	West Bengal	Sikkim	Chhattisgarh
	Maharashtra	Manipur	Madhya Pradesh
	Union territories other than Delhi		Odisha

Health System

India has a "mixed" health system, made up of both public and private health-care facilities and providers. The public (government) health infrastructure caters primarily to the rural population [38]. It is made up of community-level facilities (sub-centres that report to Primary Health Centres) and secondary level of services (Community Health Centres and sub-district hospitals), while district hospitals and medical college hospitals located in urban centres make up the tertiary level of medical care services (Table 1.2C). Like other LMICs, the underlying philosophy of public health services in India is based on primary healthcare, which emphasizes the prevention of diseases and the promotion of healthy behaviours. Basic health services, maternal and child health services, and services for endemic diseases and selected non-communicable diseases are offered through "vertical" programmes. Medical services are limited in availability and offered free of charge. The public health system caters to around 30% of the population, although nearly 50% of maternity services are accessed through government health services [37].

Table 1.5 shows the magnitude of the public health functions in India. For example, in the year 2017–18, the government health system conducted 2070 million haemoglobin tests [38]. There were 1603.5 million outpatient consultations across the country. India has one of the lowest public investments in health, being as low as 1% of the GDP. Less than a quarter of the population have any form of health insurance (Table 1.2D). A recent national insurance scheme (Pradhan Mantri Jan Arogya Yojana) provides insurance coverage for secondary and tertiary hospitalizations for 100 million families, representing the lowest 40% of the Indian population. Each family is eligible for a health cover of 500,000 Indian rupees per year [42].

The large population, low investment and quality issues are reasons for a flourishing private healthcare sector in India, where services are obtained on payment [43]. The private healthcare sector is marked by its heterogeneity. It has large, specialized accredited hospitals, as well as smaller hospitals, maternity hospitals and a large

Table 1.5 Government
health service performance
indicators, India (provisional
figures as on 6 January 2020)
[38]

Pregnant women registered for ANC	29 million
Pregnant women receiving four ANCs	180.4 million
Pregnant women receiving 180 iron and folic acid tablets	22.4 million
Live births	20.8 million
Outpatient consultations	1603.5 million
Admitted patients	75 million
Major surgeries	0.43 million
Minor surgeries	11.5 million
Haemoglobin tests conducted	2070 million

number of private providers. The complexity of the health system is enhanced by plural systems of medical practice, referred to as the AYUSH (ayurveda, yoga and naturopathy, unani, siddha and homeopathy) systems of medicine. While government health services and programmes undergo periodic review, the quality and types of private sector services are rarely reviewed. Furthermore, these hospitals are non-compliant to data reporting. Thus, health data of India are either those reported among users of public facilities (representing the lower socio-economic strata) or those obtained from national surveys or ad hoc studies.

The mixed health system structure of India has important implications for birth defects surveillance (Chap. 4). As use of private hospitals is through personal expenditure, there is considerable population mobility for choice of hospital for delivery. The catchment area of hospitals is very difficult to define. It is also difficult to draw out a random sample of private hospitals, as currently, there is no centralized database of all hospitals other than large accredited hospitals. The limitation in data is most acute in smaller towns and villages.

Maternal and Child Health Services

Table 1.5 identifies the enormity of the maternal and child health services in India. In 2017–18, 29 million pregnant women registered for antenatal care (ANC) across the country, and there were over twenty million births. With congenital anomalies affecting 2–3% of births, over 500,000 pregnancies are likely to be affected each year. Table 1.2E shows that 50% of births occur at government health facilities (where data are recorded and reported), while the remaining births take place at private hospitals and maternity homes (which do not report data). There are few home births in India (3.2%) [37], indicating improving services. The majority of child immunization (82%) takes place at government health centres [37]. In addition to immunization services, other child services include routine monitoring of child growth parameters (there is no developmental monitoring through government health

Table 1.6 Estimated number of deaths in the neonatal period and in children in the age group of 1–59 months in India (2015) (from [44])

Neonatal deaths		Post-natal deaths	
Neonatal deaths	695,852	Post-natal deaths	505,146
Preterm birth complications	296,177	Pneumonia	143,228
Intrapartum-related events	135,303	Diarrhoea	112,277
Sepsis/meningitis	100,451	Injuries	39,478
Congenital	76,470	Congenital	30,253
Pneumonia	36,497	Preterm birth complications	24,851
Tetanus	5973	Measles	22,897
Diarrhoea	5112	Meningitis	21,106
Other disorders	39,870	Malaria	6627
		Intrapartum-related events	6213
		Pertussis	4164
		AIDS	4036
		Other disorders	90,016

services as of now), feeding programmes and free of cost treatment for common childhood ailments.

In terms of a birth defects service, training of nurse midwives at government health facilities to identify a newborn with a birth defect is imperative, as half of all births occur in public facilities. Referral services for care are needed, as well as a surveillance system for recording these births. It should be noted that recording data on birth defects is likely to be more systematic at government facilities. For the remaining population accessing services from the private health sector, establishing surveillance is a more challenging prospect.

Table 1.6 shows that nearly half of the deaths in children under five years of age occur during the neonatal period. Birth defects are the fourth leading cause of mortality, but the proportion of deaths caused by preterm birth complications and intrapartum-related events are considerably higher [44]. Pneumonia and diarrhoea are the major causes of child deaths. Health service attention is focussed on the reduction of these major causes of mortality, as a result of which birth defects are less prioritized than the other causes of mortality.

Several services for the care of children with birth defects and developmental disabilities are already in place in India, including the community-based child screening and early intervention services of the Rashtriya Bal Swasthya Karyakram (RBSK) [45] (Chap. 12). Birth defects prevention services are not yet a component of the maternal and child health service (Chap. 9), and a systematic birth defects surveillance system has not yet been implemented. Current data collection systems do not stringently investigate neonatal and child mortality due to congenital causes. Of greater relevance is that there are scarce data on the numbers of children surviving with congenital disabilities and the quality of life of families.

For genetic disorders, guidelines for a community-based genetic programme for haemoglobinopathies have been announced [46] (Chap. 11), and free clotting factor concentrate is being provided to patients with haemophilia. A rare disease policy has been drafted but not implemented [47] (Chap. 5).

A large but poorly implemented service for persons with disabilities is available, but there is no referral linkage with the RBSK programme. Although scattered, these services have the potential to be developed into a cohesive programme of services for birth defects and developmental disabilities in India.

References

1. Christianson A, Howson CP, Modell B (2006) March of Dimes Global Report on Birth Defects: the hidden toll of dying and disabled children. March of Dimes Birth Defects Foundation, White Plains, New York
2. World Health Organization (2015) Birth defects surveillance: a manual for programme managers. In: Birth defects surveillance: a manual for programme managers. Available via https://www.who.int/nutrition/publications/birthdefects_manual/en/
3. Boyle CA, Cordero JF (2005) Birth defects and disabilities: a public health issue for the 21st century. Am J Public Health 95:1884–1886
4. McPherson M, Arango P, Fox H et al (1998) A new definition of children with special health care needs. Pediatrics 102(1):137–139
5. Lawn JE, Blencowe H, Pattinson R, Committee LSSS et al (2011) Stillbirths: Where? When? Why? How to make the data count? Lancet 377(9775):1448–1463
6. Lemacks J, Fowles K, Mateus A, Thomas K (2013) Insights from parents about caring for a child with birth defects. Int J Environ Res Public Health 10(8):3465–3482
7. Farrell M, Corrin K (2005) The stigma of congenital abnormalities. In: Mason T, Carlisle C, Watkins C, Whitehead E (eds) Stigma and social exclusion in healthcare. Routledge, pp 69–80
8. Murphy NA, Christian B, Caplin DA, Young PC (2007) The health of caregivers for children with disabilities: caregiver perspectives. Child Care Health Dev 33(2):180–187
9. Neely-Barnes SL, Dia DA (2008) Families of children with disabilities: a review of literature and recommendations for interventions. J Early Intensive Behav Interv 5(3):93–107
10. Zuurmond M, Nyante G, Baltussen M et al (2019) A support programme for caregivers of children with disabilities in Ghana: understanding the impact on the wellbeing of caregivers. Child Care Health Dev 45(1):45–53
11. Whitney DG, Shapiro DN, Peterson MD, Warschausky SA (2019) Factors associated with depression and anxiety in children with intellectual disabilities. J Intellect Disabil Res 63(5):408–417
12. Farr SL, Oster ME, Simeone RM, Gilboa SM, Honein MA (2016) Limitations, depressive symptoms, and quality of life among a population-based sample of young adults with congenital heart defects. Birth Defects Res A Clin Mol Teratol 106(7):580–586
13. Wehby GL, Cassell CH (2010) The impact of orofacial clefts on quality of life and healthcare use and costs. Oral Dis 16(1):3–10
14. Jones L, Bellis MA, Wood S et al (2012) Prevalence and risk of violence against children with disabilities: a systematic review and meta-analysis of observational studies. Lancet 380(9845):899–907
15. Cruz-Inigo AE, Ladizinski B, Sethi A (2011) Albinism in Africa: stigma, slaughter and awareness campaigns. Dermatol Clin 29(1):79–87
16. Dharmarajan S, Phadnis S, Gund P, Kar A (2014) Out-of-pocket and catastrophic expenditure on treatment of haemophilia by Indian families. Haemophilia 20(3):382–387

17. Dharmarajan S, Phadnis S, Gund P, Kar A (2012) Treatment decisions and usage of clotting factor concentrate by a cohort of Indian haemophilia patients. Haemophilia 18:27–29
18. Health Resources and Service Administration Maternal and Child Health Bureau NS-CSHN. Chartbook 2009–2010 illustrated findings from the national survey of children with special health care needs. Available at https://mchb.hrsa.gov/cshcn0910/population/pages/pc/pcih.html. Accessed Aug 2018
19. National Birth Defects Prevention Network (2004) NDPN guidelines for conducting birth defects surveillance article 3 case definition rev. 06/04. Available via https://www.nbdpn.org/current/resources/sgm/Ch_3_Case_Definition6-04%20no%20app.pdf. Accessed 21 Mar 2018
20. EUROCAT European congenital anomalies registry? Available via https://eu-rd-platform.jrc.ec.europa.eu/eurocat. Accessed Mar 2018
21. Latin American collaborative study of congenital malformations (ECLAMC)? Available via https://www.eclamc.org/eng/index.php. Accessed Mar 2018
22. World Health Organization Regional Office for South-East Asia (2015) Neonatal–perinatal database and birth defects surveillance report of the regional review meeting, New Delhi, 19–21 Aug 2014. Available via https://apps.searo.who.int/PDS_DOCS/B5227.pdf. Accessed Mar 2018
23. International clearinghouse for birth defects surveillance and research? Available via https://www.icbdsr.org/. Accessed Mar 2018
24. World Health Organization congenital anomalies facts sheets? Available via https://www.who.int/en/news-room/fact-sheets/detail/congenital-anomalies. Accessed Mar 2018
25. Modell B, Darlison MW, Malherbe H, Moorthie S, Blencowe H, Mahaini R, El-Adawy M (2018) Congenital disorders: epidemiological methods for answering calls for action. J Community Genet 9:335–340
26. Darmstadt GL, Howson CP, Walraven G et al (2016) Prevention of congenital disorders and care of affected children: a consensus statement. JAMA Pediatr 170(8):790–793
27. Bhide P, Gund P, Kar A (2016) Prevalence of congenital anomalies in an Indian maternal cohort: healthcare, prevention, and surveillance implications. PLoS ONE 11(11):e0166408. https://doi.org/10.1371/journal.pone.0166408
28. Christianson A, Modell B (2004) Medical genetics in developing countries. Annu Rev Genomics Hum Genet 5:219–265
29. Liu L, Oza S, Hogan D et al (2016) Global, regional, and national causes of under-5 mortality in 2000–15: an updated systematic analysis with implications for the Sustainable Development Goals. Lancet 388(10063):3027–3035
30. Ujagare D, Kar A (2020) Birth defect mortality in India 1990–2017: estimates from the Global Burden of Disease data. J Community Genet 12:81–90
31. Hug L, Alexander M, You D et al (2019) National, regional, and global levels and trends in neonatal mortality between 1990 and 2017, with scenario-based projections to 2030: a systematic analysis. Lancet Glob Health 7(6):e710–e720
32. Czeizel AE, Intody Z, Modell B (1993) What proportion of congenital abnormalities can be prevented? BMJ 306(6876):499–503
33. Czeizel AE (2005) Birth defects are preventable. Int J Med Sci 2(3):91
34. Sixty-third World Health Assembly WHA63.17 agenda item 11.7 21 May 2010 birth defects. Available via https://apps.who.int/gb/ebwha/pdf_files/WHA63/A63_R17-en.pdf?ua=1. Accessed Mar 2018
35. UNICEF country profile. India Available via https://data.unicef.org/country/ind/. Accessed Mar 2018
36. Central Bureau of Health Intelligence, Directorate General of Health Services, Ministry of Health and Family Welfare, Government of India. Source national health profile 2018. Available via file:///Users/anita/Downloads/NHP%202018.pdf. Accessed Mar 2018
37. International Institute for Population Sciences (IIPS) and ICF (2017) National family health survey (NFHS-4), 2015–16: India. IIPS, Mumbai. Available via https://rchiips.org/nfhs/pdf/NFHS4/India.pdf. Accessed Mar 2018

38. Ministry of Health and Family Welfare. Health management information system. Available via https://nrhm-mis.nic.in/SitePages/Home.aspx. Accessed Feb 2020
39. National Accreditation Board for hospitals and healthcare providers? Available via https://www.nabh.co/frmViewAccreditedHosp.aspx. Accessed Mar 2018
40. Sagar R, Dandona R, Gururaj G et al (2020) The burden of mental disorders across the states of India: the Global Burden of Disease Study 1990–2017. Lancet Psychiatry 7(2):148–161
41. Dandona L, Dandona R, Kumar GA et al (2017) Nations within a nation: variations in epidemiological transition across the states of India, 1990–2016 in the Global Burden of Disease Study. Lancet 390(10111):2437–2460
42. National Health Authority. Ayushman Bharat Pradhan Mantri Jan Arogya Yojana (PM-JAY)? Available via https://pmjay.gov.in/about/pmjay. Accessed Mar 2018
43. Mackintosh M, Channon A, Karan A, Selvaraj S, Cavagnero E, Zhao H (2016) What is the private sector? Understanding private provision in the health systems of low-income and middle-income countries. Lancet 388(10044):596–605
44. Liu L, Chu Y, Oza S et al (2019) National, regional, and state-level all-cause and cause-specific under-5 mortality in India in 2000–15: a systematic analysis with implications for the Sustainable Development Goals. Lancet Glob Health 7(6):e721–e734
45. Government of India, Ministry of Health and Family Welfare, National Health Mission (2013) Rashtriya Bal Swasthya Karyakram. A child health screening and early intervention services under NRHM. Ministry of Health and Family Welfare. Available at https://nhm.gov.in/images/pdf/programmes/RBSK/Operational_Guidelines/Operational%20Guidelines_RBSK.pdf. Accessed 6 Aug 2019
46. Ministry of Health and Family Welfare, Government of India. Guidelines on haemoglobinopathies in India, 2016. Available via https://nhm.gov.in/images/pdf/programmes/RBSK/Resource_Documents/Guidelines_on_Hemoglobinopathies_in%20India.pdf. Accessed Mar 2020
47. Ministry of Health and Family Welfare, Government of India. National policy for treatment of rare diseases? Available via https://main.mohfw.gov.in/sites/default/files/Rare%20Diseases%20Policy%20FINAL.pdf. Accessed Mar 2020

Chapter 2
Some Common Birth Defects

Anita Kar

Abstract This article provides brief descriptions of some common birth defects, genetic disorders and developmental disabilities, their prevalence and health and rehabilitation implications. The conditions have been described in three broad groups. The first group includes some common congenital anomalies, which are structural birth defects caused by improper development of one or more body structures of the fetus. Congenital heart defects, orofacial clefts, neural tube defects, congenital talipes equinovarus, developmental dysplasia of the hip and limb defects are described. Some of these conditions can be treated, but others have cosmetic implications and cause disability. The second group of conditions include common developmental disabilities that cause intellectual impairment, sensory deficits, and impairment in socio-behavioral functioning. Down syndrome, intellectual disability disorders, congenital hearing and vision impairments, autism and attention-deficit hyperactivity disorders are illustrated. Physical rehabilitation and social welfare services are imperative. The third group of conditions is the single gene disorders (hemoglobinopathies, hemophilia, muscular dystrophy and achondroplasia). The severity of these conditions and lifelong dependence on medical care for survival is described. The article illustrates that although birth defects services need to have disorder-specific medical services, physical rehabilitation and social welfare activities would be a common need for many children with disabilities and disabling conditions.

Keywords Birth defects · Congenital malformations · Developmental disabilities · Single gene disorders · Public health · Epidemiology · Rehabilitation

Congenital and developmental disorders (congenital anomalies, genetic disorders and developmental disabilities) are lifelong and involve higher than normal usage of medical, surgical and rehabilitation services. This article illustrates the health care and rehabilitation needs of children with these disorders, by providing overviews of some common conditions. Child health services of most low- and middle-income countries (LMICs) do not have dedicated services for these disorders. The focus is on

A. Kar (✉)
Birth Defects and Childhood Disability Research Centre, Pune 411020, India

© Springer Nature Singapore Pte Ltd. 2021
A. Kar (ed.), *Birth Defects in India*,
https://doi.org/10.1007/978-981-16-1554-2_2

31

providing immunization, monitoring child growth, addressing nutritional deficiencies and providing services for common acute, treatable childhood diseases. These include oral rehydration therapy for diarrhea, anti-malarial treatment, de-worming activities, and routine treatment for common childhood diseases (Table 2.1). These services are also required by children with birth defects, who make up a special health service group (children with special healthcare needs, CSHCN) CSHCN have chronic medical, physical, developmental, behavioral or emotional conditions, and therefore require specific health and rehabilitation services [1].

This article illustrates these special health care needs for children with some common birth defects, genetic disorders and developmental disabilities. The first section describes five common congenital malformations, which are structural defects that arise during fetal development. The second section describes some common developmental disabilities. The final section describes some common single gene (genetic) disorders that require specialized medical management and may involve frequent hospitalization. Table 2.2 summarizes the prevalence, risk factors, recurrence risk and medical and rehabilitation care needs for these conditions.

Table 2.1 Child health services in India

Newborn care	Essential newborn care, emergency medical care
Immunization services	At birth OPV zero dose, BCG, Hep-B birth dose, OPV 1, 2 and 3 at 6, 10, 14 weeks, Pentavalent 1, 2, 3 at 6, 10, 14 weeks, Rotavirus vaccine at 6, 10, 14 weeks, fIPV at 6, 14 weeks Measles/MR 1st at 9 completed—12 months First dose of Vitamin A at 9 completed months, together with measles/MR Booster dose of DPT, second measles/MR and OPV booster between 16 and 24 months, Vitamin A doses every six months till 5 years of age Second DT booster between 5 and 6 years TT 10–16 years
Nutrition services	Mid-day meal programme, supplementary calories for children in government play schools and schools
Other	De-worming, iron and folic acid supplementation
Medical care	At community-level primary health centers and at government hospitals for conditions that cannot be managed at the primary care level Community-based screening and early intervention for common diseases, nutritional deficiencies, birth defects and developmental disabilities (Rashtriya Bal Swasthya Karyakram, RBSK) Medical care for common childhood conditions, Nutrition Rehabilitation Centres and District Early Intervention Centres for children with deficiencies, birth defects and developmental delays

OPV = oral polio vaccine, BCG = Bacillus Calmette Guerin, Hep-B = hepatitis B, fIPV = fractional inactivated poliomyelitis vaccine; pentavalent is a combined vaccine against diptheria, pertussis, tetanus, hepatitis B and Haemophilus influenzae; measles-rubella is being introduced in a phased manner across the country; pneumococcal conjugate vaccine (PCV) is being provided in selected states

Table 2.2 Characteristics of some common birth defects[a]

	Prevalence	Consequence of suboptimal or no treatment	Diagnosis[b]	Medical/surgical services	Rehabilitation needs	Prevention	Recurrence risk
1 Congenital heart defect	9.1 per 1000 live births	Mortality, poor growth and development, repeated episodes of morbidity, chest pain, dyspnea on exertion, syncope and fainting, death	At least chest X-ray, echo-cardiography	Surgery, medical management	No, but some data suggest neuromotor deficits in patients	Risk factors include maternal smoking, overweight or obesity, maternal illness, teratogenic drug use	2–5% in isolated cases, 50% for autosomal dominant inheritance, associated with several chromosomal anomalies such as Down syndrome
2 Orofacial cleft (isolated)	1 in 700 births, ethnic variation	Facial disfigurement, emotional and psychosocial distress, difficulty in speech and feeding, inner ear infections, regurgitation	Oral examination of the newborn	Surgical, in two or more stages, beginning between 3 and 6 months of age	Speech therapy, counseling and support	Maternal smoking before pregnancy increases risk by 1.3 times. Folate in the preconception period and during first trimester decreases risk	Increased 32 times for cleft lip, 56 times for cleft palate among first degree relatives

(continued)

Table 2.2 (continued)

	Prevalence	Consequence of suboptimal or no treatment	Diagnosis[b]	Medical/surgical services	Rehabilitation needs	Prevention	Recurrence risk
3	Spina bifida (myelomeningocele)	1 in 1000 births, ethnic variation	Locomotor impairment, weakness and paralysis of the legs, pressure sores, urinary and bowel incontinence, repeated urinary tract infection or constipation	Prenatal ultrasound or visible at birth	Sometimes, surgery in the neonatal period	Physical therapy and other interventions for locomotor impairment	Dietary folate deficiency, maternal overweight and obesity	Recurrence increased by almost 20–50% above the general population
4	Congenital talipes equinovarus	1 in 1000 live births, ethnic variations	Locomotor disability	Physical examination	Ponseti method of foot manipulation, casting	Physiotherapy	Maternal smoking during pregnancy	Recurrence risk increases by 2–3% when one parent is affected, 15% when both parents are affected

(continued)

Table 2.2 (continued)

	Prevalence	Consequence of suboptimal or no treatment	Diagnosis[b]	Medical/surgical services	Rehabilitation needs	Prevention	Recurrence risk
5	Developmental dysplasia of the hip	1–2 per 1000 through physical examination, increases to 5–30 per 1000 with ultrasound screening of hips	Locomotor disability, rarely causes pain in childhood, waddling gait; in adulthood chronic knee and backpain, limp because of shortened leg length	Physical examination at birth, Ortolani and Barlow maneuvers, hip ultrasonography for high risk infants, screening of children till walking is established	90% of unstable hips resolve without treatment by 12 weeks, intervention, i.e., Pavlik harness. Delayed treatment causes complications, surgery for severe cases	Assistive devices if required	Careful screening of all children till walking is established	Positive family history increases the risk 12-fold, risk to sibling of affected child is 6%, 12% if a parent has DDH and 36% (1 in 3) for a subsequent pregnancy if both parent and child have DDH
6	Congenital limb defects	4 and 2 per 10,000 for upper and lower limbs, respectively	Disabling, require assistance in activities of daily living, limitation in participation in some types of activities, psychosocial impact	Physical	Prosthetics (if required) and medical management if associated with other anomalies	Prosthetics	Genetic, teratogenic drugs like thalidomide, retinoic acid	Recurrent risk if of monogenic origin

(continued)

Table 2.2 (continued)

	Prevalence	Consequence of suboptimal or no treatment	Diagnosis[b]	Medical/surgical services	Rehabilitation needs	Prevention	Recurrence risk
7	Down syndrome	1 in 800, risk increases with increasing maternal age	Intellectual disability, frequently congenital heart defects	Physical characteristics, karyotyping	Rehabilitation therapies, medical management of comorbidities	Early intervention, with appropriate physical, occupational and speech therapies, special education, social service support	Awareness on association with maternal age and family history, access to family planning services, triple/quadruple test, ultrasonography	Older women at increased risk
8	Attention-deficit hyperactivity disorder	7.2% prevalence	Disruptive behavior, poor social skills and peer relationships	Inattention, hyperactivity and impulsivity	Medication and behavioral therapy	Behavioral therapy	History of affected sibling or parent, environmental toxins like lead, maternal smoking, alcohol, non-therapeutic drugs, prematurity	First degree relatives at five times increased risk

(continued)

Table 2.2 (continued)

		Prevalence	Consequence of suboptimal or no treatment	Diagnosis[b]	Medical/surgical services	Rehabilitation needs	Prevention	Recurrence risk
9	Autism spectrum disorders	1–2%, 3–4 times higher in boys	Impaired social behavior with characteristic signs	Severe disability in social behavior	Medication and behavioral therapy	Behavioral, speech, occupational therapies and social skills training	Prenatal factors (teratogenic drug exposures, viruses like rubella), perinatal factors (low birth weight, abnormally short gestation, birth asphyxia) and post-natal factors (auto-immune disease, viral infections, hypoxia)	High, 50–100 times higher for subsequent pregnancies
10	Cerebral palsy	1.5–2.5 per 1000 live births, 3 per 1000 live births in 4–48 year age group	Progressive disability, early mortality	Movement disorder, may be accompanied with intellectual disability, behavior disorders, visual/hearing impairment and seizure disorders	Medical and surgical interventions, concurrent with rehabilitation therapies	Physical, occupational, speech therapy	Preterm birth, intra-uterine disorders, neonatal encephalopathy, kernicterus, perinatal asphyxia, stroke and CNS infections	

(continued)

Table 2.2 (continued)

	Prevalence	Consequence of suboptimal or no treatment	Diagnosis[b]	Medical/surgical services	Rehabilitation needs	Prevention	Recurrence risk
11 Congenital hearing loss	1.33 per 1000 live births for congenital bilateral permanent hearing loss	Poor quality of life, inability to reach individual potential	Inability to respond to sound		Appropriate hearing technology (hearing aids or cochlear implants), auditory-verbal therapy, sign language, family counseling and support	Multifactorial etiology, single gene disorders, prematurity, low birth weight, developmental delay, craniofacial anomalies, admission to neonatal intensive care unit, pregnancy acquired infections, CMV, rubella, syphilis, herpes simplex virus, *Toxoplasmosis gondii*	18% probability for deafness in children for hearing couple with one deaf child and no family history
12 Congenital cataract	0.63–9.74 per 10,000 births	Visual impairment/blindness	Visual impairment	Early detection followed by intraocular lens implant		Monogenic conditions, genetic syndromes or rubella and CMV infections	

(continued)

Table 2.2 (continued)

		Prevalence	Consequence of suboptimal or no treatment	Diagnosis[b]	Medical/surgical services	Rehabilitation needs	Prevention	Recurrence risk
13	Sickle cell disease	1–5% of the global population, variation in prevalence by regions and ethnicity	Mortality in childhood	Failure to thrive, with feeding problems, frequent bouts of fever, diarrhea and distended abdomen caused by splenomegaly	For thalassemia, transfusion, iron chelation, management of comorbidities. For sickle cell disease, management of infections, hydroxyurea, pain relief		Consanguinity, endogamous communities, ethnicity	25% risk of affected birth for each pregnancy, 50% risk of carrier
	Duchenne and Becker muscular dystrophies	1/3300 (DMD) 1/18,000–1/31,000 (BMD) males	Progressive degeneration in muscle function	Clinical, Gower's sign, mutation analysis	Supportive treatment	Physical therapy		X-linked recessive inheritance[c]
	Achondroplasia	1 in 25,000–30,000 live births	Psychosocial, musculoskeletal complications	Disproportionate size, short limbs, macrocephaly	For complications	Psychosocial support	FGFR3 gene mutation, older paternal age, radiation, chemotherapy	Autosomal dominant, 0.23% (that is 1 recurrence per 443 in siblings of a sporadic case)

(continued)

Table 2.2 (continued)

	Prevalence	Consequence of suboptimal or no treatment	Diagnosis[b]	Medical/surgical services	Rehabilitation needs	Prevention	Recurrence risk
Hemophilia	1 in 5000 (hemophilia A) 1 in 30,000 (hemophilia B) male births	Mortality in the first decade of life	Hemorrhagic episodes, hematological measures	Replacement therapy with clotting factor concentrate	Physiotherapy to restore joint function	Genetic	X-linked recessive inheritance[c]

[a]The numbers may be specific for ethnic groups. The reader is referred to the text and cited references for further details
[b]Refers to the minimum first step in identifying the condition
[c]In X-linked recessive inheritance, there is a 50% risk of the male child being affected; 50% of female children will be carriers of the disorder

Section I : Common Congenital Anomalies/Malformations

Congenital anomalies (malformations) are caused by anomalous organogenesis during fetal development. Several malformations can be surgically corrected or corrected through appropriate treatment. Many children (such as those with congenital heart defects, CHDs) might require lifelong supervision by medical specialists. Others such as congenital talipes equinovarus (CTEV) and developmental dysplasia of the hip (DDH) require orthopedic management. Availability and treatment compliance can ensure disability-free life. Yet other structural birth defects, like orofacial clefts, cause cosmetic disfigurement and psychological distress if left untreated. Congenital limb defects where upper and/or lower limbs may be reduced or missing cannot be corrected. Prosthetic devices, counseling and social rehabilitation measures can improve functioning.

Congenital Heart Defects

Congenital heart defects (CHD) are malformations in the structure of the heart and/or great vessels (that is the major arteries bringing and carrying away blood from the heart) [2]. CHDs arise during fetal development and affect cardiac functioning. They may be detected before or at birth, or later in childhood [2]. CHDs are the most frequently encountered birth defect, affecting nearly 1% of births, and responsible for 30% of all birth defect mortality. Globally, the birth prevalence of CHDs is 9.1 per 1000 live births (95% CI 9.0–9.2), with significant geographical variations [3]. The highest CHD prevalence is reported from Asia (9.3 per 1000 live births). Several types of CHDs are treatable, but treatment is costly [4]. It requires specialists like cardiologists and pediatric surgeons. These skills may not be available in LMICs. Most CHD deaths occur in LMICs [5], possibly due to lack of availability and access to care. Where free-of-charge diagnostic and treatment facilities are unavailable, caregivers may experience considerable out-of-pocket expenditure in availing treatment from private health care facilities.

Congenital heart defects result in either   inter-mixing of deoxygenated and oxygenated blood, or reduction in the flow of blood. Based on the type of defect, CHDs can be broadly divided into acyanotic and cyanotic defects [2]. Acyanotic defects are of two main types, namely septal defects (like ventricular or atrial septal defects) or obstructive cardiac anomalies (such as aortic or pulmonary stenosis and coarctation of the aorta). Septal defects are caused by incomplete closure of the septum (the partitioning wall) between the upper (atrial) or the lower (ventricular) sections of the heart. Ventricular septal defects (VSD) are common and account for 20–25% of CHDs. Atrial septal defects (ASD) account for 5–10% of CHDs. The second group of acyanotic defects is obstructive cardiac anomalies. These defects arise from narrowing (stenosis) or blockage (atresia) of the great vessels (the major veins and arteries and associated structures) of the heart. Common obstructive cardiac anomalies are aortic stenosis, caused by narrowing of the aortic valve. Aortic stenosis accounts for 5% of CHDs. Pulmonary stenosis, found in 5–8% of CHDs, is caused by narrowing of the pulmonary valve. Coarctation of the aorta, found in around 8% of CHDs, is caused by narrowing of the aorta that blocks blood flow.

Cyanotic heart defects are complex structural defects of the heart. Due to poor oxygenation, children present with a blue discoloration of the skin (cyanosis) that is especially visible around the lips. There are several types of defects such as Tetralogy of Fallot (occurring in 10% of CHDs), transposition of great vessels (found in 5% of CHDs), less common conditions like pulmonary and tricuspid atresia, hypoplastic left heart syndrome and other complex structural defects of the heart.

The time of diagnosis of CHDs varies. Some defects can be detected prenatally, or they may be detected after birth. In LMICs, where health care providers are insufficiently trained, symptoms may be overlooked, and the condition may remain undiagnosed for a considerable period of time. Common diagnostic investigations involve echo-cardiography and chest X-ray. Limited access to such facilities in LMICs may also account for under-diagnosis and under-reporting of the true prevalence of CHDs. Treatment is based on the type of the defect and its severity. Small VSDs, especially muscular defects, may close spontaneously during infancy. Around 10–15% of neonates with critical congenital heart disease need surgery or intensive medical management in early life. Untreated neonates have a high risk of mortality. Open heart surgery is the effective method for repair of heart defects. Surgery may be delayed as per medical judgement, and children may be put on treatment such as diuretics that remove excess fluids from the body and lungs (to reduce the functional load on the heart), digoxin (which enhances the pumping action of the heart) and angiotensin-converting enzyme (ACE) inhibitors which relax blood vessels and assist the heart to pump blood with less effort. For obstructive cardiac defects, cardiac catheterization is used to open narrow vessels or open valves, avoiding the need for open heart surgery [2].

The major public health role is to make available free-of-charge treatment for CHDs, as several studies have shown that CHDs not only affect the child, but also affect the quality of life (QoL) of caregivers [6, 7]. Parenting stress initiates from diagnosis and continues throughout the treatment process. Parents require to understand the diagnosis and arrange appropriate treatment. Treatment of children till surgery is also expensive, as it involves several medications. Dietary habits (such as low salt intake, limiting liquid intake) may be necessary. Medication may be lifelong, even after surgery. There is evidence that children with CHDs are at an increased chance of neurodevelopmental impairments (mild cognitive impairment, fine and gross motor skills, functioning, visual, communication and social skills, [8]). Many adults survive with functional deficits of cognition, attention and behavior that limit educational achievement and employment opportunities.

Although prevention is an important public health activity, CHDs have a multifactorial etiology, where several low penetrant genes interact with environmental exposures to cause the defect [9, 10]. Non-genetic risk factors that increase the risk of

CHDs include maternal illness (e.g., diabetes mellitus, rubella, systemic lupus erythematosus and preterm preeclampsia), teratogenic drug use during pregnancy (anticonvulsants, retinoic acid), and older maternal age [11]. Maternal smoking during the first trimester is a strong risk factor for selected CHDs, with the risk increasing proportionately with maternal age and number of cigarettes smoked [12]. The risk of congenital heart defects increases proportionately with increasing body mass index, with the risk being higher among overweight and obese mothers [13]. The evidence for maternal peri-conceptional multivitamin supplementation containing folic acid and reduction in birth prevalence of CHDs is debated [14]. CHDs have a strong heritable basis, with genetic etiology attributable to 20% of cases. CHD is found in several genetic disorders such as Down syndrome, trisomies 13, 18 and Turner syndrome. The recurrence risk in isolated cases is 2–5%, but may be as much as 50% where the genetic determinant is transmitted in an autosomal dominant manner [2]. Universal screening for critical congenital heart disease using pulse oximetry has been recommended for newborns prior to discharge. This public health activity is only possible when the screening can be accompanied by treatment services, which are yet to be put in place in many LMICs.

Orofacial Clefts

Several types of birth defects are caused by abnormal growth of the bones of the head and/or face of the developing fetus [15]. The most common among these defects is orofacial clefts (OC) (Fig. 2.1). OCs are caused by improper development of only the lip (cleft lip), lip and palate (cleft lip and palate) or only palate (cleft palate) [16]. OCs may be syndromic that is they are caused by chromosomal abnormalities,

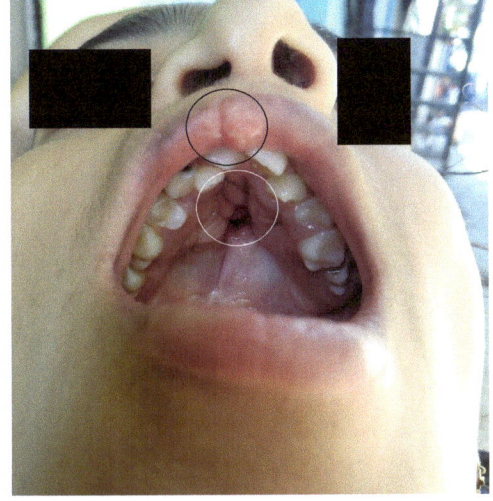

Fig. 2.1 Orofacial cleft. The patient has had one surgery to repair the cleft lip (marked with a black circle). The child was waiting for the second surgery to repair the cleft palate (marked with a white circle)

monogenic syndromes, or associated with other congenital syndromes or multiple congenital anomalies. Almost 70% of OCs are non-syndromic/isolated cases, that is, they appear without any other anomaly.

In addition to their severe cosmetic effects, OCs have several functional consequences. At birth, neonates with OCs have difficulty in closing their lips around the nipple during feeding. Special bottles are available to assist in feeding the infant. Insertion of artificial dental plates to cover the cleft palate facilitates better suckling. As the child grows, unrepaired cleft palate causes difficulty in eating and speech. As the cleft palate connects the roof of the mouth to the nasal passage, infants are predisposed to repeated risk of infections, especially ear infections. Treatment for OCs involves surgical closure of the defect. The type of defect and the health of the infant determines the timing of surgery. Usually, surgery is completed in two stages. The cleft lip, nose and soft palate is repaired between 3 and 6 months of age. The hard palate cleft is repaired after the first year of life, by 15–18 months of age.

Although surgery significantly improves the cosmetic aspects of the birth defect, in severe cases, the facial appearance might still be affected. Patients may have a nasal voice, and there may be a tendency to regurgitate. Dental and orthodontic treatment and speech therapy are usually recommended. The QoL of children with OCs is affected. Poorer QoL is associated with severity of speech problems, facial appearance, age of the child, access to surgery and number of surgeries [17]. In LMIC settings, where surgical skills and infrastructure are unavailable, children survive with the disfiguring cosmetic effects of OC. When services are available in the private sector, affording parents undergo major out-of-pocket expenditure on treatment.

OCs are familiar birth defects, having a prevalence of 1 in 700 live births [16]. There is variation in prevalence based on geographic and ethnic characteristics. The prevalence of OCs is highest among Asian and American Indian populations with rates as high as 1/500. European-derived populations have rates of 1 per 1000, while African populations have lower rates of 1 per 2500. The frequency varies by gender, with males having a higher risk of cleft lip and females having a higher risk of cleft palate. Among unilateral cleft lip, the left side is more affected than the right side [16]. The prevalence of cleft lip/palate was 1.38 per 1000 births in a systematic review and meta-analysis of published studies from LMICs. The authors concluded that every one in 730 children was born with an OC, but there was high heterogeneity between studies due to lack of standard classification of OCs [18].

The variation in the presentation of OCs makes their classification difficult. As such, epidemiological studies to identify risk factors for OCs are also challenging. Environmental and genetic factors have been implicated in the development of OCs [19]. Prenatal tobacco and alcohol use are associated with increased risk. Maternal smoking before pregnancy elevates the risk by 1.3 times. Folate in the preconception period and during the first trimester is protective and decreases the risk of OC. The genetic component of OCs is evident from the increased risk of a second affected birth. In a Norwegian study, the relative risk of occurrence of OC among first degree relatives was 32 (95% CI 24.6–40.3) for cleft lip and 56 (95% CI 37.2–84.8) for cleft palate [20].

Spina Bifida and Neural Tube Defects

Neural tube defects (NTDs) are irreversible defects in the development of the brain and spinal chord that cannot be corrected. The symptoms are variable, ranging from intellectual disability, to paralysis and incontinence. NTDs occur early in pregnancy, caused by improper closure of the neural tube in the fetus [21]. The most severe form of NTD is anencephaly, where the brain fails to form. Anencephaly is nearly always fatal. Spina bifida (Fig. 2.2) is a condition where the vertebrae fail to close completely, so that the spinal chord may protrude through this gap [21, 22]. The most serious form of spina bifida is those where the spinal chord and/or the meninges protrudes through vertebral openings. Collectively referred to as spina bifida cystica, it can include meningocele (only the meninges protrude), meningoencephalocele (meninges and brain tissue protrudes), meningomyelocele (meninges and spinal chord tissues protrude), encephalocele (brain tissue protrudes) or meningocele (spinal chord tissue protrudes). Spina bifida occulta is the least severe form of the condition, where the spinal chord remains covered with the meninges, the membrane covering the spinal chord and the skin. The only visible effect may be a tuft of hair or different skin coloration in the affected region. Most spina bifida occulta are asymptomatic, but in severe cases, children may have leg weakness or incontinence. Occult spinal dysraphism is a more severe form of spina bifida, where abnormalities (birthmarks, tuft of hair, lumps or openings in the skin) are visible. Abnormalities in the underlying spinal chord can lead to nerve damage.

The most familiar form of spina bifida is myelomeningocele, where the uncovered neural tube degenerates in utero, resulting in neurological deficits. Occurring

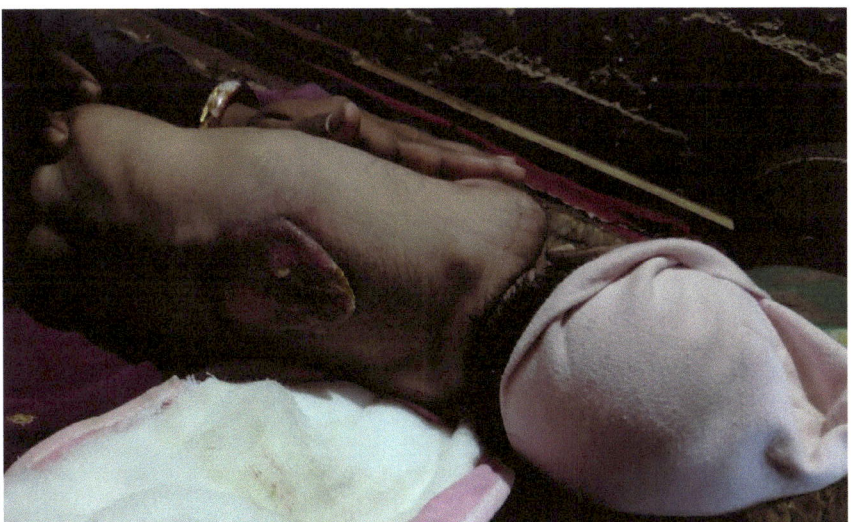

Fig. 2.2 Spina bifida showing the lesion in the lumbar region of the spine

in 1 per 1000 births globally, significant variations in prevalence across geographic regions and ethnic groups have been reported. The condition causes mortality. For surviving children, spinal chord problems may result in weakness and paralysis of the legs. Lack of sensation may lead to the development of pressure sores. Children may have urinary and bowel incontinence, repeated urinary tract infection or constipation. Newborns require immediate surgery. In some cases, (such as Chiari II malformation, where there is a herniation of the hindbrain), spina bifida is associated with hydrocephalus which requires shunting to relieve the intracranial pressure. Majority of spina bifida are sporadic and non-syndromic, but they may also be associated with multiple malformations, such as clubfoot, arthrogryposis (contractures), dislocated hip or scoliosis [21, 22].

Spina bifida can be suspected through prenatal triple marker tests that measure alpha fetoprotein levels in maternal blood in the second trimester of pregnancy. Elevated alpha fetoprotein levels are indicative of increased risk of spina bifida. Prenatal ultrasound offers the opportunity for elective termination of pregnancy.

Spina bifida has a multifactorial etiology with both genetic and environmental risk factors [21]. Among the latter, dietary folate deficiency is one the most vital modifiable risk factors and has been the major focus of birth defect prevention programmes [23]. Beginning with a trial conducted by the Medical Research Council in 1991, there is unequivocal evidence that 400 μg of peri-conceptional folic acid (and 4 mg for those with a prior history of a NTD-affected birth) is associated with a reduction in NTD prevalence [24]. (Around 28% of NTDs are however folate-resistant, [24].) It is imperative that folate supplementation/fortification programmes target women prior to pregnancy, so that women enter pregnancy with normal physiological folate levels. Preconception supplementation benefits are however not fully realized as there is poor uptake by women, even in industrialized countries [25]. Food (flour) fortification with folate has been mandated in many countries as a population-based measure. The effectiveness of this measure has been demonstrated by the fact that spina bifida is more common in countries without mandatory folic acid fortification of the food supply [26]. In addition to folate insufficiency, another non-genetic risk factor is maternal overweight and obesity. Almost 60–70% of spina bifida has a genetic component. The recurrence of NTDs in siblings is increased by almost 20–50% as compared to siblings of unaffected children.

A systematic review and meta-analysis reported wide variations in NTD prevalence across the world [27]. Estimated median prevalence and range was 11.7 (5.2– 75.4) per 10,000 births for African countries and 21.9 (2.1–124.1) per 10,000 births in Eastern Mediterranean countries. It was 9.0 (1.3–35.9) per 10,000 births for European countries, 11.5 (3.3–27.9) per 10,000 births for Americas, 15.8 (1.9–66.2) per 10,000 births for Southeast Asian countries and 6.9 (0.3–199.4) per 10,000 births for countries of the Western Pacific region. The variation in prevalence could be attributed to differences in the methodology of data collection, including case ascertainment methods and quality of data. The birth prevalence of NTDs in India has been reported through two systematic reviews which have reported similar estimates of 4.1 per 1000 [28, 29]. Several research studies have reported higher prevalence rates of NTDs in India, but the study results need to be considered with caution due to methodological issues [30]. Folic acid supplementation/fortification remains one of the most important tools to prevent NTDs in India. Reduction in folic acid preventable spina bifida and anencephaly would contribute significantly in assisting the country in achieving SDG 3 child health targets [31]. NTDs and serum folate deficiency are further discussed in Chap. 10.

Spina bifida affects the quality of life of children with the condition, caused by locomotor impairment and associated complications. Spina bifida impacts education and independence. It impairs the QoL of caregivers [32]. The lifetime cost of children born with myelomeningocele has been estimated at USD 600,000 with 37% being direct medical costs, and the remaining required for special education, caregiver needs and financial costs from loss of employment [33]. Appropriate medical and disability services can to some extent ameliorate the QoL of children, adults and their families.

Hydrocephalus, Macrocephaly, Microcephaly

Hydrocephalus (Fig. 2.3a), macrocephaly and microcephaly (Fig. 2.3b) are defects in the brain and spinal chord, but unlike NTDs, these develop later in pregnancy [20], when elective termination of the pregnancy is not possible. Hydrocephalus is a condition where the malformation causes buildup of cerebrospinal fluid in the ventricles of the brain or in the subarachnoid space [34]. Fluid accumulation results

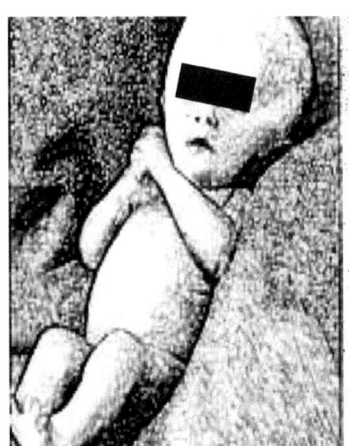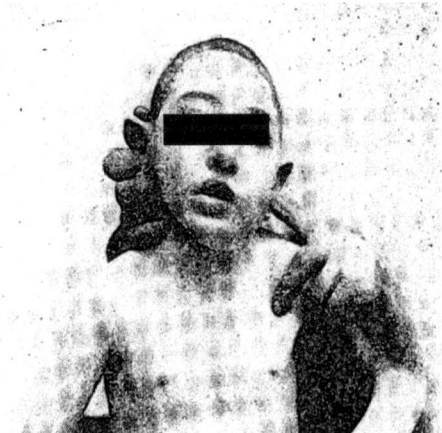

Hydrocephalus image from page 142 of "Modern diagnosis and treatment of diseases of children; a treatise on the medical and surgical diseases of infancy and childhood" (1911) Sheffield, Herman Bernard. Microcephaly image from Gordon Alfred Diseases of the nervous system: for the general practitioner and student 1913

Fig. 2.3 Hydrocephalus (left) and **microcephaly** (right)

in a larger than normal head size. Hydrocephalus may be acquired, caused by conditions such as prematurity. Congenital hydrocephalus is associated in nearly 70% of cases with other anomalies, the most common being mylomeningocele (spina bifida). Hydrocephalus affects 1 in 2000 live births, affecting more males than females.

The diagnosis of congenital hydrocephalus is relatively straightforward, recognized at birth by an abnormally large head size. Unrelieved pressure within the brain causes irritability and listlessness, accompanied by vomiting and seizures. In older children, poor growth and development are accompanied by severe headaches and vision disturbances. Congenital hydrocephalus is usually associated with intellectual disabilities. Treatment attempts to address the cause of the accumulation of fluid in the brain. A shunt is inserted to drain the accumulating fluid. Surgical complications occur, with the frequency increasing from 30% in the first year of life, to 60% by ten years of age. Untreated or poorly treated hydrocephalus is associated with severe disability. The impairment, the frequent need for medical care, out-of-pocket expenditure and the dependence of the child on caregivers, severely affects the QoL of families.

Macrocephaly is another birth defect where the individual has a larger than normal head circumference, as compared to others of the same gender, sex and ethnic group [35]. Macrocephaly can be acquired or be caused by genetic disorders. Microcephaly (Fig. 2.3b) is an abnormally small head, caused by abnormal development of the brain [36]. Microcephaly has a prevalence of 2–12 per 10,000 births, occurs due to genetic conditions (chromosomal or single gene disorders), teratogenic drugs, infections during pregnancy, like rubella, syphilis, toxoplasmosis, cytomegalovirus, Zika virus infections and brain malformations. Severe microcephaly is a debilitating condition, with children experiencing frequent seizures, developmental delay, visual and hearing impairment, locomotor difficulties and intellectual impairment. Hydrocephalus, macrocephaly and microcephaly can be detected with ultrasound in late pregnancy, requiring adequate medical preparations for the birth. The detection of these conditions in late pregnancy causes severe distress to parents. Counseling and support for parents and perinatal palliative care services are required.

Clubfoot (Congenital Talipes Equinovarus)

Congenital talipes equinovarus (CTEV) (clubfoot) (Fig. 2.4) is a common defect of the lower limb, in which the foot is fixed in adduction (inclined inwards), in supination (axially rotated outwards) and in varus (that is pointing downwards) [37, 38]. It is the most common congenital musculoskeletal disorder, having an incidence of 1 in 1000 births, but with population specific variation. CTEV affects both feet (bilateral) in 50% of cases. It is nearly two-fold more common in males than females. Majority, that is nearly 80% of clubfoot, presents as isolated defects, while the rest are associated with other malformations. Clubfoot has a multifactorial etiology, which includes both genetic and environmental factors. The recurrence risk of clubfeet increases by 2–3% when one parent is affected, but the risk increases by 15% when both parents are affected. Certain single nucleotide polymorphisms linked to muscle development

Fig. 2.4 Congenital talipes equinovarus

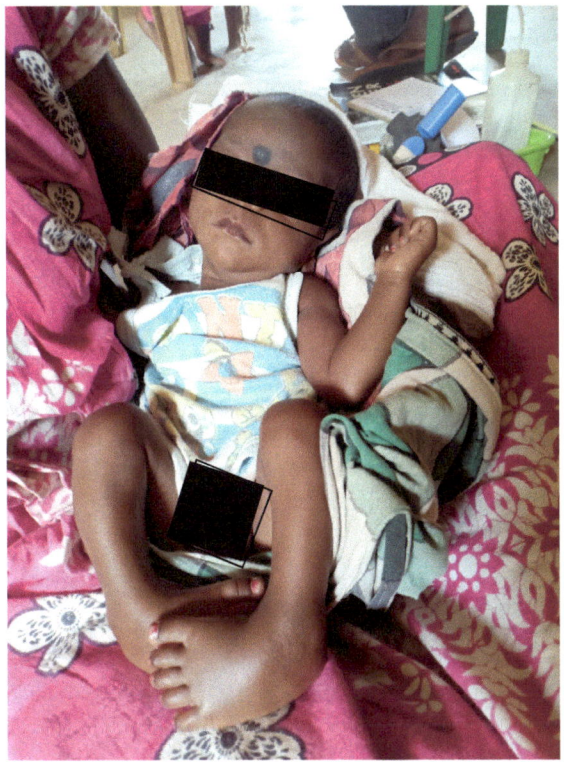

Acknowledgement. Parent for permission, and Dr Ajay Meher

have been associated with CTEV. Environmental factors such as maternal smoking during pregnancy have been associated with elevated risk.

Treatment is mostly non-operative. The Ponseti method of foot manipulation followed by foot to groin casting is the standard care. The leg is gently manipulated with stretching of the tendons and ligaments of the leg. A cast is applied to retain the foot in its new position. This procedure is conducted every 4–7 days, and usually, five to six sessions are needed to bring the foot back to its normal position. Foot ankle orthotics are needed for two to three months, followed by foot braces at night till 4 years of age. Surgery may be required in some cases. The Ponseti method has a success rate of 90%, and successfully treated children show no signs of locomotor disability [39]. In some, the condition may reappear after treatment. Treatment compliance, which includes weekly casting appointments and compliance to instructions for home care, can differentiate between lifelong disability and normal mobility for the child. Although treatable, data indicate that the quality of life and physical functioning scores may be lower in individuals treated for clubfoot as compared to controls [37].

Developmental Dysplasia of the Hip

One of the most common birth defects of the hip joint is developmental dysplasia of the hip (DDH). DDH occurs from abnormal development of the hip bones, so that the head of the femur (thigh bone) does not articulate appropriately with the acetabular socket [40, 41]. DDH presents as a spectrum, from stable (femur cannot be dislocated from the socket), to less frequent dislocatable forms. DDH may affect one (unilateral) or both (bilateral) hip joints. DDH causes shortening of the affected leg and movement difficulty. Untreated DDH is a common cause of locomotor disability. Training medical care providers to identify DDH in the newborn is important, as it can prevent lifelong difficulty in locomotion.

Newborn examination at birth can detect neonatal hip instability. Physical examination in the neonatal period and early infancy uses the Ortolani and the Barlow maneuvers. DDH can be suspected from asymmetric skin creases in the groin and thigh, limited hip abduction (movement away from the midline of the body), a waddling (Trendelenburg gait), walking on toes on the affected side or a limp and Galeazzi sign (unequal knee height). Recognizing these signs is important. In case of suspicion of DDH, infants younger than 4 months of age are recommended ultrasonography of the hips. Early diagnosis is important as duration of treatment is reduced, and outcomes are more favorable. The recommended treatment is the Pavlik harness which stabilizes the hip joint and keeps the infant's hips spread outwards and upwards toward the chest. Pavlik harness needs to be used continuously for 12 weeks. Surgical options are needed for severe cases [40].

The incidence of DDH through physical examination is 1–2 per 1000 [41]. Family history increases the risk 12-fold, the risk to a sibling of a child with DDH is 6% (1 in 17), 12% if a parent has DDH (1 in 8), and if both parent and child have DDH, the risk for a subsequent pregnancy is increased to 36% (that is 1 in 3). Girls with positive family history are at an enhanced risk, although approximately 75% of DDH occur in females without any family history or other risk factors. Breech position is another strong risk factor for DDH. Breech position occurs in 2–3% of births and has an incidence of 84/1000 in females with DDH (as compared to 18 per 1000 in males). DDH is associated with other skeletal deformities of the spine such as torticollis and congenital foot deformities. Other risk factors for DDH are primiparity and oligohydramnios, birth weight, prematurity, multiple pregnancy, mode of delivery and presence of foot deformities. Some DDH occur postnatally and have been associated with swaddling methods. Healthy hip positioning is an important prevention measure, especially in at-risk infants. Poorly treated DDH is associated with pain, movement difficulties and secondary arthritis in adults. Delayed detection involves higher financial costs [42]. Disability affects self-esteem, confidence and body image. Hip pain and the impairment in general, significantly affect quality of life and psychosocial functioning [43].

Congenital Limb Defects

Congenital limb defects are structural birth defects, where a part of or the complete upper or lower limb does not form. The limb is missing or reduced from its normal size [44]. The extent of the defect varies. The defect may involve the lower arm or lower leg (longitudinal deficiencies), or the arm may appear amputated (transverse deficiencies). Upper limb defects are more common than lower limb reduction defects. Limb reduction defects are common birth defects, occurring in 1 in 2000 neonates, but prevalence varies with the classification adopted. The rates of upper and lower limb reduction defects are 4 and 2 per 10,000 live births [45].

Congenital limb defects may be isolated defects or may be associated with other anomalies or conditions unrelated to the limb. Limb defects are frequently associated with congenital syndromes such as Adams–Oliver syndrome, Holt–Oram syndrome, Fanconi anemia and VACTERL (that is vertebral anomalies, anal atresia, cardiac malformation, tracheoesophageal fistula, renal anomalies, radial aplasia and limb anomalies syndrome). Limb deformities may also be caused by amniotic bands, where strands of tissues from the amniotic sac intertwine with the limb of the developing fetus, restricting further development of the structure.

In addition to genetic syndromes, limb defects may be caused by teratogenic medications like thalidomide and retinoic acid. Historically, thalidomide exposure is the best known example of a teratogen causing limb reduction defects. Between 1964 and 1968, the prevalence of these conditions was as high as 30 per 10,000 live births in Scotland. With the goal of monitoring the prevalence and risk factors of limb reduction defects and preventing further tragedies due to teratogenic exposures, several international birth defects registries were established, including the 37 country European Surveillance of Congenital Anomalies (EUROCAT) and the International Clearinghouse of Birth Defects Surveillance and Research (ICBDSR). Presence of a limb defect requires a thorough examination for other anomalies, a family history and possible teratogen exposure. Genetic testing may be needed.

Limb reduction defects are potentially severely disabling, but with proper management, most children are able to lead productive lives [46]. Children may require assistance in activities of daily living or in sport and school activities. Emotional acceptance of the disfigurement by the child is a key psychosocial concern. Therapy may involve surgery, provision of prosthetics (artificial limbs) or orthotics (such as splints or braces) and physical and occupational therapy to enhance functionality. Prosthetics need to be accepted by the child and therefore are useful only if fitted early and become accepted as a part of the child's body. Children with other medical conditions may have difficulties in normal development and in acquiring motor skills.

Polydactyly

Less disabling limb defects include polydactyly, syndactyly, split hand and split foot [47]. They may be isolated or associated with other defects or deformities.

Polydactyly is characterized by complete or partial supernumerary digits. Syndactyly is characterized by fusion of digits. Polydactyly can range from extra thumb or toe (preaxial polydactyly), central polydactyly, where the ring, middle and index finger may be duplicated, or post-axial polydactyly where there is an extra digit on the ulnar or fibular side of the limb. Syndactyly is webbing or fusion of fingers or toes, and there are many different variations. Polydactyly and syndactyly may be diagnosed by X-rays. Treatment is surgery and prosthetic devices, if needed.

Section II : Developmental (Neuromotor) Disabilities

Developmental disabilities are characterized by motor, cognitive and/or language delays. The diversity of clinical types, but similarity in terms of rehabilitation needs are illustrated through brief descriptions of Down syndrome, intellectual disabilities, cerebral palsy, behavior disorders, (autism and attention-deficit/hyperactivity disorder, ADHD), congenital vision impairment/blindness and congenital deafness (sensory impairments). Children with these conditions benefit from early intervention services (Chap. 13) that assist the development of functioning.

Intellectual Disability

Intellectual disability (ID, referred to as mental retardation in the International Classification of Diseases, 10th revision, ICD-10) is characterized by "deficits in cognition, a summary term for thought processes used in interactions with other humans and the environment" [48]. ID may be the outcome of different clinical conditions. It is characterized by sub-average intellectual functioning with intelligence quotient <70–75, which is 2 standard deviations below the IQ of 100 in the general population. Children have delayed intellectual development, impaired functional and adaptive skills, that is, lack the ability to carry out age appropriate activities of daily living [49]. The American Association of Intellectual Developmental Disabilities (AAIDD) describes intelligence as a general mental capacity that involves reasoning, planning, solving problems, thinking abstractly, comprehending complex ideas, learning efficiently and learning from experience [49]. Children with ID lack or have limited communication ability, immature behavior and social skills, and have limited or lack self-care skills.

Severe and moderate ID may be detected early, especially if accompanied by physical abnormalities such as dysmorphic physical features, other disabilities or medical complications. Mild ID is detected by preschool age as low IQ and poor adaptive skills. The AAIDD classifies the severity of ID based on the intensity of support needed [49], while the *Diagnostic and Statistical Manual of Mental Disorders, 5th Edition* (DSM-5), which is published by the American Psychiatric Association, classifies severity based on the capability of independent living [48–50] (Table 2.3).

Table 2.3 Level of support based on severity[a]

Level of severity	DSM-5 (severity classified on level of daily skills)	AAIDD (severity classified based on intensity of needed support)
Mild	Can live independently with minimum levels of support	Intermittent support
Moderate	Independent living possible with moderate support, such as those available in group homes	Limited support for daily living activities
Severe	Needs daily assistance with self-care and safety supervision	Extensive support needed for daily activities
Profound	Requires 24 h care	Pervasive support for all activities

[a]Modified from [50]

Health problems among people with ID are not dissimilar from that of the general population, but are complicated by the fact that in case of more severe disabilities, children may be unable to verbalize their complaints. Complications and comorbidities frequently ecountered among children with ID are motor deficits, epilepsy and seizure disorders, allergies, otitis media, gastroesophageal reflux disease (GERD), dysmenorrhea, sleep disturbances, mental illness, vision and hearing impairments, oral health problems and constipation [51]. Health conditions can contribute to problem behavior. Children with ID have above usual need for medical services. Other service needs include special education, family counseling and other social welfare measures.

Down Syndrome

Down syndrome (Fig. 2.5), a condition caused by duplication of whole or part of chromosome 21, is characterized by intellectual disability, which may be mild (IQ 50–75), moderate (IQ 35–50) and less frequently, severe (IQ 20–35) [52]. Children with Down syndrome have typical physical features, characterized by short stature, small head, flat facial features with upwardly slanting eyes, epicanthal folds, flat nasal bridge, small ears, protruding tongue, extra skin on the nape of the neck, single transverse palmar crease, short fifth finger and wide space between first and second toes. Nearly 50% of children with Down syndrome are at an increased risk of comorbidities such as congenital heart defects (50%), thyroid disease (15%), gastrointestinal atresia (12%), leukemia (1%), Hirschsprung disease (1%), acquired hip dislocation (6%), obstructive sleep apnea (50–75%), seizure disorders and obesity. Sensory impairments and complications are also common. Among these, 75% risk hearing loss, 50–70% have increased episodes of otitis media, eye disease (60%), including cataract (15%) and severe refractive errors (50%). Some children with Down syndrome may present with behavior suggestive of attention-deficit hyperactivity disorder in childhood. Children with profound intellectual disability may show autistic behavior.

Fig. 2.5 Down syndrome

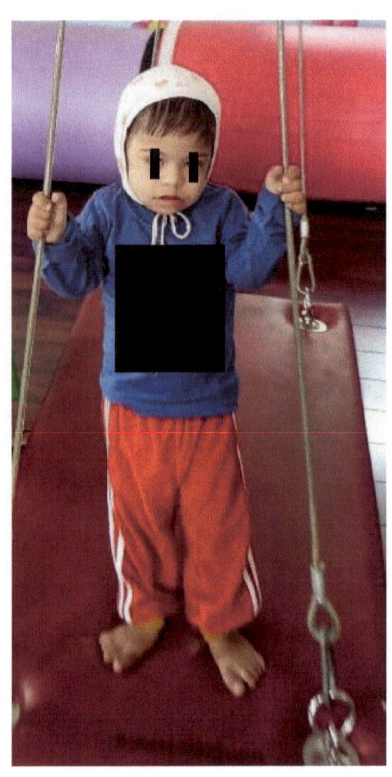

Adults and children with Down syndrome frequently suffer from depression. Down syndrome babies have low muscle tone (hypotonia) and have delayed milestones. Caregivers encounter problems in infancy such as sucking, feeding, constipation and toilet issues. Speech delay, and delay in activities of daily living are frequent. Learning difficulties are dependent on the extent of intellectual impairment. Median age of survival is till the fifth decade of life [52].

In nearly 95% of cases, Down syndrome is not familial but is caused by a maternally derived extra chromosome 21. Familial cases are 3–4%, caused by a translocation between chromosome 21 and chromosome 14. In families with chromosomal translocations, the risk is 1: 20 when the father is a carrier. In about 1–2% of cases, Down syndrome may be mosaic, with karyotypes showing both normal and trisomy 21 cells. The incidence of Down syndrome is 1 in 800, but there is a well-established relationship with maternal age [53]. The risk of Down syndrome is 1 in 2000 at age 20, which increases to 1 in 1000 at age 30, to 1 in 365 by age 35 and 1 in 100 by age 40.

Down syndrome can be detected using a number of tests conducted early in pregnancy [54]. Nuchal translucency test conducted between the 11th and 14th week of pregnancy identifies a larger and fluid-filled nuchal space (fold of skin at the back of the neck) of the fetus. This measure, together with maternal age and gestational age, is used to calculate the risk of a Down syndrome affected pregnancy. Maternal triple and quadruple screening test measures three and four maternal serum biomarkers between the 15th and 18th week of pregnancy, respectively. Down syndrome may be suspected by elevated plasma protein A levels in late first trimester and increased levels of alpha fetoprotein, beta-HCG (human chorionic gonadotropin) and unconjugated estradiol and inhibin by the 15th/16th week of gestation. Ultrasound at 18–20 weeks of gestation, amniocentesis at 15–20 weeks or chorionic villus sampling between 10 and 12 weeks with karyotyping are other prenatal tests [54]. One of the most promising prenatal tests is the non-invasive prenatal test (NIPT) that is based on identification of cell free DNA in maternal plasma [55]. It is a highly effective screening test for Down syndrome, with 99% sensitivity and with a false positive rate of 0.1%. The mean pregnancy termination rate after detection of Down syndrome varied in the USA, ranging from 67% (61–93%) in population-based studies from three states to 85% (60–90%) in hospital-based studies from other states [56].

The quality of life of children and their caregivers is dependent on medical management of the child and the counseling and support received by families [57]. While the underlying intellectual disability cannot be treated, comorbidities can be treated. Diet and exercise programmes are needed to maintain appropriate weight. Early intervention before the age of three years is recommended with appropriate physical, occupational and speech therapies. Special education programmes, appropriate for the intellectual level of the child, are needed.

Child and family support are required throughout the life-course. Socialization and self-help skills, addressing issues related to sexual development and ensuring that the child is protected from abuse, are of special concern. Psychosocial support services are required to help parents overcome guilt and loss, assist sibling adjustment and help intrafamily relationships. As appropriate, genetic counseling services need to be provided. Supplementary income, support for long-term financial planning and independent living opportunities need to be decided.

Attention-Deficit/Hyperactivity Disorder

Attention-deficit hyperactivity disorder (ADHD) is a neurobehavioral and neurobiological disorder that severely affects the quality of life of families [58]. The condition is marked by inattention, hyperactivity and impulsivity. A review of 175 prevalence studies (most conducted in Europe and the USA) estimated ADHD prevalence at 7.2% [59]. Hyperactivity-impulsivity disorders are 2–9 times more frequent in boys. ADHD is characterized by temper tantrums, aggressiveness, disobedience, impulsiveness and attention demanding behavior. It is marked by learning disability, poor social skills and poor peer relationships. Sleep may be affected, and there may be signs of anxiety and depression. Adolescents and adults may be at a higher risk of substance abuse. School performance is affected by difficulty in sustaining attention, inability to follow instructions, easy distraction, forgetfulness and restlessness. ADHD has a multifactorial etiology, with a strong genetic component. The risk to first degree relatives is five times, and it is over 1–2% of the risk in the general population for second degree relatives.

Physical hyperactivity in the home environment affects family life. Managing ADHD requires efforts and self-control on part of parents in order to guide appropriate and disciplined behavior in the child [60]. Parenting is a physically and emotionally challenging task. Parental difficulty in coping with and accepting the behavior leads to high parental anxiety and stress and experience of guilt after rebuking the child. Sibling quality of life is affected. ADHD is lifelong, but can be treated with both medications and behavioral therapy [58]. Behavioral therapy requires professional cognitive-behavioral therapists. Studies suggest that behavioral therapy is more effective when combined with drug therapy.

Autism Spectrum Disorders

Autism spectrum disorders (ASD) are disabling, neurodevelopmental conditions that are diagnosed in early childhood [61, 62]. ASD is characterized by a severe disability in social behavior, identified by inability for reciprocal social skills, reduced eye contact, reduced interest in sharing emotions and interests, lack of verbal and non-verbal communication, stereotypic repetitive behavior such as hand flapping, toe walking, playing in a different way and speaking in a unique way. Other behaviors are need for predictable routine, intense dislike to specific smells, tastes, colors or textures. ASD severity including intellectual development varies. Symptoms range from mild to severe and can change over time. Common comorbidities are seizures, mental illness and sleep problems. ASD is usually diagnosed around 2–3 years of age when delayed language development, lack of typical play and lack of bonding with parents become evident. Some children develop normally till this age and subsequently stop acquiring and loose previously acquired skills. The diagnosis of ASD is made clinically based on DSM-5. Behavioral therapy that provides speech, occupational therapies and social skills training is the main form of treatment.

The prevalence of ASD is 1–2% [63] with prevalence being higher by three or four fold in boys than in girls. ASD symptoms in girls are not as marked as that in boys. ASD is associated with anomalies of brain structure and function and has multifactorial etiology with both genetic and non-genetic risk factors. The genetic etiology is suggested by observations of high recurrence risk (50–100 times higher for subsequent pregnancies, 90% in monozygotic and nearly 20% for dizygotic twins). It is also associated with genetic disorders like Fragile X and Rett syndrome. Several genes associated with ASD have been identified, but understanding the molecular pathology of ASD is encumbered by the heterogeneity of the clinical presentation and in the spectrum of identified variants [64]. Among environmental factors, several prenatal factors (teratogenic drug exposures, viruses like rubella), perinatal factors (low birth weight, abnormally short gestation, birth asphyxia) and post-natal factors (auto-immune disease, viral infections, hypoxia) have been associated with ASD.

Children with ASD and their caregivers experience a poor quality of life [65]. Children experience more health problems than typical children. The major problems faced by caregivers are in communication as children have limitations in expressing needs and are easily frustrated by routine changes. Caregivers quality of life is profoundly impacted with parents reporting physical symptoms, in addition to anxiety and depression.

Cerebral Palsy

Cerebral palsy is a movement disorder that may be accompanied in some cases with intellectual disability, behavior disorders, visual/hearing impairment and seizure disorders [66, 67]. Cerebral palsy occurs from brain malformations during development, or brain damage occurring before, during or shortly after birth. Cerebral palsy impairs voluntary movements or postures and is marked by spasticity (increased, involuntary muscle contracture impeding movement), ataxia (lack of voluntary coordination) or involuntary movements. Spastic syndromes account for 70% of cases. The spasticity is marked by muscle stiffness and weakness that affects motor functions. Spasticity may be unilateral (monoplegia where one limb is affected), hemiplegia (where one side of the body is affected) or be bilateral (diplegia where all limbs are affected but spasticity is more marked for lower limbs). Triplegia is unilateral upper and bilateral lower limb spasticity, paraplegia is where lower part of the body including both lower limbs are affected and quadriplegia where all four limbs and trunk are affected. Scissor gait and toe walking are typical. Children with quadriplegia are the most severely affected. Dysarthria (slow or slurred speech) and dysphagia (difficulty in swallowing) are common.

Athetoid or dyskinetic cerebral palsy represents 20% of cases. Athetoid cerebral palsy is marked by slow, writhing, involuntary movements. Abrupt, jerking (choreic) movements may also occur, which increases with excitement but disappears during sleep. Dysarthria is severe. The third group is ataxic syndromes, which occur in less than 5% of cases. Cerebral palsy of this type is marked by weakness, tremors, poor coordination, wide-based gait and difficulty in rapid or fine movements. The severity of motor impairment is classified using the gross motor function classification [68].

Cerebral palsy is one of the most common disorders of childhood with a prevalence of 1.5–2.5 per 1000 live births, but increasing in prevalence to 3 per 1000 live births in populations in the 4–48 year age group [66]. This increase in prevalence is attributed to better survival rates of very premature infants. Cerebral palsy has a multifactorial etiology, involving prenatal and perinatal factors. Preterm birth is the strongest risk factor for cerebral palsy, with the risk at 28 weeks of gestation being increased 50 times that of risk of full-term births. Intra-uterine disorders, neonatal encephalopathy, kernicterus (brain damage caused by high bilirubin levels) and perinatal factors such as perinatal asphyxia, stroke and central nervous system infections are strongly associated factors. Cerebral palsy has a strong familial risk. The relative risk of occurrence of cerebral palsy in a sibling is 3.0 (95% CI 1.1–8.6). Among twins, the risk that the other twin is affected is 15.6 (95% CI 9.8–25). Cerebral palsy in a parent increases the risk by 6.5 times [69].

Cerebral palsy is usually recognized by caregivers as involuntary or jerky movements. Other signs of cerebral palsy could be the lack of typical developmental milestones in the infant such as lack of smile, inability to hold up neck by three months of age, inability to reach for objects, roll over, kick legs or appearing rigid or limp. Cerebral palsy cannot be cured. Treatment goals are aimed at overcoming disability, ensuring independence and preventing complications. The main treatment modalities are physical therapy, occupational therapy and speech therapy that are concurrent with medical and surgical interventions [67]. Physical therapy aims at improving muscle tone and overall mobility, improving sensory perceptions of touch and depth, which improve movement. Appropriate assistive mobility aids, such as braces, crutches and wheelchairs, are suggested to enhance mobility. The goal of occupational therapy is to enhance independence. Occupational therapy assists in improving the activities of daily living (feeding, bathing, toileting, dressing, transferring and maintaining continence), developing social relationships, and learning to play and concentrate on activities. Speech therapy aids speech, eating, breathing, language development and vocabulary [67]. Medications are used to treat seizures, gastric reflux, incontinence and drooling. Nearly one fifth of cases of cerebral palsy may have attention-deficit hyperactivity disorder. Botulinum toxin injections and other drugs are used to address spasticity. Surgery aims at improving the alignment of joints, muscles and tendons, especially in cases where there is muscle contracture. Children with cerebral palsy survive upto adulthood.

As with other chronic conditions of childhood, cerebral palsy affects children, caregivers, especially mothers and family life [70]. In a study conducted across seven European countries, the quality of life of children with cerebral palsy was strongly related to the severity of impairment, with a proportional relationship between the severity of gross motor function impairment and intellectual quotient levels. Children with more severe motor function were reported to have poorer physical well-being and autonomy. Children with lower IQ had lower scores of social support and were unable to develop and maintain relationships. Poorer quality of life was associated with chronic pain. Children of better educated parents had poorer quality of life in the domains of parental relations, while those in single parent household had lower scores in the mood and emotion domains. There was a relationship between parental stress and QoL of the child, as parents reporting higher levels of stress were more likely to report lower QoL of the child. Like other caregivers of children with chronic conditions, parental QoL was severely impacted in terms of psychological and physical health. The QoL was related to the severity of the impairment. Parental physical and social well-being were affected, as were freedom and independence, family well-being and financial stability. Social support was an important determinant in the QoL of families [70].

Congenital Hearing Impairment

Congenital hearing loss occurs due to the inability of the ear to convert vibratory mechanical energy of sound into the electrical energy of nerve impulses. It is defined as hearing loss ≥40 dB in the better hearing ear, averaged over the frequency range important for speech recognitions, that is at 500, 1000, 2000 and 4000 Hz [71]. Congenital hearing loss may be conductive hearing loss that affects the functioning of the outer and middle ear, sensorineural hearing loss that impairs functioning of the inner ear, auditory nerve or the central auditory pathway, or mixed hearing loss that is impairment caused by improper functioning of the conductive and sensorineural pathways [71].

The prevalence of congenital bilateral permanent hearing loss is 1.33 per 1000 live births. The prevalence increases to 2.83 per 1000 in primary school children, and 3.5 per 1000 in adolescents. Prevalence from low-income countries varies with the use of different diagnostic methods or use of different criteria for hearing loss threshold [72]. Congenital hearing loss has a multifactorial etiology. Genetic factors are responsible for 40% of childhood hearing loss with higher prevalence among consanguineous marriages. Nearly 31% hearing loss is ascribed to infections, 17% due to pre- or peri-natal factors and 4% due to the use of ototoxic drugs. Congenital hearing loss caused by single gene disorders may be syndromic (associated with conditions such as Usher and Waardenburg syndromes). Environmental risk factors associated with congenital hearing loss are prematurity, low birth weight, developmental delay, craniofacial anomalies, admission to neonatal intensive care unit, neonatal medical interventions and duration of hospitalization [72]. Pregnancy acquired infections including rubella, syphilis, herpes simplex virus and *Toxoplasmosis gondii* are implicated in congenital hearing loss. Congenital cytomegalovirus (CMV) infections are the most important congenital infection, being causative for 10–20% of cases of hearing impairment in children [73]. The prevalence of congenital CMV in developed countries was estimated at 0.58%, with 1 out of 3 symptomatic children likely to experience bilateral, severe-to-profound hearing loss. Maternal CMV infection prevalence is likely to be considerably higher in low-income countries. Only 1.43% of children with a positive family history have hearing loss. The recurrence risk of deafness for a hearing couple with one deaf child and no family history is 18% [73].

Early detection of hearing loss is critical for child development, as children identified prior to 2 weeks of age through neonatal screening programmes had a better quality of hearing amplification and significantly better quality of life as compared to children who are diagnosed at a later age of 9 months. Delay in diagnosis increases the risk of impaired speech. Universal neonatal hearing screening programmes are a part of newborn screening in the industrialized countries. Management involves early intervention like appropriate hearing technology (such as hearing aids or cochlear implants), professional support for communication, learning and education (auditory-verbal therapy, use of non-verbal communication, that is sign language) and a family-centered approach (that is counseling and support of parents). Best effects of hearing technology are observed when the device is fitted prior to 6 months of age and used continuously. Cochlear implants are available for patients with severe-to-profound hearing loss. The World Health Organization suggests a six pronged strategy for prevention and care which include (i) strengthening relevant programmes for reducing risk factors for hearing loss and strengthening organizations working for the hearing impaired, (ii) implementing screening and appropriate interventions for affected or high risk infants, (iii) training primary-level health care providers, (iv) making available appropriate technologies, (v) regulation of ototoxic medicines, regulation and monitoring of noise levels and (vi) raising awareness [74].

Vision Impairment/Blindness

Childhood blindness and vision impairment affect psychomotor, social and emotional development and can compromise opportunities for education and employment. Congenital cataract, congenital glaucoma, congenital cloudy retina and retinal degeneration (such as Leber's congenital amaurosis), optic atrophy and eye malformations like micropthalmos, anopthalmos, coloboma are common congenital causes of blindness/vision impairment [75, 76]. Congenital ocular anomalies can be caused by genetic, gene-environmental and environmental factors such as infections and exposure to teratogenic drugs during the prenatal period. Congenital ocular anomalies may also arise during the perinatal period, caused by birth asphyxia, opthalmia neonatorum (neonatal conjunctivitis), retinopathy of prematurity (ROP) and retinitis pigmentosa. Some conditions (such as retinal dystrophies, Leber's congenital amaurosis, retinoblastoma and others) are caused by single gene disorders. Although more than 60% of blindness among infants is likely to be caused by genetic factors or intrauterine infections, in majority of cases of congenital blindness, the exact etiology cannot be determined [76].

Congenital cataract is one of the most frequent causes of treatable childhood blindness, responsible for 5–20% of childhood blindness worldwide [77]. A systematic review from five geographical regions estimated that the prevalence in low-income countries was 0.42–2.05 per 10,000, 0.32–8.49 per 10,000 for low middle-income countries, 0.74–22.7 per 10,000 in upper middle-income countries and 0.63–13.6 per 10,000 in high-income countries. The estimated pooled prevalence of congenital cataract ranged between 0.63 and 9.74/10,000 (median 1.71) per 10,000 births [77]. Congenital cataract can be caused by monogenic conditions, genetic syndromes or rubella and CMV infections. Only 18% of congenital cataract cases report family history. Early detection followed by intraocular lens implant can restore vision in congenital cataract cases. Parental counseling to ensure compliance to post-operative advice is extremely important.

Retinopathy of prematurity (ROP) is another condition that can cause severe vision impairment or blindness. Important risk factors for ROP are low gestational age, low birth weight and oxygen therapy following delivery [78]. Alerting and training providers is the best strategy for reducing ROP.

Early detection of congenital visual deficits is imperative, as eye development occurs through the first year of life, and interventions during infancy lead to better outcomes. Untreated, visual deficits lead to amblyopia (lazy eyes, blurred vision) and blindness. In many cases, surgery, followed by therapeutic and low-vision aids, can be used, but in several cases, the blindness cannot be treated. Habilitation including special education remains the main strategy for children with low vision or blindness. Early detection of congenital blindness is challenged by the inability of infants to verbalize their complaints, and opthalmologists may lack appropriate training to detect ocular conditions in very young children. Genetic counseling is required when a sibling is also affected.

Vision 2020 is a global initiative launched by the World Health Organization to eliminate avoidable blindness [79]. Its core strategies include control and treatment of diseases known to cause blindness, supporting training of opthalmologists and other eye care providers in the provision of eye care, and support the development of infrastructure and technology that can make eye care services more accessible to communities.

Section III : Common Single Gene Disorders

Hemoglobinopathies

The hemoglobinopathies, thalassemia and sickle cell anemia are the most prevalent single gene disorders worldwide [80]. Thalassemias are functional defects of hemoglobin synthesis. They are caused by reduced or lack of synthesis of alpha- and beta-globin chains of hemoglobin, causing alpha- and beta-thalassemia, respectively [81, 82]. They are autosomal recessive conditions. Both parents are heterozygous carriers of the gene mutation. Every pregnancy carries a 25% risk of an affected birth and 50% risk of being a carrier (Fig. 2.6a).

The disruption in the ratio of alpha to beta-globin chains causes chronic hemolytic anemia, accompanied by ineffective erythropoiesis and compensatory hemopoietic expansion [80]. The severity of alpha thalassemia depends on the underlying mutation. HbH disease, for example, causes chronic anemia, severity of which varies between individuals [81]. The HbH-Constant Spring variant is a more severe manifestation of alpha thalassemia. Hemoglobin (Hb) Bart's hydrops fetalis is a form of alpha thalassemia that is fatal during gestation or in the perinatal period. The severity of beta-thalassemia increases from the carrier state (beta-thalassemia minor, beta-thalassemia trait) to the disease state (beta-thalassemia intermedia and beta-thalassemia major, Cooley's anemia) [82].

Beta-thalassemia is usually diagnosed when infants present with complaints of poor growth, feeding problems, frequent bouts of fever, diarrhea and distended abdomen caused by splenomegaly. Hematological parameters reveal microcytic, hypochromic anemia, with reduced mean corpuscular volume and mean corpuscular

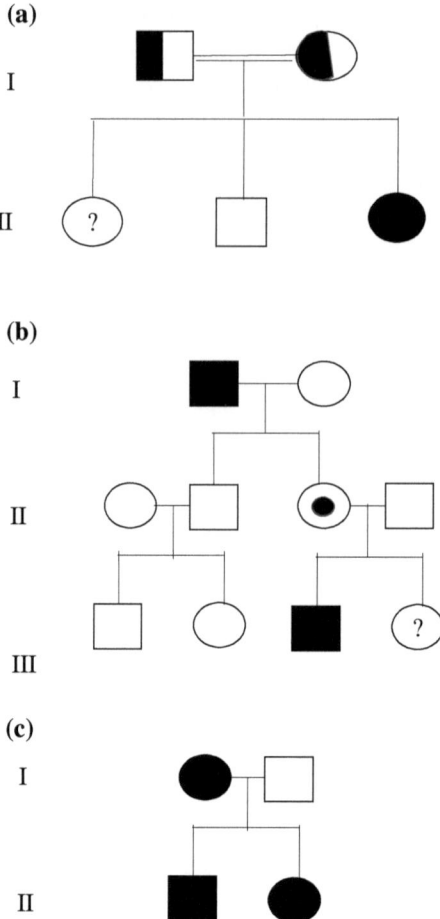

Fig. 2.6 Inheritance patterns of single gene disorders. a The mode of inheritance of autosomal recessive conditions like thalassemia and sickle cell anemia. Parents are heterozygous carriers. Each pregnancy has a 25% risk of an affected conceptus, 50% risk of heterozygous carrier state, and 25% are normal due to non-inheritance of the mutation. **b** Mode of inheritance of X-linked recessive conditions like hemophilia and Duchenne muscular dystrophy. The mutation shows a crisscross pattern of transmission, from an affected male via a carrier daughter to the grandson. **c** Autosomal dominant inheritance, for example, achondroplasia, where the condition is transmitted from parent to children. As the person with achondroplasia has a normal chromosome, not all children have achondroplasia. Square boxes represent males, circles females, shaded symbols indicate affected individuals, and half-shaded symbols and circles with a smaller black circle represent carriers. Double horizontal line in **a** indicates a consanguineous marriage, that is marriage between close blood relatives

hemoglobin and elevated HbA2 (Hemoglobin A2) levels [82]. Chronic anemia causes growth impairment. The clinical manifestations of chronic anemia are fatigue and weakness, shortness of breath, pallor, hepatomegaly and splenomegaly. Leg ulcers, jaundice and gall stones are also common. Standard treatment is routine transfusion every two to five weeks, with the goal of maintaining hemoglobin levels around 9–10.5 g/dl. Hepatitis B immunization is recommended to reduce the risk of this transfusion-transmitted pathogen. Iron overload (transfusional hemosiderosis) is the unavoidable consequence of repeated transfusions, which is managed through appropriate iron chelation therapy. Desferrioxamine B, a non-oral chelator, is administered through slow transfusion, which requires indoor admission, and has low compliance due to the mode of administration and time required for administration. Deferiprone and deferasirox are available as oral monotherapy. Bone marrow transplantation from a HLA matched donor remains the cure for this condition.

Like the thalassemias, sickle cell disease is also prevalent globally [83]. Sickle cell anemia is caused by a structural variation in the beta chain of hemoglobin. The altered hemoglobin molecule, HbS, causes sickling and enhanced fragility of the erythrocytes. Sickle cell disease refers to individuals who are homozygous that is they have two copies of sickle hemoglobin (HbS). Sickle cell trait refers to heterozygous carriers of the HbS allele. Sickled erythrocytes cause vaso-occlusion of small blood vessels of the microcirculation, resulting in lack of oxygenated blood to the tissues. Sickle cell disease is characterized by periodic episodes of pain (crises), organ ischemia and severe systemic complications. Hemolysis causes chronic hemolytic anemia. Anemia is usually severe, and mild jaundice and pallor are present. Patients experience debilitating pain in the long bones, in the hip joint, back, hands and feet which may persist for hours to weeks. Painful swelling of hands and feet occurs due to vaso-occlusion. Sudden onset of fever and chest pain caused by bacterial pneumonia may cause pulmonary complications. Sequestration of sickled cells in the spleen damages the organ, predisposing patients to repeated infections. Repeated infarctions and fibrosis result in atrophy of the spleen in adults. Damage to organ systems manifests as stroke, pulmonary complications, heart failure, seizures, chronic kidney disease and retinopathy [83, 84]. Leg ulcers are a common sign of sickle cell anemia. Hemolysis and increased bilirubin can cause gallstones. Priapism, an erectile dysfunction, is a painful condition affecting younger men. Recurrent oxygen deprivation delays growth in infants and young children. Infections are common in childhood, furthering spleen damage. Hepatosplenomegaly is common in children. In well-managed patients, life expectancy has now reached the fifth decade of life.

The hemoglobinopathies primarily affect populations in developing countries, being prevalent in sub-Saharan Africa, the Mediterranean and Middle East, the Indian sub-continent and east and southeast Asia. Migration from these regions to Europe and the USA has resulted in these conditions being distributed more globally [85]. The slave trade has been responsible for distribution of sickle cell anemia to the USA where it is a known cause of morbidity and premature mortality among the African–American population. An estimated 1–5% of the global population are carriers of hemoglobinopathies. Hemoglobinopathies reduce the risk of severe malaria and are associated with high level of resistance to *Plasmodium falciparum* malaria [86].

Thalassemia and sickle cell anemia affect the quality of life of patients and their caregivers, especially in countries where programmes to provide free-of-charge treatment for patients has not yet been instituted [87–89]. Studies on the QoL of thalassemia patients indicate that the physical suffering associated with the disorder, the need for repeated transfusions and the physical appearance in poorly transfused patients affect psychosocial well-being. Studies report symptoms of depression, emotional problems and poor self-image. The repeated nature of transfusions, direct and indirect costs of treatment, lack of information and financial costs of transfusion and chelation therapy cause emotional and economic distress among caregivers. The quality of life of patients with sickle cell anemia is similarly compromised, even in settings where the best of care is available [83]. Due to the psychosocial impact, medical therapy needs to be supported with psychological support and counseling. Genetic counseling forms an integral part of a service, especially as a large number of underlying mutations have been identified, providing the opportunity for prenatal diagnosis [90].

Hemophilia

Hemophilia A and B are inherited X-linked gene disorders (Fig. 2.6b) that affect 1 in 5000 and 1 in 30,000 male births, respectively [91, 92]. Hemophilia is a bleeding disorder, characterized by repeated bleeding in the joints, causing hemarthroses (Fig. 2.7) or in the deep or superficial tissues, visible as hematomas. Repeated episodes of hematuria (blood in urine) are also a characteristic of hemophilia. Bleeding is accompanied by severe pain and inflammation. The frequency and severity of bleeding episodes depend on residual clotting factor levels. Patients with severe hemophilia retain 1% of normal clotting factor activity, experiencing repeated, spontaneous bleeding episodes. Patients with severely moderate hemophilia retain 1–5% of clotting factor activity. Spontaneous bleeding episodes are rare, although

Fig. 2.7 Hemophilia. Knee joint hemarthroses and hematoma. Image from page 791 of Koplik, Henry "The diseases of infancy and childhood: designed for the use of students and practitioners of medicine" (1918) New York and Philadelphia, Lea and Febiger

minor trauma results in unusual bleeding. Patients with mild hemophilia retain 5–40% of normal clotting factor activity. Unusual bleeding is identified only if the patient undergoes surgery or a tooth extraction. Severe and moderately severe hemophilia are fatal within the first decade of life if untreated. Untreated or poorly treated hemophilia causes intense pain, progressive joint damage, leading to locomotor disability. Uncontrolled hemorrhage, especially intracranial hemorrhage, is fatal [91, 92].

Treatment is by replacement therapy with plasma derived or recombinant (genetically engineered) clotting factor concentrate. The major complication of treatment is the development of allo-antibodies (inhibitors) against the treatment product, which requires specialized management involving the use of larger amounts of clotting factor concentrate and other treatment products. Another complication is the risk of transfusion-transmitted infections. The life expectancy and quality of life of patients in industrialized countries have greatly improved with better understanding of the condition and their management, prophylaxis and the universal availability of improved treatment product. In low-income countries, access to treatment product remains limited due to prohibitive costs [93]. Patients have a poor quality of life, with repeated painful bleeding episodes, that interrupt education and employment. Due to poor treatment, disability sets in early, further compromising the quality of life.

Hemophilia is caused by mutations in the clotting factor VIII gene (hemophilia A) and factor IX (hemophilia B) genes, respectively. The mutation may appear de novo, or there may be maternal family history of hemophilia. As a sex-linked recessive condition, the mutation is transmitted through the maternal line to the patient. Every pregnancy carries a 50% risk of an affected birth among male offsprings [92]. Hemophilia has gender implications, with a high risk of blame and stigmatization of the mother. In low-income countries where access to free-of-cost or subsidized treatment product is limited, guilt and distress severely affect the maternal quality of life.

Muscular Dystrophy

Muscular dystrophies are a group of inherited muscle disorders where there is a defect in normal muscle structure and function [94, 95]. The most common form of muscular dystrophy is facio-scapulohumeral muscular dystrophy, occurring in 7 per 1000 people. It appears in childhood as weakness of the muscles of the face and shoulder girdle and becomes evident by the age of 20 years. The condition progresses slowly, evidenced by the inability to whistle, close eyes or raise arms. The weakness develops variably, progressing slowly in some and rapidly in others. Visual problems due to retinal complications and sensorineural hearing loss are other complications [96, 97].

Duchenne (DMD) and Becker (BMD) muscular dystrophies are characterized by progressive muscular wasting and weakness due to degeneration of skeletal, smooth and cardiac muscles [96, 97]. DMD is typically identified at 2 or 3 years of age. Weakness of the lower limbs manifests as toe walking, waddling and lordosis. Children have difficulty in running, walking, rising from the floor or climbing stairs. Falls are frequent. One of the characteristics features is the development of pseudo-hypertrophy, where fatty and fibrous deposits in some key muscles, such as the calf muscles, make them appear enlarged. Weakness of the hip muscles makes it difficult for the child to rise from sitting or supine position. The child needs to support himself in a characteristic fashion to become upright (Gowers' sign) (Fig. 2.8). Weakness progresses steadily. Children become wheelchair bound by twelve years of age. Progressive degeneration of cardiac and respiratory muscles occurs. Death occurs due to respiratory complications by the end of the second decade of life. Around one third of children may have mild intellectual impairment that affects verbal abilities. BMD is a less severe form, where disease progression is slower. Children remain ambulatory till adolescence. Death occurs in the third or fourth decade of life. Treatment is mainly supportive [95].

Diagnosis is based on clinical signs and symptoms, measurement of creatine kinase levels (which is elevated due to muscle damage) and mutation analysis. DMD and BMD are X-linked recessive disorders, transmitted from the mother to the male offspring. Nearly one third of cases are sporadic in nature. The mutation causing DMD or BMD affects the dystrophin gene, one of the largest human genes, with 79 exons. Most DMD mutations are caused by deletions or duplications. The dystrophin gene encodes a protein that supports the sarcolemma (sheath covering the muscle cell) and links it to actin, a key molecule required for muscle contraction. The difference in the severity of DMD and BMD is due to the mutation. While mutations causing DMD cause loss or reduced levels of dystrophin, in the case of BMD, the mutation affects the structure of dystrophin. This allows some level of function, so that the age of onset of the condition is delayed. DMD affects 1/3300, and BMD affects 1/18,000–1/31,000 males [94, 95].

Fig. 2.8 Gowers' sign. From William Richard Gowers (1845–1915)—Gowers WR. Clinical lecture on pseudohypertrophic muscular paralysis. Lancet 1879; ii, 73–75

Achondroplasia

Achondroplasia (dwarfism) is an inherited single gene disorder where affected individuals have disproportionate short stature, with average height of 131 ± 5.6 cm for men and 124 ± 5.9 cm for women [98, 99] (Fig. 2.9). Short limbs with shortening of the proximal limb, macrocephaly (enlarged head), underdeveloped midface with depressed nasal bridge, short stubby trident hands with limited elbow extension, small chest, lordotic lumbar spine, enlarged buttocks and protrudent abdomen are other characteristic features. Newborns with achondroplasia are at a risk of hydrocephalus. Infants show delayed motor milestones due to hypotonia, but psychomotor

Fig. 2.9 Achondroplasia.
Image from page 515 of
Miller, Reginald "The
medical diseases of children"
(1911)

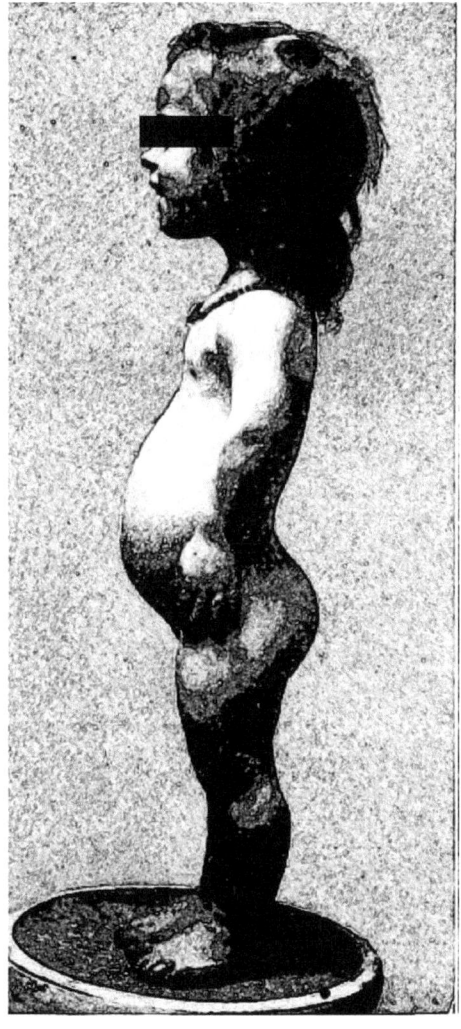

development is normal. Due to under-development of the midface, recurrent otitis media is common as is speech and articulation delay. There is a risk of obesity with age, which enhances joint problems. Leg and lower back pain are reported in nearly half of adults. Cardiorespiratory complications and sleep apnea are also common.

Most persons with achondroplasia have normal life expectancy, but there is an enhanced cardio-vascular mortality risk in midlife. Mortality risk in infancy is caused by abnormal development of the cranio-cervical junction and increased risk of apnea, leading to death. Small chest size is also responsible for restrictive pulmonary disease in some infants. To date, there is no treatment to enhance growth in people with achondroplasia.

Achondroplasia affects 1 in 25,000–30,000 live births worldwide [96, 97]. It is caused by a mutation in the fibroblast growth factor receptor (*FGF3*) gene that encodes a transmembrane protein that regulates linear bone growth. Although inherited in an autosomal dominant pattern, 90% of cases are sporadic that is there is no family history of this condition. Nearly, all mutations involve a glycine to arginine change at position 380 of *FGFR3*. The presence of this mutation in majority of individuals reduces the cost of DNA testing. In families without a history of the disorder, achondroplasia can be detected during the third trimester by shortened limbs, increased biparietal diameter and depressed nasal bridge. Prenatal diagnosis at 30 weeks brings in the difficult issue of late termination of the pregnancy. The recurrence risk of achondroplasia among siblings of a sporadic case was estimated at 0.23% (that is 1 recurrence per 443 siblings) [97]. Older paternal age is a risk factor for achondroplasia. Environmental factors that may increase the risk of achondroplasia have been speculated as agents that cause mutations in the paternal germ cells such as radiation, chemotherapy or genetic variations that affect DNA repair mechanisms.

The quality of life of nearly two thirds of adults with achondroplasia is affected by pain [99]. Reduction in ability to walk is responsible for loss of independence in the fourth or fifth decade of life, as the activities of daily living are affected. One of the major consequences of achondroplasia is the psychosocial impact of short stature, which causes reduced self-esteem, increased depression and withdrawal from social activities. The socioeconomic consequences of the disability are evident in that individuals with achondroplasia are more likely to have lower levels of education, have higher unemployment rates and have lower income.

Conclusion

The description of common birth defects, genetic disorders and developmental disabilities illustrates the chronic nature of these conditions and the spectrum of medical and rehabilitation needs. Hospitalizations and medical consultations are higher and required throughout life, either for correction of defects and deformities, for symptom control and for prevention of comorbidities. Rehabilitation services are essential for those birth defects that affect functioning and cause activity limitation. Early diagnosis and skilled management can lead to improvement of symptoms, reduction of complications and improved quality of life. Putting in place free-of-charge medical and rehabilitation services is the first step toward providing care. Certain types of birth defects, especially those that require surgery or expensive therapeutics, are high cost activities. Estimates from the USA show that although birth defects were responsible for only 3% of hospital admissions, they accounted for 5.2% of total hospital costs [100].

The disabling nature of many birth defects, however, highlights the necessity for assistance, not only for health and rehabilitation, but also for education and employment opportunities. The vulnerability caused by disability indicates the need to introduce and implement regulations to protect children against abuse. Estimates report that there is a 3.5 times and 2.88 times increased risk of physical and sexual violence against children with disabilities [101]. A difficult but essential service is the need to introduce palliative care services [102].

When not diagnosed, or not optimally treated, as is the case in many LMICs, birth defects affect growth and development, enhance morbidity episodes, lead to progressive disability and functional impairment and cause premature mortality. Empowerment of communities and community-based providers with knowledge on common birth defects and putting in place a medical and rehabilitation service with a well-defined referral system is the first step in addressing the needs of children and their caregivers. Providing knowledge on caring and management at home, identifying warning signs and understanding the steps to address medical emergencies are important. The need to locate rehabilitation services close to communities is important. Discontinuation of treatment, and non-compliance to medical advice is associated with limited knowledge about the condition and inability to sustain the financial costs of transportation [103]. Chapter 3 further elaborates the essential components of a birth defects service.

References

1. Bradshaw S, Bem D, Shaw K et al (2019) Improving health, wellbeing and parenting skills in parents of children with special health care needs and medical complexity—a scoping review. BMC Pediatr 19(1):301. https://doi.org/10.1186/s12887-019-1648-7
2. Baffa JM (2019) Overview of heart defects. Available at https://www.msdmanuals.com/home/children-s-health-issues/birth-defects-of-the-heart/overview-of-heart-defects?query=Overview%20of%20Congenital%20Cardiovascular%20Anomalies. Accessed 16 Aug 2019
3. Liu Y, Chen S, Zühlke L et al (2019) Global birth prevalence of congenital heart defects 1970–2017: updated systematic review and meta-analysis of 260 studies. Int J Epidemiol 48(2):455–463
4. Botto LD (2015) Epidemiology and prevention of congenital heart defects. In: Muenke M, Kruszka PS, Sable CA, Belmont JW (eds) Congenital heart disease: molecular genetics, principles of diagnosis and treatment. Karger Medical and Scientific Publishers, Basel, pp 28–45
5. Zimmerman MS, Smith AGC, Sable CA et al (2020) Global, regional, and national burden of congenital heart disease, 1990–2017: a systematic analysis for the Global Burden of Disease Study 2017. Lancet Child Adolesc Health 4(3):185–200

6. Gregory MRB, Prouhet PM, Russell CL et al (2018) Quality of life for parents of children with congenital heart defect: a systematic review. J Cardiovasc Nurs 33(4):363–371
7. Woolf-King SE, Anger A, Arnold EA, Weiss SJ, Teitel D (2017) Mental health among parents of children with critical congenital heart defects: a systematic review. J Am Heart Assoc 6(2):e004862
8. Mebius MJ, Kooi EM, Bilardo CM, Bos AF (2017) Brain injury and neurodevelopmental outcome in congenital heart disease: a systematic review. Pediatrics 140(1):e20164055
9. Edwards JJ, Gelb BD (2016) Genetics of congenital heart disease. Curr Opin Cardiol 31(3):235–241
10. Russell MW, Chung WK, Kaltman JR, Miller TA (2018) Advances in the understanding of the genetic determinants of congenital heart disease and their impact on clinical outcomes. J Am Heart Assoc 7(6):e006906
11. Jenkins KJ, Correa A, Feinstein JA et al (2007) Non-inherited risk factors and congenital cardiovascular defects: current knowledge: a scientific statement from the American Heart Association Council on Cardiovascular Disease in the Young: endorsed by the American Academy of Pediatrics. Circulation 115(23):2995–3014
12. Sullivan PM, Dervan LA, Reiger S, Buddhe S, Schwartz SM (2015) Risk of congenital heart defects in the offspring of smoking mothers: a population-based study. J Pediatr 166(4):978–984
13. Persson M, Cnattingius S, Villamor E et al (2017) Risk of major congenital malformations in relation to maternal overweight and obesity severity: cohort study of 1.2 million singletons. Br Med J 357:j2563.
14. Øyen N, Olsen SF, Basit S et al (2019) Association between maternal folic acid supplementation and congenital heart defects in offspring in birth cohorts from Denmark and Norway. J Am Heart Assoc 8(6):e011615
15. Boyd SAB (2020) Introduction to birth defects of the face, bones, joints and muscles. https://www.msdmanuals.com/professional/pediatrics/congenital-craniofacial-and-musculoskeletal-abnormalities/introduction-to-congenital-craniofacial-and-musculoskeletal-abnormalities. Accessed 18 Aug 2019
16. Mossey PA, Little J, Munger RG, Dixon MJ, Shaw WC (2009) Cleft lip and palate. Lancet 374(9703):1773–1785
17. Wehby GL, Cassell CH (2010) The impact of orofacial clefts on quality of life and healthcare use and costs. Oral Dis 16(1):3–10
18. Kadir A, Mossey PA, Orth M et al (2017) Systematic review and meta-analysis of the birth prevalence of orofacial clefts in low- and middle-income countries. Cleft Palate Craniofac J 54(5):571–581
19. Dixon MJ, Marazita ML, Beaty TH, Murray JC (2011) Cleft lip and palate: understanding genetic and environmental influences. Nat Rev Genet 12(3):167–178
20. Sivertsen Å, Wilcox AJ, Skjærven R et al (2008) Familial risk of oral clefts by morphological type and severity: population based cohort study of first degree relatives. BMJ 336(7641):432–434
21. Copp AJ, Adzick NS, Chitty LS et al (2015) Spina bifida. Nat Rev Dis Primers 1:15007. https://doi.org/10.1038/nrdp.2015.7
22. Falchek SJ (2018) Spina bifida. Manual Professional Version. Available at https://www.merckmanuals.com/en-ca/professional/pediatrics/congenital-neurologic-anomalies/spina-bifida?query=Neural%20Tube%20Defects%20and%20Spina%20Bifida. Accessed 16 Aug 2019
23. Williams J, Mai CT, Mulinare J, Isenburg J et al (2015) Updated estimates of neural tube defects prevented by mandatory folic acid fortification—United States, 1995–2011. MMWR Morb Mortal Wkly Rep 64(1):1–5
24. MRC Vitamin Study Research Group (1991) Prevention of neural tube defects: results of the Medical Research Council Vitamin Study. Lancet 338(8760):131–137
25. Bestwick JP, Huttly WJ, Morris JK, Wald NJ (2014) Prevention of neural tube defects: a cross-sectional study of the uptake of folic acid supplementation in nearly half a million women. PLoS ONE 9(2):e89354. https://doi.org/10.1371/journal.pone.0089354

26. Atta CA, Fiest KM, Frolkis AD et al (2016) Global birth prevalence of spina bifida by folic acid fortification status: a systematic review and meta-analysis. Am J Public Health 106(1):e24–e34
27. Zaganjor I, Sekkarie A, Tsang BL et al (2016) Describing the prevalence of neural tube defects worldwide: a systematic literature review. PLoS ONE 11(4):e0151586. https://doi.org/10.1371/journal.pone.0151586
28. Bhide P, Sagoo GS, Moorthie S et al (2013) Systematic review of birth prevalence of neural tube defects in India. Birth Defects Res A 97(7):437–443
29. Allagh KP, Shamanna BR, Murthy GV et al (2015) Birth prevalence of neural tube defects and orofacial clefts in India: a systematic review and meta-analysis. PLoS ONE 10(3):e0118961. https://doi.org/10.1371/journal.pone.0118961. Published 13 Mar 2015
30. Bhide P, Kar A (2018) A national estimate of the birth prevalence of congenital anomalies in India: systematic review and meta-analysis. BMC Pediatr 18(1):175
31. Kancherla V, Oakley Jr GP (2018) Total prevention of folic acid-preventable spina bifida and anencephaly would reduce child mortality in India: implications in achieving Target 3.2 of the Sustainable Development Goals. Birth Defects Res 110(5):421–428
32. Pit-ten Cate IM, Kennedy C, Stevenson J (2002) Disability and quality of life in spina bifida and hydrocephalus. Dev Med Child Neurol 44(5):317–322
33. Yi Y, Lindemann M, Colligs A, Snowball C (2011) Economic burden of neural tube defects and impact of prevention with folic acid: a literature review. Eur J Pediatr 170(11):1391–1400
34. Kahle KT, Kulkarni AV, Limbrick DD Jr, Warf BC (2016) Hydrocephalus in children. Lancet 387(10020):788–799
35. Boyd SAB. Macrocephaly. Merck Manual Professional Version. Available at https://www.merckmanuals.com/en-ca/professional/pediatrics/congenital-craniofacial-and-musculoskeletal-abnormalities/macrocephaly?query=Macrocephaly. Accessed 18 Aug 2019
36. Boyd SAB. Microcephaly. Merck Manual Professional Version. Available at https://www.merckmanuals.com/en-ca/professional/pediatrics/congenital-craniofacial-and-musculoskeletal-abnormalities/microcephaly. Accessed 16 Aug 2019
37. Miedzybrodzka Z (2003) Congenital talipes equinovarus (clubfoot): a disorder of the foot but not the hand. J Anat 202(1):37–42
38. O'Shea RM, Sabatini CS (2016) What is new in idiopathic clubfoot? Curr Rev Musculoskelet Med 9(4):470–477
39. Smith PA, Kuo KN, Graf AN et al (2014) Long-term results of comprehensive clubfoot release versus the Ponseti method: which is better? Clin Orthop Relat Res 472(4):1281–1290
40. Sewell MD, Rosendahl K, Eastwood DM (2009) Developmental dysplasia of the hip. BMJ 339:b4454
41. Woodacre T, Ball T, Cox P (2016) Epidemiology of developmental dysplasia of the hip within the UK: refining the risk factors. J Child Orthop 10(6):633–642
42. Woodacre T, Dhadwal A, Ball T, Edwards C, Cox PJA (2014) The costs of late detection of developmental dysplasia of the hip. J Child Orthop 8(4):325–332
43. Gambling TS, Long A (2019) Psycho-social impact of developmental dysplasia of the hip and of differential access to early diagnosis and treatment: a narrative study of young adults. SAGE Open Med 7:2050312119836010
44. Wilcox WR, Coulter CP, Schmitz ML (2015) Congenital limb deficiency disorders. Clin Perinatol 42(2):281–300
45. Vasluian E, van der Sluis CK, van Essen AJ et al (2013) Birth prevalence for congenital limb defects in the northern Netherlands: a 30-year population-based study. BMC Musculoskelet Disord 14(1):323
46. Johansen H, Østlie K, Andersen LØ, Rand-Hendriksen S (2016) Health-related quality of life in adults with congenital unilateral upper limb deficiency in Norway. A cross-sectional study. Disabil Rehabil 38(23):2305–2314
47. Boyd SAB. Polydactyly. Merck Manual Professional Version. Available at https://www.merckmanuals.com/professional/pediatrics/congenital-craniofacial-and-musculoskeletal-abnormalities/congenital-limb-abnormalities#v37841809. Accessed 19 Aug 2019

48. Boat TF, Wu JT et al (2015) Clinical characteristics of intellectual disabilities. In: Boat TF, Wu JT (eds) Mental disorders and disabilities among low-income children. Committee to Evaluate the Supplemental Security Income Disability Program for Children with Mental Disorders; Board on the Health of Select Populations; Board on Children, Youth, and Families; Institute of Medicine; Division of Behavioral and Social Sciences and Education; The National Academies of Sciences, Engineering, and Medicine. National Academies Press (US), Washington
49. AAIDD (American Association on Intellectual Developmental Disabilities) (2010) Intellectual disability: definition, classification, and systems of supports. AAIDD, Washington
50. Sachdev P, Blacker D, Blazer DG et al (2014) Classifying neurocognitive disorders: the DSM-5 approach. Nat Rev Neurol 10:634–642
51. May ME, Kennedy CH (2010) Health and problem behavior among people with intellectual disabilities. Behav Anal Pract 3(2):4–12
52. Roizen NJ, Patterson D (2003) Down's syndrome. Lancet 361(9365):1281–1289
53. Morris JK, Mutton DE, Alberman E (2002) Revised estimates of the maternal age specific live birth prevalence of Down's syndrome. J Med Screen 9(1):2–6
54. Public Health England. Public health functions to be exercised by NHS England. Service specification No. 16. NHS Down syndrome screening (trisomy 21) programme. Available at https://assets.publishing.service.gov.uk/government/uploads/system/uploads/attachment_data/file/256467/16_nhs_downs_syndrome_screening_trisomy_21.pdf. Accessed 19 Aug 2019
55. Mackie FL, Allen S, Morris RK, Kilby MD (2017) Cell-free fetal DNA-based noninvasive prenatal testing of aneuploidy. Obstet Gynecol 19(3):211–218
56. Natoli JL, Ackerman DL, McDermott S, Edwards JG (2012) Prenatal diagnosis of Down syndrome: a systematic review of termination rates (1995–2011). Prenat Diagn 32(2):142–153
57. Haddad F, Bourke J, Wong K, Leonard H (2018) An investigation of the determinants of quality of life in adolescents and young adults with Down syndrome. PLoS ONE 13(6):e0197394
58. Singh A, Yeh CJ, Verma N, Das AK (2015) Overview of attention deficit hyperactivity disorder in young children. Health Psychol Res 3(2):2115
59. Thomas R, Sanders S, Doust J, Beller E, Glasziou P (2015) Prevalence of attention-deficit/hyperactivity disorder: a systematic review and meta-analysis. Pediatrics 135(4):e994–e1001
60. Sikirica V, Flood E, Dietrich CN et al (2015) Unmet needs associated with attention-deficit/hyperactivity disorder in eight European countries as reported by caregivers and adolescents: results from qualitative research. Patient 8(3):269–281
61. Park HR, Lee JM, Moon HE et al (2016) A short review on the current understanding of autism spectrum disorders. Exp Neurobiol 25(1):1–13
62. Christensen DL, Braun K, Baio J et al (2018) Prevalence and characteristics of autism spectrum disorder among children aged 8 years—autism and developmental disabilities monitoring network, 11 sites, United States, 2012. Morb Mortal Wkly Rep 65(13):1–23
63. Baxter AJ, Brugha TS, Erskine HE, Scheurer RW, Vos T, Scott JG (2015) The epidemiology and global burden of autism spectrum disorders. Psychol Med 45(3):601–613
64. Ziats MN, Rennert OM (2016) The evolving diagnostic and genetic landscapes of autism spectrum disorder. Front Genet 7:65
65. Ten Hoopen LW, de Nijs PFA, Duvekot J et al (2020) Children with an autism spectrum disorder and their caregivers: capturing health-related and care-related quality of life. J Autism Dev Disord 50(1):263–277
66. Graham H, Rosenbaum P, Paneth N et al (2016) Cerebral palsy. Nat Rev Dis Primers 2:15082. https://doi.org/10.1038/nrdp.2015.82
67. O'Shea TM (2008) Diagnosis, treatment, and prevention of cerebral palsy in near-term/term infants. Clin Obstet Gynecol 51(4):816–828
68. Palisano RJ, Rosenbaum P, Bartlett D, Livingston MH (2008) Content validity of the expanded and revised gross motor function classification system. Dev Med Child Neurol 50(10):744–750

69. Tollånes MC, Wilcox AJ, Lie RT, Moster D (2014) Familial risk of cerebral palsy: population based cohort study. BMJ 349:g4294
70. Pousada M, Guillamón N, Hernández-Encuentra E et al (2013) Impact of caring for a child with cerebral palsy on the quality of life of parents: a systematic review of the literature. J Dev Phys Disabil 25(5):545–577
71. Korver AM, Smith RJ, Van Camp G et al (2017) Congenital hearing loss. Nat Rev Dis Primers 3:16094
72. Shearer AE, Hildebrand MS, Smith RJ (2017) Hereditary hearing loss and deafness overview. In: GeneReviews® (Internet). University of Washington, Seattle. Available at https://www.ncbi.nlm.nih.gov/books/NBK1434/. Accessed 1 Sept 2019
73. Goderis J, De Leenheer E, Smets K, Van Hoecke H, Keymeulen A, Dhooge I (2014) Hearing loss and congenital CMV infection: a systematic review. Pediatrics 134(5):972–982
74. World Health Organization (2016) Childhood hearing loss: strategies for prevention and care. Available at https://www.who.int/docs/default-source/imported2/childhood-hearing-loss--strategies-for-prevention-and-care.pdf?sfvrsn=cbbbb3cc_0. Accessed 1 Sept 2019
75. Gilbert C, Awan H (2003) Blindness in children. BMJ 327(7418):760–761
76. Gogate P, Gilbert C, Zin A (2011) Severe visual impairment and blindness in infants: causes and opportunities for control. Middle East Afr J Ophthalmol 18(2):109–114
77. Sheeladevi S, Lawrenson JG, Fielder AR, Suttle CM (2016) Global prevalence of childhood cataract: a systematic review. Eye 30(9):1160–1169
78. Wheatley CM, Dickinson JL, Mackey DA, Craig JE, Sale MM (2002) Retinopathy of prematurity: recent advances in our understanding. Br J Ophthalmol 86(6):696–700
79. Pizzarello L, Abiose A, Ffytche T et al (2004) VISION 2020: the right to sight: a global initiative to eliminate avoidable blindness. Arch Ophthalmol 122(4):615–620
80. Taher AT, Weatherall DJ, Cappellini MD (2018) Thalassaemia. Lancet 391(10116):155–167
81. National organization for rare disorders alpha thalassemia. Available at https://rarediseases.org/rare-diseases/alpha-thalassemia/. Accessed 1 Sept 2019
82. Origa R (2017) β-thalassemia. Genet Med 19(6):609–619
83. Kato GJ, Piel FB, Reid CD et al (2018) Sickle cell disease. Nat Rev Dis Primers 4:18010
84. Azar S, Wong TE (2017) Sickle cell disease: a brief update. Med Clin 101(2):375–393
85. Williams TN, Weatherall DJ (2012) World distribution, population genetics, and health burden of the hemoglobinopathies. Cold Spring Harb Perspect Med 2(9):a011692
86. Taylor SM, Parobek CM, Fairhurst RM (2012) Haemoglobinopathies and the clinical epidemiology of malaria: a systematic review and meta-analysis. Lancet Infect Dis 12(6):457–468
87. Aydinok Y, Erermis S, Bukusoglu N, Yilmaz D, Solak U (2005) Psychosocial implications of thalassemia major. Pediatr Int 47(1):84–89
88. Nahalla CK, FitzGerald M (2003) The impact of regular hospitalization of children living with thalassaemia on their parents in Sri Lanka: a phenomenological study. Int J Nurs Pract 9(3):131–139
89. McClish DK, Penberthy LT, Bovbjerg VE et al (2005) Health related quality of life in sickle cell patients: the PiSCES project. Health Qual Life Outcomes 3(1):50
90. Sabath DE (2017) Molecular diagnosis of thalassemias and hemoglobinopathies: an ACLPS critical review. Am J Clin Pathol 148(1):6–15
91. Franchini M, Mannucci PM (2012) Past, present and future of hemophilia: a narrative review. Orphanet J Rare Dis 7(1):24
92. Pr Claude Negrier (1997) Hemophilia. Orphanet: an online database of rare diseases and orphan drugs. Available at https://www.orpha.net/consor/cgi-bin/OC_Exp.php?Lng=EN&Expert=448. Accessed 26 Sept 2019
93. Kar A, Phadnis S, Dharmarajan S, Nakade J (2014) Epidemiology and social costs of haemophilia in India. Indian J Med Res 140(1):19–31
94. Pula S, Qunilivan R (2020) Duchenne and Becker muscular dystrophy. Orphanet: an online database of rare diseases and orphan drugs. Copyright INSERM 1997. Available at https://www.orpha.net/consor/cgi-bin/OC_Exp.php?Lng=EN&Expert=448. Accessed 26 Sept 2019

95. Rubin M. Duchenne and Becker muscular dystrophy. MSD Manual Professional Version. Available at https://www.msdmanuals.com/en-in/professional/pediatrics/inherited-muscular-disorders/duchenne-muscular-dystrophy-and-becker-muscular-dystrophy?query=Duchenne%20Muscular%20Dystrophy%20and%20Becker%20Muscular%20Dystrophy. Accessed 29 Sept 2019

96. Mettler G, Fraser FC (2000) Recurrence risk for sibs of children with "sporadic" achondroplasia. Am J Med Genet 90(3):250–251

97. Waller DK, Correa A, Vo TM et al (2008) The population-based prevalence of achondroplasia and thanatophoric dysplasia in selected regions of the US. Am J Med Genet A 146(18):2385–2389

98. Baujat G, Legeai-Mallet L, Finidori G, Cormier-Daire V, Le Merrer M (2008) Achondroplasia. Best Pract Res Clin Rheumatol 22(1):3–18

99. Pauli RM (2019) Achondroplasia: a comprehensive clinical review. Orphanet J Rare Dis 14(1):1

100. Arth AC, Tinker SC, Simeone RM, Ailes EC, Cragan JD, Grosse SD (2017) Inpatient hospitalization costs associated with birth defects among persons of all ages—United States, 2013. MMWR Morb Mortal Wkly Rep 66(2):41–46

101. Jones L, Bellis MA, Wood S et al (2012) Prevalence and risk of violence against children with disabilities: a systematic review and meta-analysis of observational studies. Lancet 380(9845):899–907

102. Meiring MA (2011) Caring for children with life-limiting and life-threatening illnesses: what the GP should know. Contin Med Educ 29(7):286–230

103. Kar A, Radhakrishnan B, Girase T, Ujagare D, Patil A (2020) Community-based screening and early intervention for birth defects and developmental disabilities: lessons from the RBSK programme in India. Disabil CBR Inclus Dev 31(1):30–46

Chapter 3
Thalidomide: Understanding the Responsibilities of a Birth Defects Service

Anita Kar

Abstract In the late 1950s and early 1960s, the teratogenic drug thalidomide caused an epidemic of serious birth defects. Thalidomide embryopathy in its severest form produced limb reduction defects and other malformations, causing severe disability in survivors. This article uses the thalidomide incident as a case study to illustrate the health and social welfare responses that were put into place to address the needs of children born with disabilities and their families. These health and social service activities represent the core functions of a birth defects service. While thalidomide left its impact on industrialized countries, fifty years later, another teratogen, the Zika virus, threatened low- and middle-income countries. Congenital Zika virus syndrome, another disabling birth defect, is marked by severe microcephaly and intellectual disability in survivors. The article briefly compares the global health response to this teratogen five decades after the thalidomide episode, identifying that health systems of low- and middle-income countries still lack the preparedness to address severe birth defects caused by an unexpected teratogen.

Keywords Congenital · Thalidomide · Disability · Zika virus · Congenital Zika virus syndrome · Public health · Health service · LMIC

The world witnessed the catastrophic consequences of the teratogenic drug thalidomide in the late 1950s and early 1960s. Thalidomide was used by pregnant women across 46 countries to manage symptoms of morning sickness. It caused an epidemic of severe birth defects that left children severely disabled. Birth defects (congenital anomalies, congenital malformations) are a group of clinically heterogeneous conditions that are of prenatal origin and that manifest as structural or functional abnormalities in the newborn [1]. Examples of some frequently encountered birth defects are congenital heart defects (CHDs), spina bifida, clubfoot (congenital talipes equinovarus, CTEV), cleft lip/palate and hydrocephalus. Severe birth defects are fatal, but non-fatal birth defects cause a deficiency or deformity in a body part, causing disability and medical complications. For example, untreated CTEV or developmental dysplasia of the hip or spina bifida leads to lifelong locomotor disability.

A. Kar (✉)
Birth Defects and Childhood Disability Research Centre, Pune, India

© Springer Nature Singapore Pte Ltd. 2021
A. Kar (ed.), *Birth Defects in India*,
https://doi.org/10.1007/978-981-16-1554-2_3

Isolated cleft lip/palate has severe cosmetic effects and may affect both functional (speech difficulty, difficulty in eating, repeated ear infections) and emotional well-being. CHDs are the most frequently encountered birth defect. Majority of CHDs require specialized medical management, and several CHDs require surgical correction. Intellectual disability, congenital hearing or visual impairments, and behavioural disorders like autism are complications associated with a number of birth defects. Rehabilitation services are needed for management of these conditions.

The need to provide public health services for children with such special health-care needs underlines the crux of the public health challenge of birth defects in low- and middle-income countries (LMICs). LMIC child health services are focused on treatment of acute conditions, and provision of supplemental nutrition and immunization. Services for children with chronic conditions and disabilities are either absent or rudimentary [2]. In addition to providing medical care and rehabilitation, other public health functions like surveillance and prevention are needed. The purpose of this article is to illustrate the services required to address birth defects. Using the thalidomide incident as a case study, this article describes how health and social welfare services responded to the large number of children with disabilities. The activities put in place by several countries constituted the essential public health activities to address birth defects. Over five decades later, a teratogen of a different kind, the Zika virus, a mosquito-borne infectious agent, caused a similar epidemic of severe birth defects in Brazil and few other countries. The numbers of children born with microcephaly and neurological complications underscored the reality of unexpected teratogens. They emphasized the need for health service preparedness for addressing the consequences of unforeseen teratogenic agents, especially in LMICs, where long-term care for children with disabilities may be undeveloped and health services may be unable to meet all needs of children and caregivers.

Thalidomide

Thalidomide (alpha-phthalimido-glutarimide) (Fig. 3.1) is a synthetic derivative of glutamic acid [3, 4]. The drug had been first synthesized and tested by the pharmaceutical company Ciba, but it was discarded as it failed to show any obvious effects

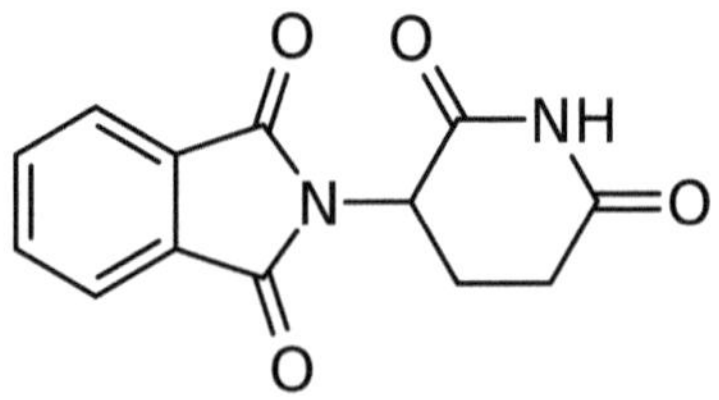

Fig. 3.1 **Structure of thalidomide**. From https://commons.wikimedia.org/wiki/File:Thalidomide_structure.svg

in experimental animals. The drug was investigated once again by a German pharmaceutical company, Chemie Grünenthal. Initial clinical trials failed to show any significant activity. However, the drug appeared to have a unique sedative property. Unlike barbiturates, thalidomide induced prolonged sleep but did not cause death, even after an overdose. This observation led to the drug being marketed as a safe, non-lethal, non-addictive, non-barbiturate sedative [5]. The drug was first introduced in Germany in May 1957 in an anti-flu medication, Grippex, and then, in October of the same year, as a sedative, Contergan. It was available over the counter without a prescription, dispensed widely in hospitals and mental institutions and even used to "quiet a child for an electroencephalogram" [5]. Thalidomide also found its way into a variety of medical compounds, such as Asmaval for asthma, Tensival for hypertension, Entero-Sediv for dysentery and Valgraine for migraine. Furthermore, free samples were distributed to physicians to distribute to patients. Sometimes shortly after its launch, the drug was used off-label as an anti-emetic for morning sickness.

The teratogenicity of the drug was unknown to doctors and the public at that time [6]. Thalidomide has two enantiomeric states. The dextro form (the S isomer) is more strongly teratogenic than the levo form (R isomer), which also harbours the sedative property of the drug. At physiological pH, the S and R forms rapidly interconvert (racemize), making it difficult to make a stable, non-teratogenic form of thalidomide. Furthermore, within its half-life of 8–12 h in the body, the drug undergoes non-enzymatic hydrolysis, resulting in multiple metabolites which also have teratogenic properties [7]. Drug testing norms were not stringent in the mid-1950s, so that scientific evidence on drug safety was limited [7]. Teratogenicity tests were not mandatory as a part of clinical trials for drugs during this time. Additionally, thalidomide exhibited species-specific differences in teratogenicity. For example, the teratogenic effect on rodents was less severe than that in rabbits [8]. In studies conducted subsequently, a median lethal dose (LD_{50}) could not be established in rodent models [9].

Even before the teratogenic effects of the drug were reported, side effects of thalidomide were being observed by medical practitioners [5]. The toxicity of the drug was evident as irreversible peripheral neuritis after prolonged use. Other neurological effects such as dizziness, loosing balance, loss of memory, trembling and systemic effects such as constipation, petechial haemorrhage and hypotension were being reported. The manufacturer ignored these reports. In April 1961, following several complaints from physicians, the drug was placed under prescription in Germany.

Thalidomide Distribution Across the World

After its popularity in Germany, the drug was licensed for marketing/manufacturing to 17 different companies. The distribution of thalidomide spanned across 46 countries. It included eleven European, seven African, seventeen Asian and eleven North and South American countries [10]. Available literature on the impact of thalidomide is available from seven countries like Australia, Canada, Japan, Germany and the UK [11]. During its short market life between 1958 and 1961/62, the drug was marketed under different names. The drug was sold as Distaval in UK, Australia and New Zealand by the Distillers Company (now Diageo). It was sold as Isomin in Japan, and in Canada it was marketed as two brands, Kevadon and Talimol. It was sold as Sedalis in Brazil and as Softenon in Europe [3]. The total number of women who used thalidomide, before or during pregnancy, was never known.

There were at least two instances where the drug was not licensed for marketing, as regulators were dissatisfied with the available safety data. In the US, Frances Kelsey, a U.S.Food and Drug Administration (FDA) regulator was cautioned by early reports of irreversible peripheral neuropathy after thalidomide use. She demanded additional data from the manufacturer, especially about the effect of the drug on pregnant animals. Even though the licensing application was submitted several times, thalidomide was never licensed in the USA. Despite this, the manufacturer, through 1200 medical practitioners, conducted unregulated clinical trials, and at least 20,000 doses of thalidomide were distributed to trial participants that included pregnant women. Approximately 40 children were born with thalidomide embryopathy/syndrome in the USA. Turkey also prevented a similar disaster. Professor ST Aygun, a virologist at Ankara University, was unconvinced about the safety of the drug and prevented imports of thalidomide into Turkey between 1958 and 1961 [10].

Thalidomide Embryopathy/Syndrome

Soon after the drug was marketed in Germany, it found off-label use in the treatment of morning sickness. The first birth defect, reported in December of 1956, was that of a baby born with anotia (ear malformation). Ironically, the mother was the wife of a Grünenthal employee [5]. From the 1960s, there were reports of increase in the numbers of a rare but severe birth defect, characterized by a unique limb deformity [12]. The first case series of 33 children was presented in 1961 by German clinicians [5]. A number of German clinicians attempted to identify the teratogen. The association between the birth defect and thalidomide use by mothers in early pregnancy was noted by Widukind Lenz, a German paediatrician [5, 13, 14]. By November 1961, Lenz was convinced that these severe birth defects were caused by thalidomide. The drug manufacturers were alerted, and the drug was withdrawn in Germany in 1961. In the meantime, similar birth defects had been observed in other countries. In Australia, McBride in a communication to the medical journal Lancet reported a 20% rise in this unique birth defect [15]. McBride's concerns and communication with the manufacturer led to Distaval being removed from Australia within a few days after it was taken off the market in Germany. The sale of thalidomide was stopped in the UK, and the Swedish manufacturer Astra withdrew the drug from the market on the same day.

Although reports of severe birth defects had started emerging from as early as 1956/57, thalidomide was withdrawn from the market in 1961/62. The withdrawal of the drug was not uniform across the world. For example, the drug was banned in Japan nearly 9 months later than that of European countries. Another challenge was that thalidomide was present in different formulations, and marketed under different names. As such, physicians unaware of the composition continued its prescription. Thus, the teratogenic effects of thalidomide were observed till more than two years after the drug had been withdrawn [4].

The true numbers of thalidomide affected were unknown, as there was no surveillance for either the numbers of women who received the drug or the number of affected births. The numbers of cases of thalidomide embryopathy are reported to be over 10,000. Germany had the largest numbers of survivors at 2700. In the UK, 2000 babies were born of whom around 466 survived. Thalidomide was responsible for reproductive loss, as foetuses with severe defects would have died in utero or been miscarried or been stillborn due to severe internal defects of the heart and the kidneys. Without a birth defects surveillance, data on the reproductive loss associated with thalidomide remains known from small observational studies [16]. The combined stillbirth and neonatal mortality rate among 29 affected pregnancies was 45%. In another report from Australia, of 50 women who had received thalidomide in the first trimester of pregnancy, 20% reported miscarriage, 30% had babies with thalidomide syndrome, while 50% escaped the teratogenic effects of the drug [17].

Thalidomide embryopathy/thalidomide syndrome refers to the spectrum of anomalies observed in infants exposed in utero to thalidomide. In mild conditions, only digits of the hand were affected, but the most severe consequences of thalidomide were drastic limb anomalies [12]. In its severest form, the birth defect manifested as complete absence of limbs (amelia) and as phocomelia where the long bones were affected. Severe phocomelia presented as a flipper-like structure, with the hand plate articulated to the body. Less severe forms of phocomelia were more common, characterized by missing or reduced long bones. Lower limb anomalies were less common and rarely seen alone. Other anomalies included damage to the ears (anotia and microtia), eyes (microphthalmia, anophthalmos, coloboma, strabismus), and internal organs including heart, kidney, gastrointestinal and genital defects [12]. Vertebral column problems and facial palsy were also presenting features of thalidomide embryopathy. Another unique feature was the appearance of a transient facial hemangioma which extended from the forehead to the lip, but which disappeared within the first two years of life. These anomalies presented either individually or as multiple malformations in the child. There were individual variations in the defects caused by thalidomide. Some defects were more frequent than others (Table 3.1), but not all defects were likely to be present in all children.

Like other teratogenic drugs, thalidomide exerted its teratogenic effects during a narrow window after conception [6] (Fig. 3.2). This sensitive window of teratogenic effects was between 20 and 36 days post-fertilization that is 34–50 days after the last menstrual period, coinciding with the period of early embryogenesis and organogenesis. Thalidomide exposure before the critical period caused miscarriage in humans and rats, while exposure after 36 days had no morphological effects on the embryo/foetus. There was a remarkable relationship between the affected organ system and the timing of thalidomide exposure. Thalidomide intake immediately

Table 3.1 Phenotypic description of thalidomide embryopathy based on frequency of occurrence [12]

Frequent presentation (30–79%)	Occasional presentation (5–29%)
Abnormality of cardiovascular system	Anotia
Abnormality of the fibula	Chronic rhinitis
Aplasia or hypoplasia of the femur	Hearing impairment
Aplasia or hypoplasia of the humerus	Insulin resistance
Aplasia or hypoplasia of the thumb	
Aplasia or hypoplasia of the ulna	
Preaxial hand polydactyly	
Radial club hand	
Short stature	
Triphalangeal thumb	
Upper limb phocomelia	

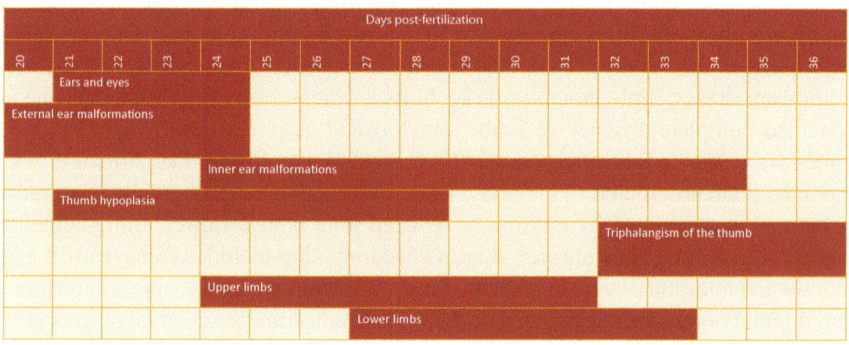

Fig. 3.2 Timing of thalidomide exposure and affected organ system. Numbers represent days post-fertilization

into the critical period (day 21–24) affected the development of the ears and eyes, and caused external ear malformations (day 20–24), inner ear malformations (day 24–34), thumb hypoplasia (day 21–28) and triphalangism of the thumb (day 32–36). Use of thalidomide between day 24 and 31 affected upper limbs, while day 27–33 of exposure affected lower limb development. The teratogenic potency of thalidomide was exemplified by the fact that as little as a single 50 mg capsule was sufficient to result in a malformation. Like other teratogens, there were instances where there was no foetal harm after thalidomide exposure. A dose–response relationship has however not been reported for thalidomide. A large body of research has investigated the mode of action of thalidomide, with data suggesting that there may be multiple pathways by which the drug could affect embryogenesis [3, 4, 6, 7, 9].

Some of the presenting features of thalidomide embryopathy were not unique to this teratogen, but also associated with other genetic syndromes. Roberts syndrome (pseudo-thalidomide syndrome) has a striking similarity to thalidomide embryopathy, with tetraphocomelia being the main feature of the syndrome [3, 4]. Other defects such as the Okihiro syndrome or the Holt–Oram syndrome also cause limb deficiency. As thalidomide can phenocopy these single gene disorders, establishing causation was difficult in thalidomide-related litigation.

Health and Social Impact of Thalidomide

Clinical Needs of Children

In terms of medical needs, the severity of the malformation required immediate surgical interventions, for example, correcting congenital heart defects and urogenital malformations. In other instances, the severity of the malformation resulted in early neonatal death. A major focus was on correcting the physical deformity [18]. Evaluation of children was done by a multidisciplinary team consisting of paediatrician and orthopaedic surgeon, physical medicine and rehabilitation specialist, supported by radiological and laboratory investigations. Other specialities were involved in addressing other problems, such as vision problems or anotia. The latter was frequently associated with severe deafness to total hearing loss.

A considerable effort was dedicated to the development of orthotics and prosthetics [19]. For children with phocomelia, the emphasis was placed on prosthetics for upper arms, although it was believed that it would be "quite unrealistic to compare their performance with normal upper limb functions" [20]. Prosthetics for lower limbs were found not to be very effective as children could only cover small distances and that too very slowly. Staircases were not manageable, and children required help in putting on and getting out of the prosthetics. Basic activities like feeding, dressing and toileting were dependent on assistance. Despite this intensive effort, some thalidomide survivors accepted and continued using these devices [20].

Impact of Disability: Schooling

The severe physical impairment affected children in several ways. In one of the earliest studies, Gingras et al. identified the need for counselling and support, as children were affected by the consequences of the disability [21]. Toddlers appeared to be aware about their disability. Early adolescence heralded the child's anxiety and emotional conflicts, and the deformity contributed to a sense of self-devaluation. The sense of inferiority caused by the disability led adolescents to limit their social contacts and conceal the disability. Other studies identified that even when there was no major emotional disorder, mild emotional conflicts were always present due to the deformity.

The physical impairment caused by thalidomide affected schooling and social integration of children. Limb malformations affected daily functions such as dressing and eating, and made school work difficult [22]. Lower limb malformations limited play and social interactions with other children. Hearing difficulties affected articulation. Facial paresis in many children gave a blank facial expression and conveyed the impression of the child being slow. As Smithells noted, "a combination of deafness and bilateral facial palsy makes it difficult to know what is going on in the child's mind" [16]. Several studies conducted at that time, showed that thalidomide survivors had normal intelligence. Significantly, the physical limitation of the child was not a factor in adaptation to the school environment. Rather, the child's abilities were determined by the attitude of the mother [23]. Gingras et al. noted that the emotional adjustment and acceptance of the impairment by the mother had a direct effect in the integration of children in school [21].

Impact on Parents

Early observations noted that parents were acutely affected by the child's disability. A range of psychosocial responses were observed. These ranged from grief, and inability to accept the child's disabling deformity, to acceptance and over-protection. One of the reactions was parental rejection, so that children were reared in care homes and denied a home environment [21]. There were reports of infanticide [24]. The severe emotional impact of the deformity on parents identified the need for counselling and support to assist them to cope and accept the anomaly. Most parents learnt to accept the situation over time, and this was dependent on the education of parents [18]. The child's disability however continued to affect normal family life in several ways. It resulted in marital discord and break-up of families. Repeated visits to hospitals, sometimes for surgery to permit fitting of assistive devices, disrupted family life and affected siblings. Family counselling and support was integral to tide over the initial shock associated with the birth of a child with a disability.

Health and Welfare Needs with Ageing

Reports and studies identified that the impairments at birth were compounded with age in thalidomide survivors (reviewed in [11]). One of the major consequences of ageing was musculoskeletal disorders, especially joint-related complications, associated with limb deformities, and overuse of the functional limb. Musculoskeletal problems involving the neck, shoulders and spine reduced work capacity, leading to early retirement. Increased physical disability and pain affected independence and self-esteem, and increased dependence on family members for personal assistance. The impact of the impairment and dependence was starkly highlighted in a description on challenges of correcting worsening eyesight of ageing thalidomide survivors. "... the problem with thalidomiders is how to correct their sight when they have no ears to hold their glasses on and no arms or hands to put in contact lenses ..." [23]. Other issues that arose with age included growing obesity and continence issues.

Newbronner and Atkin [11] conducted a scoping review of the literature on needs of ageing thalidomide survivors and observed that the largest number of studies was biomedical and most focused on musculoskeletal problems. Several studies addressed chronic pain and neuropathic symptoms among survivors. A small number of studies examined dental status and oral functions of thalidomide survivors, alluding poor dental status to, among other causes, the use of teeth as a tool. Other studies identified deterioration of eyesight and hearing among survivors who were born with these impairments. The increased prevalence of hypertension and obesity was reported, as was the challenge in measuring hypertension in persons with limb deformity. Studies identified mental health issues among thalidomide survivors, identifying a twofold increased risk of lifetime prevalence of depressive disorders [25]. The study identified that depression was associated with poor social networks and unemployment.

The Response to the Thalidomide Tragedy

Health and Social Sector Service Response

The epidemic of thalidomide-disabled children required appropriate responses from the health and social welfare services. In Canada, for example, 93 "thalidomide-deformed" children were identified by 1963, half of whom had severe disability. Free hospital and diagnostic services were available through the publicly funded insurance plan introduced in Canada in 1957, but the federal government allotted an additional half a million dollars for thalidomide victims and their families [26]. This funding was available for payment of doctor's fees, surgery, drugs, prosthetic devices and psychological assessment and to support special education, vocational training and custodial care when children were given up for adoption. As parents had to forsake employment for care of children, these funds also provided for income supplements.

With the understanding that the medical, social and rehabilitation requirements of the children "pose problems of no small magnitude and complexity", a programme of rehabilitation care for children was devised by a committee of experts [26]. Teams made up of paediatrician, orthopaedic surgeon/and or physiatrist, an occupational therapist, a physiotherapist, a prosthetist and a social worker were drawn from all Canadian provinces and trained in rehabilitation of children with severe limb defects.

Three special research and training units for prosthetic services and rehabilitation were set up, although none of the prosthetics ultimately assisted the thalidomide victims. In fact, upper limbs were found to be more functional without prosthetics because the weight and complexity of the device. Thus, in addition to medical services and resources to support caregivers, physical and social rehabilitation made up essential components of services to assist children with congenital disabilities.

Birth Defects Surveillance

Another significant development was the need for birth defects surveillance, as there was a need "to improve the quantity, quality and availability of information on the occurrence of congenital anomalies", arising from a need for "more detailed, complete and reliable information on congenital anomalies in general" [26]. Within a few years, several countries established birth defects surveillance programmes to monitor new teratogens. The rationale was to use vital records, as birth defects appeared to show an "undramatic endemicity"; that is, the rates remained constant over time [17]. As such, outbreaks, recognized as atypical increased occurrence of birth defects, could be rapidly detected. Birth defects surveillance would therefore create the opportunity to investigate and identify the teratogen and introduce appropriate measures to prevent exposure. The task was not simple, as there were challenges associated with classification, investigation and statistical interpretation of data. The thalidomide tragedy led to establishment of birth defects surveillance systems. In the USA, the Centres for Disease Control and Prevention started the Metropolitan Atlanta Congenital Defects Programme in 1967 [27]. The International Clearinghouse for Birth Defects Surveillance and Research was founded in 1974 as an aftermath of the thalidomide tragedy [28].

Drug Regulation

One of the major impacts of the thalidomide tragedy was stringent regulation on the licensing of drugs [29]. The tragedy highlighted the need for rigorous testing, in more than one species, as teratogenicity was species sensitive. The Australian Drug Evaluation Committee was established in June 1963 as a response to the thalidomide tragedy. Even though not majorly affected by thalidomide, a significant number of changes were introduced in the USA [30, 31]. An outdated Food, Drug and Cosmetics Act of 1938 was amended in 1962. The Kefauver Harris Amendments, as it was termed, mandated that the sponsor of a drug had to provide the FDA with detailed protocol and data on pre-clinical findings. Sponsors (usually pharmaceutical companies) were required to monitor and report the progress of the trial, and investigators were to maintain complete records including data on whom the drugs were being dispensed to. It was necessary to disclose the qualifications of the investigators. Subsequently, pharmacovigilance, "the science and activities relating to the detection, assessment, understanding and prevention of adverse events or any other drug-related problem", evolved with a global impetus to establish a system to monitor the side effects of drugs [31]. Regulations regarding drug labelling were first introduced in the USA. In 1979, the FDA introduced labelling for prescription drugs used during pregnancy. This labelling permitted the classification of drugs based on the risk to the foetus during pregnancy. It served to alert physicians about the risk and side effects of the drug, especially in terms of causing birth defects [32].

Abortion Legislation

Another consequence of thalidomide was a wide public, political and religious debate on abortion. Till this time, abortion was associated with a sense of immorality, and abortion services were not very visible. Widespread media coverage of thalidomide-affected children, and the predicaments of parents, influenced public sentiments about abortion. It instigated a rethinking on abortion under specific instances, such as the man-made catastrophe of thalidomide. It helped in removing the stigma of immorality associated with abortion. The impact of thalidomide, and a rubella epidemic of the 1940s and 1960s moved the abortion agenda to the political sphere. It paved the way for the enactments that led to legalizing abortion in several countries, including the UK in 1967 and in France in 1975 [33].

Elective termination of pregnancy for foetal malformation was however not an easy discussion [33]. An early proponent of termination of pregnancy for foetal anomalies was Dr. Norman Gregg, who had described the teratogenic effects of German measles during pregnancy. Babies who survived the lethal effects of the infection had a high possibility of being born with hearing and visual impairments, and congenital heart defects. Gregg and other physicians considered these births as "ruinous" to mothers, being emotionally painful, traumatic, overwhelmingly exhausting and leading to social isolation. While these were the arguments that justified therapeutic abortions, debates about this difficult issue have continued, with thinking encompassing the rights of the woman and the unborn child.

Compensation and Economic Support

There was initially no financial and community-based care for thalidomide-affected families, who were left to deal with the physical, social and financial impact of caring for a child with a disability. Thalidomide survivors did receive compensation from pharmaceutical companies, but this was after prolonged litigation [34]. Thalidomide injury set up the precedence for litigation against drug companies by individuals experiencing drug-related injury. Often, the funds provided by companies were insufficient and needed to be supplemented by government funds. In Germany, for instance, thalidomide compensation payments included in addition to a one-off compensation payment, lifetime monthly thalidomide pension. In the UK, thalido-mide survivors receive economic support from the Thalidomide Trust (established by the pharmaceutical company), supplemented by the UK Health Grant. The latter grant, implemented through the National Health Service, promoted independent living among thalidomide survivors. The grant permitted expenditure for indepen-dent mobility (such as wheelchairs or taxi rides), home adaptations that encouraged independent living, communication technology to protect against harm and vulner-ability, medical costs that were not covered under the National Health Service and respite, that is, therapeutic breaks for individuals and their caregivers. A Wellbeing Service, a psychological therapy service and social care support, was also available for thalidomide survivors and their caregivers.

Regulated Reintroduction of Thalidomide

Thalidomide was licensed for use in the treatment of erythematous nodosum leprosy in 1998. Because of the severe teratogenic effects of the drug, marketing was and is permitted in industrialized nations under a restricted distribution programme, where only registered physicians and pharmacists are allowed to prescribe and distribute the drug. Patients are educated on the risk of severe birth defects and the need to comply with the regulations. Such regulations were important, as identification of the efficacy of thalidomide in the treatment of erythematous nodosum leprosy and multiple myeloma resulted in several clinical investigations to test the potential of thalidomide for a range of immunosuppressive disorders [9].

Thalidomide in Developing Countries

During the peak year of thalidomide sales in 1961, 25% of the drug was exported to developing countries including seven African countries: Angola, Ghana, Guinea, Mozambique, Somalia, Sudan and Western Africa. Smaller amounts entered Uganda and Kenya through informal routes [10]. There is very little literature on how thalido-mide affected these and other developing countries where the drug may have been prescribed or used by women.

The consequence of unregulated distribution of thalidomide was reported from Brazil, where thalidomide was available for the treatment of leprosy [35]. Over 30 cases of thalidomide syndrome were reported, underlining the need for stringent drug regulations in LMICs. Thalidomide was not licensed in India till 2002. But since the early 1970s, it was brought into the country by international non-governmental orga-nizations that were involved in providing care to leprosy patients. With limited data, it is difficult to understand whether India, like Brazil, had cases of thalidomide embry-opathy. The cases from Brazil, however, highlighted that drug regulations, stringent implementation, and education of healthcare professionals and the community on teratogenic medications were critical in preventing birth defects. Awareness was of utmost importance, as medical, physical and social rehabilitation services are not in place for thalidomide survivors [36].

Lessons for a Birth Defects Service

Thalidomide affected thousands of families, illustrating the catastrophic potential of teratogens. An Editorial in the Canadian Journal of Medical Research in 1963 hoped that "… in the long term, something of value may be salvaged from the thalidomide tragedy", and that the activities to support children and caregivers might in the long run assist other children born with birth defects [26]. The thalidomide case study illustrated some key public health functions and responsibilities of a birth defects service.

- Medical and physical rehabilitation services

Like thalidomide embryopathy, neonates with malformations require medical services, primarily surgical and orthopaedic services for correcting the defect, either immediately after birth in the neonatal period or early in childhood. Several birth defects require care for existing medical conditions, complications and co-morbidities. As observed with thalidomide survivors, medical needs are lifelong and increase with age. Public health resource allocation has to be carefully estimated to accommodate cumulative cohorts of birth defect survivors.

Providing services for children with disabilities is an important responsibility of the birth defects service. Organizing physical rehabilitation therapies and ensuring that referral pathways are in place to link rehabilitation and medical services are essential to reduce/ameliorate the consequences of disabilities. Ensuring referral to organizations providing assistive devices is also a key function of a birth defects service. Keeping in mind the need to respect the capabilities and wishes of children and adults with disabilities is equally important. For thalidomide survivors, most of the fitted prosthetic devices were subsequently not used, as children and adults found alternative ways to overcome functional impairment.

- Social rehabilitation and economic support

Equally important is the need for social welfare services, ensuring that parents can obtain information on education and employment opportunities, and information on rights of children and adults with disabilities. Providing respite services for caregivers is another important social welfare service. Respite care provides an opportunity to parents to de-stress and rest from the physically and emotionally difficult task of caregiving. The need for economic support is equally important. For thalidomide-affected, initially families were left to care for children without any economic support, but subsequently, financial support was provided through social welfare services (and from long drawn-out compensation from thalidomide manufacturers).

- Psychosocial support

The distress experienced by parents underlines the importance of psychosocial support as another essential function of a birth defects service. Like parents of thalidomide survivors, parents of children with disabilities experience emotional stress and may encounter or perceive stigma. Mothers especially are frequently confronted with lack of support from the spouse and family members. Birth defects services need to develop a network of parent–patient organizations, where parents can be referred for counselling support.

- Surveillance

The thalidomide incident identifies the essential role of surveillance. Birth defects surveillance with regular collection and monitoring of data is the first step towards detecting a teratogenic exposure. Cases identified through the surveillance system can be investigated in order to identify the teratogenic exposure. The utility of birth defects surveillance was especially illustrated during the Zika virus outbreak, where surveillance data formed the basis of establishing the association between this teratogen and congenital Zika syndrome (CZS) [37].

- Prevention

Prevention activities initiate with increasing awareness about teratogens and birth defects, especially among women intending a pregnancy or among pregnant women. Teratogenic effects are most severe when the exposure occurs in the early stages of pregnancy, at a time when women may not be aware of the event. Health promotion messages targeted at women in the preconception and the inter-conception period are important. Depending on the teratogen, other preventive activities are implemented. This may range from withdrawal (or regulating) a potentially teratogenic drug, to rubella immunization, or vector control measures in case of CZS.

Underlying all these activities would be the need for developing the infrastructure for providing care and capacity building activities. Health systems of industrialized countries were capable of providing care during the thalidomide episode, but additional budgetary provisions had to be implemented. For LMICs, birth defects services would need to be carefully designed within existing resources of the maternal and child health service.

Zika Virus and the Public Health Response

Nearly five decades later, another teratogen, the Zika virus, caused a cluster of cases of severe birth defects in Brazil and some other countries. Zika virus was first reported from Uganda in 1947, and the first human case was identified in 1962–63. Spread by two mosquito species, *Aedes aegypti* and *Aedes albopictus*, the benign nature of Zika virus disease and lack of specific serological diagnostics contributed to the reporting of only 20 cases over 60 years from Africa and Asia. In 2001, the virus spread to the

Pacific, causing the first human outbreak in the Yap Islands of the Federated States of Micronesia [38].

The symptoms of Zika virus disease are mild fever accompanied by a rash, arthralgia and conjunctivitis. In 2013–14, an outbreak of 42 cases of Guillain-Barre syndrome (GBS) in French Polynesia was linked to the Zika virus [39]. In late 2014, an epidemic characterized by an exanthemic rash was reported from north-east Brazil, and by September 2015, an increase in numbers of microcephaly cases was reported [40]. A causal association between Zika virus and CZS was subsequently established [37]. A retrospective analysis identified a similar increase in numbers of microcephalic births during the Zika virus outbreak in French Polynesia. Till July 2019, autochthonous Zika virus transmission has been reported from 87 countries. The highest number of cases (over 3000) was in Brazil.

Unlike thalidomide, the implications of this pandemic were enormous. Tropical countries (where the mosquito vectors were widespread) were at a high risk of jeopardizing health system achievements with an epidemic of Zika virus-associated birth defects. In accordance with International Health Regulations, the World Health Organization (WHO) in February 2016 declared Zika virus infection associated with microcephaly and other neurological symptoms as a Public Health Emergency of International Concern (PHEIC) [41]. With the threat of continued spread of Zika virus in areas where the vector was active, the WHO encouraged member countries to develop long-term plans for tackling the spread of Zika virus [42].

Congenital Zika Virus Syndrome

CZS is a severe birth defect, characterized by severe microcephaly with partially collapsed skull, thin cerebral cortices with sub-cortical and intracranial calcifications, macular scarring and focal pigmentary retinal mottling, congenital contractures and marked early hypotonia with extra-pyramidal signs [43]. Cardiac and renal malformations are common. Zika virus infection in late pregnancy is associated with normal head circumference, but with significant brain damage. Only 5–15% women exposed to the infection experienced pregnancy complications [38]. Like thalidomide embryopathy, CZS was an extremely severe birth defect with lifelong caregiving implications, as children have intellectual deficit, epilepsy, cerebral palsy, delayed language and motor development, contracture of limbs, high muscle tone, eye abnormalities and hearing loss.

Public Health Response

In contrast to thalidomide, there was a global response to the Zika virus outbreak. In addition to the leadership role of the WHO and the declaration of Zika virus infection as a PHEIC, guidelines for controlling the outbreak, prevention of infection among pregnant women and care of pregnant women and infants were rapidly developed

[44]. Research contributed to the speedy development of laboratory test kits and reagents, and their distribution globally, which led to uniformity in case detection. Research identified sexual transmission, transmission via body fluids, as well as silent transmission, where infected women did not manifest signs of disease. A significant public health contribution was the establishment of a causal association between Zika virus infection during pregnancy and CZS [37]. The establishment of Zika virus pregnancy surveillance was invaluable in understanding the consequences of maternal infection on foetuses and infants [45].

The Zika virus outbreak identified lack of access to contraceptives, especially among women in endemic settings. Improving access to contraceptives to decrease unintended pregnancies, especially in areas of ongoing Zika virus transmission, was a significant step in prevention of the complications of Zika virus infection during pregnancy. The Zika virus outbreak generated further focus on routine and new vector control strategies [44]. Although surveillance for the virus (rash like illness during pregnancy) was established, this key indicator was challenged by silent transmission and transmission through the sexual route and through blood transfusion.

Health System Challenges

The unexpected numbers of children with microcephaly impacted the Brazilian health system [46, 47]. Providing care to babies born with microcephaly who require early stimulation, and family support was challenging. It required development of guidelines and specialized training for physical therapists, psychologists, physicians and audiologists. Quality of rehabilitation services was ensured through a process of certification of these centres. The epidemic differentially affected vulnerable societies in Brazil, with the risk being higher in more socio-economically vulnerable communities [48, 49]. A legislation was passed for monthly financial support for poor families with a child with microcephaly.

Impact on Families

Like thalidomide, CZS severely affected families, [50] especially, mothers [51, 52]. Zika virus-related microcephaly was associated with negative emotions, maternal anxiety, fear of stigma, poor self-esteem and poor body image. The WHO issued guidelines for psychosocial support of pregnant women or women (and families) with an infant with CZS [53]. The WHO guidelines reminded healthcare providers on the importance of supportive communication. Like thalidomide, the risk of birth defects caused by Zika virus infection during pregnancy reignited the issue of abortion, which is not legal or highly restricted in several South American countries. The request for abortions increased significantly in Latin American countries.

In conclusion, five decades after the thalidomide incident, the Zika virus pandemic identified that an epidemic of birth defects is still possible. The unexpected nature of the Zika virus outbreak caught health systems of LMICs unprepared to tackle the large numbers of cases of children with disabilities. Thalidomide and Zika virus thus flag the importance of birth defects surveillance and for having services for medical care, rehabilitation and social services in place. Having in place a birth defects service would ensure that health systems are not unduly stressed by an unanticipated teratogen and an epidemic of children with disabilities.

References

1. Stevenson RE, Hall JG (2015) Introduction. In: Stevenson RE, Hall JG, Everman DB, Solomon BD (eds) Human anomalies and related anomalies, No. 66. Oxford University Press
2. Were WM, Daelmans B, Bhutta Z, Duke T, Bahl R, Boschi-Pinto C, Bhan MK (2015) Children's health priorities and interventions. BMJ (Clin Res Ed) 351:h4300
3. Vargesson N (2013) Thalidomide embryopathy: an enigmatic challenge. ISRN Dev Biol 241016. https://doi.org/10.1155/2013/24101
4. Vargesson N (2015) Thalidomide-induced teratogenesis: history and mechanisms. Birth Defects Res C Embryo Today 105(2):140–156
5. Taussig HB (1962) A study of the German outbreak of phocomelia. The thalidomide syndrome. JAMA 180:1106–1114
6. Kim JH, Scialli AR (2011) Thalidomide: the tragedy of birth defects and the effective treatment of disease. Toxicol Sci 122(1):1–6
7. Ito T, Ando H, Handa H (2011) Teratogenic effects of thalidomide: molecular mechanisms. Cell Mol Life Sci 68(9):1569–1579
8. Emanuel M, Rawlins M, Duff G, Breckenridge A (2012) Thalidomide and its sequelae. Lancet 380(9844):781–783
9. Rajkumar SV (2004) Thalidomide: tragic past and promising future. Mayo Clin Proc 79(7):899–903
10. Clow B (2008) Defining disability, limiting liability: the care for thalidomide victims in Canada. In: Heaman EA, Li A, McKellar S (eds) Essays in honor of Michael Bliss: figuring the social. University of Toronto Press
11. Newbronner E, Atkin K (2018) The changing health of thalidomide survivors as they age: a scoping review. Disabil Health J 11(2):184–191
12. Fetal thalidomide syndrome. Genetic and Rare Disease Information Centre, National Centre for Advancing Translational Sciences, National Institute of Health. Available at https://rarediseases.info.nih.gov/diseases/2313/fetal-thalidomide-syndrome. Accessed 8 Nov 18
13. Smithells RW, Newman CGH (1992) Recognition of thalidomide defects. J Med Genet 29:716–723
14. Lenz W (1988) A short history of thalidomide embryopathy. Teratology 38(3):203–215
15. McBride W (1961) Thalidomide and congenital malformations. Lancet 1:358
16. Smithells RW (1973) Defects and disabilities of thalidomide children. Br Med J 1(5848):269–272
17. Ingalls TH, Klingberg M (1965) Implications of epidemic embryopathy for public health. Am J Public Health Nations Health 55(2):200–208
18. Mongeau M, Gingras G, Sherman ED, Hebert B, Hutchison J, Corriveau C (1966) Medical and psychosocial aspects of the habilitation of thalidomide children. Can Med Assoc J 95(9):390–395
19. Martin JK, Rathbun JC (1963) Habilitation of patients with congenital malformations associated with thalidomide: pediatric aspects. Can Med Assoc J 88(19):959–962
20. Bent N, Tennant A, Neumann V, Chamberlain MA (2007) Living with thalidomide: health status and quality of life at 40 years. Prosthet Orthot Int 31(2):147–156

21. Gingras G, Mongeau M, Moreasult P, Dupuis M, Herbert B, Corriveau C (1964) Congenital anomalies of the limbs II. Psychological and educational aspects. Can Med Assoc J 91:115–119

22. D'Avignon M, Hellgren K, Juhlin I, Atterbäck B (1967) Diagnostic and habilitation problems of thalidomide-traumatized children with multiple handicaps. Dev Med Child Neurol 9:707–712

23. Support for Australia's thalidomide survivors. Submission 4. The experience of UK-thalidomide damaged people (thalidomiders) and the British government's financial support. Available at file:///C:/Users/jayanta%20Pal/Downloads/sub04_UKTCT.pdf. Accessed 2 Nov 2018

24. Klausen SM, Parle J (2015) 'Are we going to stand by and let these children come into the world?': the impact of the 'thalidomide disaster' in South Africa, 1960–1977. J South Afr Stud 41(4):735–752

25. Peters KM, Albus C, Lüngen M, Niecke A, Pfaff H, Samel C (2015) Damage to health, psychosocial disorders and care requirements of thalidomide survivors in North Rhine West-phalia from a long-term perspective. Expert Opinion Commissioned by LZG.NRW Federal Health Centre North Rhine Westphalia (LZG.NRW), Westerfeldstraße. Available at https://www.thalidomidetrust.org/wp-content/uploads/2016/10/The-Cologne-Report.pdf. Accessed 8 Nov 2018

26. Editorial (1963) Rehabilitation of thalidomide-deformed children. Can Med Assoc J 83:488–489

27. Correa-Villaseñor A, Cragan J, Kucik J, O'Leary L, Siffel C, Williams L (2003) The Metropolitan Atlanta Congenital Defects Program: 35 years of birth defects surveillance at the Centers for Disease Control and Prevention. Birth Defects Res A Clin Mol Teratol 67(9):617–624

28. Botto LD, Robert-Gnansia E, Siffel C, Harris J, Borman B, Mastroiacovo P (2006) Fostering international collaboration in birth defects research and prevention: a perspective from the International Clearinghouse for Birth Defects Surveillance and Research. Am J Public Health 96(5):774–780

29. Junod SW. FDA and clinical drug trials: a short history. US Food and Drug Adminis-tration. Available at https://www.fda.gov/downloads/AboutFDA/History/ProductRegulation/UCM593494.pdf. Accessed 8 Nov 2018

30. Tantibanchachai C (2014) US regulatory response to thalidomide (1950–2000). Embryo Project Encyclopedia. ISSN: 1940-5030 https://embryo.asu.edu/handle/10776/7733

31. Fornasier G, Francescon S, Leone R, Baldo P (2018) An historical overview of pharmacovig-ilance. Int J Clin Pharm 40:744

32. Mosley JF 2nd, Smith LL, Dezan MD (2015) An overview of upcoming changes in pregnancy and lactation labeling information. Pharm Pract (Granada) 13(2):605

33. Parker C (2012) From immorality to public health: thalidomide and the debate for legal abortion. Aust Soc Hist Med 25(4):863–880

34. Lachmann PJ (2012) The penumbra of thalidomide, the litigation culture and the licensing of pharmaceuticals. QJM 105(12):1179–1189

35. Sales Luiz Vianna F, Kowalski TW, Fraga LR, Sanseverino MT, Schuler-Faccini L (2017) The impact of thalidomide use in birth defects in Brazil. Eur J Med Genet 60(1):12–15

36. Moro A, Invernizzi N (2017) The thalidomide tragedy: the struggle for victims' rights and improved pharmaceutical regulation. Hist Cienc Saude Manguinhos 24(3):603–622

37. Rasmussen SA, Jamieson DJ, Honein MA, Petersen LR (2016) Zika virus and birth defects—reviewing the evidence for causality. N Engl J Med 374(20):1981–1987

38. Baud D, Gubler DJ, Schaub B, Lanteri MC, Musso D (2017) An update on Zika virus infection. Lancet 390(10107):2099–2109

39. Sampathkumar P, Sanchez JL (2016) Zika virus in the Americas: a review for clinicians. Mayo Clin Proc 91(4):514–521

40. França GV, Schuler-Faccini L, Oliveira WK, Henriques CM, Carmo EH, Pedi VD et al (2016) Congenital Zika virus syndrome in Brazil: a case series of the first 1501 livebirths with complete investigation. Lancet 388(10047):891–897

41. World Health Organization. Zika virus and complications: 2016 public health emergency of international concern. Available at https://www.who.int/emergencies/zika-virus-tmp/en/. Accessed 8 Nov 2018
42. World Health Organization. Zika strategic response plan revised for July 2016-December 2017. World Health Organization. Available at https://apps.who.int/iris/bitstream/handle/10665/246091/WHO-Zikavirus-SRF-16.3-eng.pdf;jsessionid=7B8090D79000C866C2A2D 8BDD9ACD76F?sequence=1. Accessed 8 Nov 2018
43. Moore CA, Staples JE, Dobyns WB, Pessoa A, Ventura CV, Fonseca EB et al (2017) Characterizing the pattern of anomalies in congenital Zika syndrome for pediatric clinicians. JAMA Pediatr 171(3):288–295
44. Oussayef NL, Pillai SK, Honein MA et al (2017) Zika virus—10 public health achievements in 2016 and future priorities. MMWR Morb Mortal Wkly Rep 65:1482–1488
45. Honein MA, Dawson AL, Petersen EE et al (2017) Birth defects among fetuses and infants of US women with evidence of possible Zika virus infection during pregnancy. JAMA 317(1):59–68
46. Castro MC (2016) Zika virus and health systems in Brazil: from unknown to a menace. Health Syst Reform 2(2):119–122
47. Gómez EJ, Perez FA, Ventura D (2018) What explains the lacklustre response to Zika in Brazil? Exploring institutional, economic and health system context. BMJ Glob Health 3(5):e000862
48. Skråning S, Lindskog BV (2017) The Zika outbreak in Brazil: an unequal burden. Tidsskr Nor Laegeforen 137(22). Available at https://tidsskriftet.no/en/2017/11/global-helse/zika-out break-brazil-unequal-burden. Accessed Aug 2018
49. Souza WV, Albuquerque MFPM, Vazquez E, Bezerra LCA, Mendes ADCG, Lyra TM et al (2018) Microcephaly epidemic related to the Zika virus and living conditions in Recife, Northeast Brazil. BMC Public Health 18(1):130
50. Bailey DB Jr, Ventura LO (2018) The likely impact of congenital Zika syndrome on families: considerations for family supports and services. Pediatrics 141(Suppl 2):S180–S187
51. Dos Santos Oliveira SJG, Dos Reis CL, Cipolotti R, Gurgel RQ, Santos VS, Martins-Filho PRS (2017) Anxiety, depression, and quality of life in mothers of newborns with microcephaly and presumed congenital Zika virus infection: a follow-up study during the first year after birth. Arch Womens Ment Health 20(3):473–475
52. Diniz D (2016) Zika virus and women. Cad Saude Publica 32(5):e00046316
53. World Health Organization (2016) Psychosocial support for pregnant women and for families with microcephaly and other neurological complications in the context of Zika virus. Interim guidance for health-care providers. WHO/ZIKV/MOC/16.6. World Health Organization, Geneva. Available at https://www.who.int/csr/resources/publications/zika/psychosocial-support/en/. Accessed 8 Nov 2018

Part II
Surveillance, Registries and Magnitude

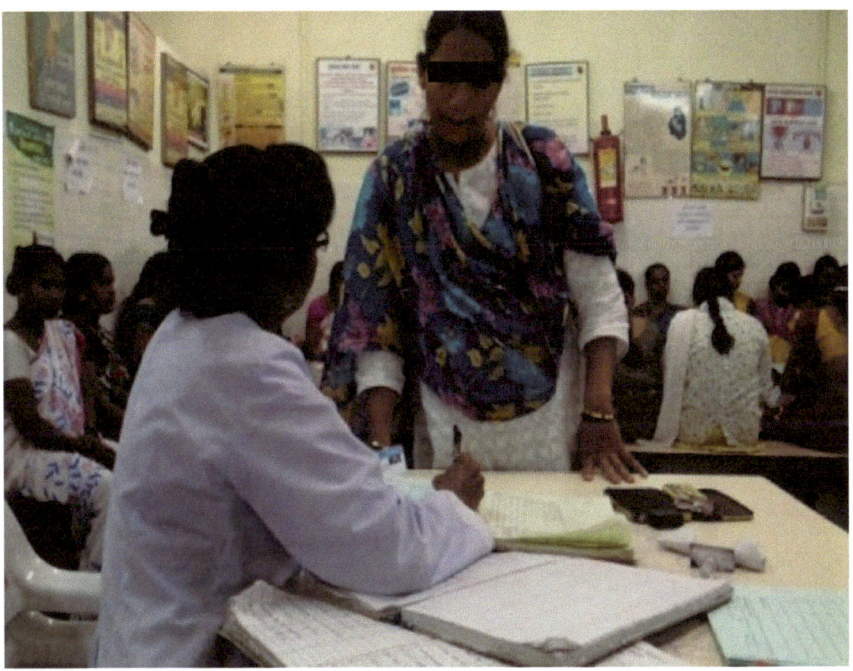

Part II explores the magnitude of birth defects and developmental disabilities in India. The first article in this part (4. "Birth Defects Surveillance in India") explains the key issues in birth defects surveillance. The article highlights the methodological problems that may be encountered while establishing birth defects surveillance in India. The second article (5. "Rare Disease Registries: A Case Study of the Haemophilia Registry in India") describes rare disease registries, and Indian initiatives. The article uses the example of the haemophilia registry in India, to illustrate how a partnership between a global organization and the national haemophilia organization has led to a national registry for this relatively uncommon single-gene disorder. Maintained by volunteers, this registry yields a wealth of data on the epidemiology of haemophilia in India. The next two articles (6. "Magnitude of Congenital Anomalies in India" and 7. "Magnitude of Developmental Disabilities in India") describe the challenge of estimating the prevalence of these conditions. The reviews however point to very large numbers of affected births, high numbers of neonatal and child deaths and significant numbers of survivors with disabilities. The final article in this part (8. "Magnitude and Characteristics of Congenital Disabilities in India") uses the only available national dataset to estimate the magnitude of survivors with congenital disabilities in India.

Chapter 4
Birth Defects Surveillance in India

Prajkta Bhide

Abstract Surveillance for birth defects is an essential public health function for informing policy. Surveillance systems are vital for understanding the magnitude and distribution of birth defects in a popualtion as well as for detecting teratogenic exposures during pregnancy. Instances like the rubella-associated outbreak of congenital cataract, severe limb defects associated with thalidomide during the early 1960s, and more recently, the Zika virus outbreak have emphasized the need for birth defects surveillance. This article introduces the key elements required for ensuring a surveillance system that yields rigorous data. It explores the challenges faced by low- and middle-income countries and discusses birth defects surveillance initiatives in India. The article concludes with an argument for establishing birth defects surveillance in order to monitor not only birth defects but also maternal and perinatal health parameters in the country.

Keywords Birth defects · Congenital anomalies · Surveillance · India

Background

The impact of congenital anomalies is dissimilarly perceived among high-income countries (HICs) and low- and middle-income countries (LMICs). In HICs, congenital anomalies are the largest contributor to deaths in children under five years [1] and are responsible for a significant burden of disability-adjusted life years (DALYs) during the neonatal period and childhood years [2]. In contrast, in LMICs dearth of data on the magnitude of congenital anomalies is responsible for their invisibility as issues of public health interest. The perception of congenital anomalies as rare self-limiting conditions, their lower proportionate contribution to mortality compared to other infectious causes and their resource intensive management, have been the key reasons leading to their neglect in LMICs [3–5]. This under prioritization as a public health issue has in turn led to lack of national agendas to address congenital anomalies, including the absence of surveillance systems. Lack of surveillance, that

P. Bhide (✉)
Birth Defects and Childhood Disability Research Centre, Pune 411020, India

is lack of a system to count the number of affected individuals cycles back to the invisibility of the issue. As LMICs undergo epidemiological transition in the rates and causes of mortality, a shift in the causes of child mortality from predominantly infectious to non-communicable conditions including congenital anomalies is now evident [6]. Assuming a low prevalence of congenital anomalies in LMICs would be misleading for various reasons. Higher birth rates in LMICs and large prevalence of risk factors like nutritional deficiencies, maternal morbidities including intrauterine infections and teratogenic exposures suggest a significant number of affected births in such settings [4, 7].

Birth Defects Surveillance

Public health surveillance has been described as 'the ongoing, systematic collection, analysis and interpretation of health-related data essential to planning, implementation and evaluation of public health practice' [8]. It forms the cornerstone of any public health practice. The objectives of surveillance are multifold. Data from surveillance systems are critical for monitoring the burden of a disease, detecting changes over time and place including outbreaks, determining risk factors and identifying at-risk populations. Quality data, generated through surveillance, are also needed for planning and advocacy for public health actions and for evaluation of prevention and control programmes. Data from surveillance systems can, in addition, guide research and policies on the broader social and economic impacts of the disease beyond the health domain.

Although many birth defects are individually uncommon, their collective impact is substantial and their human cost is steep. As such, birth defects surveillance is essential for assessing their burden with the ultimate aim of preventing these adverse health conditions and their complications [9]. While several birth defects are fatal, a good proportion of survivors suffer from lifelong disability and need continuous medical care. In the absence of adequate care, which is an unfortunate reality in many LMICs, the toll exacted on individuals and families is considerable. In 1959, the World Health Organization (WHO) planned the first international study on the epidemiology of birth defects. The intention was to determine the occurrence and types of congenital malformations and the study was carried out in 24 centres across 16 countries [10]. Despite its limitations, the study served to acknowledge the worldwide occurrence of birth defects. By 2000, it was evident that there was a paucity of reliable global epidemiological data on birth defects, especially from developing countries [11]. This did not signify that there were no studies during those years. There were many isolated studies across the world that looked at the epidemiology of birth defects, but it was not projected on the global stage. It was the March of Dimes Global Report on Birth Defects in 2006 that catapulted the problem of birth defects into the international arena for the first time [4]. The report's estimate greatly exceeded prevalent projections for genetic and partially genetic (multifactorial) birth defects and identified the five most common globally occurring birth defects as congenital heart defects

(CHDs, 1,040,835 births), neural tube defects (NTDs, 323,904 births), haemoglobin disorders (thalassaemia and sickle cell anaemia, 307,897 births), Down syndrome (trisomy 21) (217,293 births) and glucose-6-phosphate dehydrogenase deficiency (177,032 births). It highlighted the lack of quality epidemiological data from LMICs bringing it into the context of the Millennium Development Goals [4]. Subsequently, in 2010, the WHO during its sixty-third World Health Assembly (WHA) adopted a resolution that recognized the contribution of birth defects to global neonatal and child mortality. It called on member countries to prioritize birth defects, raise awareness about their effects and build in-country capacity for the prevention and care of affected individuals.

Establishing a surveillance system for birth defects was considered an integral component of the overall national health information system [12]. Firstly, knowing the relative frequencies of common birth defects can help the government prioritize services targeted to those specific defects, estimating the necessary economic impact and planning appropriate health services. An example would be talipes equinovarus (clubfoot), which when detected early can be corrected with a low-cost intervention [13], thus avoiding lifelong locomotor disability in the child. Secondly, surveillance also aids in impact evaluation. Data on the prevalence of NTDs collected from surveillance systems were used to evaluate the effectiveness of intervention, namely mandatory folic acid fortification on the decrease in occurrence of NTDs in the USA [14]. Additionally, a major goal of surveillance is to identify risk factors. Understanding the risk factors for birth defects is important as many of these risk factors are shared with other adverse pregnancy outcomes, so targeting them is likely to positively impact other maternal and child health outcomes as well. This is of special concern in resource-poor settings where maternal risk factors for birth defects are highly prevalent. Birth defects have complex etiologies varying from fully or partly genetic causes to a mix of gene-environmental interactions and teratogenic exposures like physical agents, environmental pollutants, maternal illnesses and infections and certain drugs [15]. Some established teratogens are hyperthermia [16], certain chemical endocrine disruptors [17], pregestational diabetes [18], rubella virus [19] and drugs like sodium valproate [20], thalidomide [21] and alcohol [22]. While approximately half of all congenital anomalies occur due to unknown causes [15], it is estimated that among the rest 4–8% can be attributed to teratogenic exposures [15, 23].

The earliest known infective cause for a birth defect was described in 1941 by an Australian ophthalmologist, Dr. Norman Gregg, who reported the occurrence of congenital cataracts in infants born to women with rubella infection during pregnancy [19]. Given the absence of systematic surveillance in those times, this description was based on his observations alone. Another example is of the drug thalidomide. In the 1950s, thalidomide was released as a non-addictive and non-barbiturate sedative. Soon, it was discovered to be an effective anti-emetic and began to be widely used in treating morning sickness in pregnant women. Before long reports started emerging of wide-spread occurrence of major defects (estimated at more than 10,000 affected) like severe limb reduction defects including phocomelia, which soon came to be identified as a characteristic of thalidomide embryopathy. Finally, in 1961, the association between thalidomide and severe birth defects was independently confirmed by two physicians, McBride in Australia [21] and Lenz in Germany [24]. This thalidomide epidemic demonstrated to the world the need to establish surveillance systems to monitor birth defects and identify teratogenic exposures and their harmful effects on the foetus. The importance of birth defects surveillance was further reinforced, when surveillance data indicated an increase in the number of newborns with microcephaly in the state of Pernambuco, Brazil during 2015–16 [25]. The spike in the number of cases was obvious in comparison with previous data from the region. Later, investigations into the epidemic revealed an association with Zika virus infection during pregnancy [26].

Knowledge of risk factors for birth defects can benefit birth defects prevention through targeting the removal of risk factors (e.g., avoiding maternal infections and mitigating illnesses) and the reinforcement of protective factors (e.g., folic acid supplementation) through preconception care packages. Epidemiological studies provide insight into the genetic and environmental factors contributing to birth defects. While prospective cohort studies are ideal for capturing the true association between risk factors and birth defects, these necessitate a very large sample size in order to reach the adequate number of anomaly cases required for analysis. This can be extremely troublesome, especially if the exposure is also a rare event. In this regard, birth defect registries and population-based surveillance systems can be used for identifying novel risk factors and deciphering associations. Since birth defects are typically rare conditions, using data from registries and surveillance systems is practical. This is especially pertinent for common anomalies like CHDs and NTDs. However, certain epidemiological challenges like precise exposure assessment and adequate statistical power need to be addressed.

One example of a large study based on surveillance data is the US National Birth Defects Prevention Study that includes cases identified from population-based birth defect surveillance systems in eight states [27]. The study aims to evaluate genetic and environmental factors associated with major birth defects in the country. Such a large-scale collaborative effort would have enough statistical power to detect associations as well as the capacity to generate new hypotheses. Another example is the European Surveillance of Congenital Anomalies (EUROCAT): a network of congenital anomaly registries (all population-based) in Europe [28]. An example of hospital-based surveillance is Estudio Colaborativo Latino Americano de Malformaciones Congenitas (ECLAMC), which includes a network of maternity hospitals reporting more than 200,000 births per year across 10 countries in South America [29]. Collaborative efforts, such as these with uniform methodologies, help in the identification and investigation of changes in frequency of congenital anomalies in relation to potential teratogenic exposures over large regional populations. As these systems evolved in their respective regions, it became apparent that for the work to have a true global impact, it was necessary to ensure coordination and cooperation. With this view, in 1974, representatives of birth defects surveillance programmes from 10 countries came together and established the International Clearinghouse for Birth Defects Surveillance and Research (ICBDSR). The ICBDSR network has now grown to include 42 programmes from 36 countries [30].

Birth Defects Surveillance: Factors Affecting Reporting of Prevalence Data

The magnitude of birth defects is typically reported using prevalence estimates [31]. Unlike for other diseases/health conditions, true incidence of birth defects cannot be measured. This is due to the sizable proportion of lethal anomalies resulting in unrecognized or early pregnancy loss [31]. Prevalence rates can be influenced by various factors including the characteristics of and methodologies adopted by the different surveillance systems, and the effects of these may in addition be defect specific [32]. In such a scenario, comparisons across systems on a global scale become difficult in the absence of standardized methodology and may often be misleading. The following factors need to be considered when setting up a birth defects surveillance programme.

The WHO defines congenital anomalies as structural or functional anomalies (e.g., metabolic disorders) that occur during intrauterine life and can be identified prenatally, at birth, or sometimes later in infancy [33]. Often the terms 'congenital anomalies', 'birth defects', 'congenital disorders' and 'congenital malformations' are used interchangeably. However, there exist subtle differences that have important considerations for surveillance. For instance, the International Statistical Classification of Diseases and Related Health Problems, 10th Revision (ICD-10) definition of congenital anomalies includes congenital malformations, deformations and chromosomal abnormalities but does not include inborn errors of metabolism [34]. Single-gene disorders which account for over 30% of all congenital disorders worldwide [3] are dispersed throughout ICD-10 and hence are often not part of surveillance systems.

The primary consideration when setting up surveillance is to delineate the scope, that is, the population and geographical area under surveillance. Here, there are two possible approaches: (1) population-based surveillance and (2) hospital-based surveillance. Population-based surveillance programmes compile data from the entire population in a defined geographical area. Cases include all foetuses or neonates having a congenital anomaly who are born to resident mothers living in the defined area, and this forms the numerator for calculating prevalence rates. The denominator includes all births to resident mothers for that area. Thus, all births to resident mothers are included, irrespective of whether they occurred at home or in a healthcare facility. A resident mother delivering outside the designated area is also included in the calculation of prevalence (Fig. 4.1). A non-resident mother delivering in the defined area is not included in the prevalence calculations. Anomaly data in a given area can be collected from various sources such as through periodic survey of the community by health workers or from all mapped health facilities in the designated area. Other sources of data include vital statistics records, and data from referral treatment centres in a given area, provided affected children are referred to these centres [9]. Population-based surveillance programmes are ideal for measuring the magnitude of congenital anomalies in a given geographical region as they have well-defined catchment areas. Such programmes yield accurate estimates of magnitude and data that can be easily

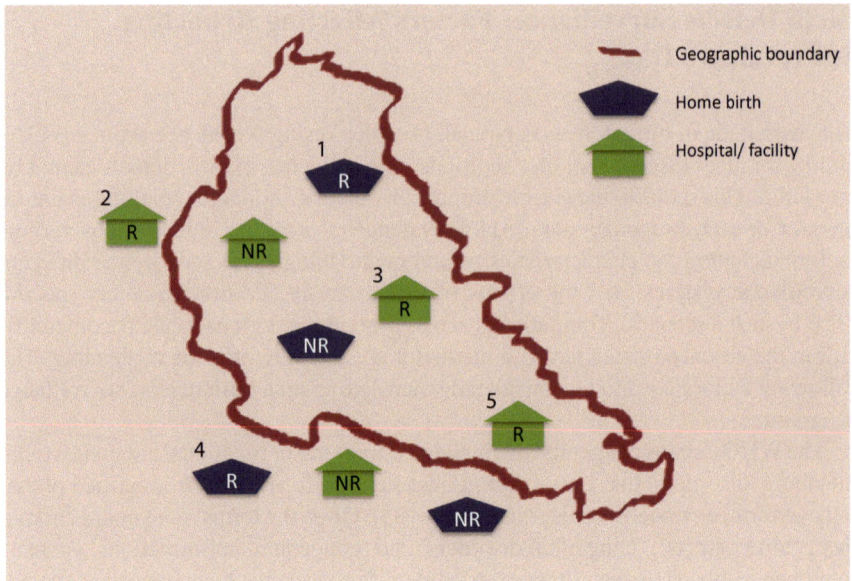

Fig. 4.1 Population-based surveillance. Five cases registered are only of resident mothers (R), even if the birth has occurred outside the geographic area under surveillance. Data of non-resident mothers (NR) are not included

generalized to the population. However, their resource-intensive nature makes them difficult to establish and sustain in developing countries.

In contrast, hospital-based surveillance programmes collect data on pregnancy outcomes occurring in certain select hospitals in a defined geographic area. The numerator includes all cases (foetuses or neonates) identified with an anomaly at any of the participating hospitals, and the denominator includes all births (both live and stillbirths) occurring at the participating hospitals. Unlike population-based surveillance, the residence status of mothers does not matter for inclusion in hospital-based surveillance (Fig. 4.2). Data on births from non-participating hospitals in the area are not included in the analysis.

Hospital-based surveillance programmes are relatively easier to sustain as they are less resource intensive. However, they suffer from some serious drawbacks as they are prone to inherent biases and may not truly represent the population under study. This is especially true for LMICs where unlike HICs, catchment areas of hospitals are not well defined. Patient mobility is high as most healthcare expenses are out-of-pocket. Inclusion of large public hospitals that serve as referral centres can overestimate the magnitude of congenital anomalies while inclusion of smaller centres will underestimate the true magnitude. The estimates of prevalence generated by such a hospital-based system can only be applied to the limited population born in those hospitals. Furthermore, only those anomalies that are detected until discharge are reported. Accurate assessment is also hampered by the limited number of qualified

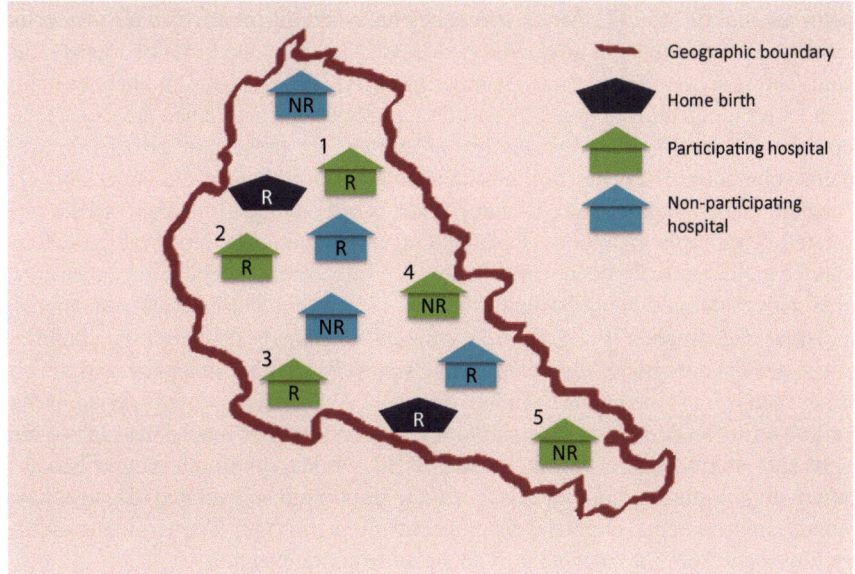

Fig. 4.2 Hospital-based surveillance. Five cases occurring only at participating hospitals are registered, irrespective of whether the mother is a resident (R) or a non-resident (NR)

healthcare personnel in such settings. Despite these issues, the comparative ease of collecting hospital-based data, makes it the feasible data source for LMICs that lack rigorous birth defects surveillance systems. A subset of hospital-based surveillance, known as sentinel surveillance, involves setting up surveillance in key facilities in order to gather data for rapid estimation of birth defects prevalence. Despite certain limitations, this approach is best suited for resource-poor settings.

Prevalence estimates are also influenced by the definitions and methods of classification used in the surveillance programme. Accurate classification and coding of congenital anomalies using standardized classification systems are important for estimating prevalence and also for ensuring comparability of data across surveillance systems. Most countries have adopted the WHO recommended ICD-10 coding system to classify congenital anomalies [9]. Congenital anomalies are included under Chapter XVII 'congenital malformations, deformations and chromosomal abnormalities' (formerly the Q chapter) of ICD-10 which lists congenital disorders with structural effects regardless of cause and excludes inborn errors of metabolism. The chapter excludes majority of single-gene disorders and disorders due to genetic risk factors. The ICD-10 codes for congenital anomalies are alphanumeric (Q00–Q99) and are distributed across 11 blocks for various organ systems [34]. Another widespread classification system in use is the British Paediatric Association coding system extension that uses a five digit code and covers all medical conditions of childhood [35]. Congenital anomalies are also classified as major or minor depending on the severity of their consequences. Most surveillance systems focus on reporting only major anomalies [36, 37]. Major anomalies have significant adverse effects on the health and development of affected individuals and require medical or surgical treatment. They may also have serious cosmetic impacts [36, 37]. Minor anomalies usually have lesser medical, functional or cosmetic implications and hence are usually not reported in surveillance programmes unless they are present in association with major anomalies. They may at times indicate other problems in the infant [36, 37]. Congenital anomalies can also be categorized based on their clinical presentation as isolated or multiple anomalies. Isolated congenital anomalies are those conditions where a major anomaly occurs in isolation in a foetus/baby, without any other unrelated major anomaly being present. Majority (75%) of major congenital anomalies are isolated anomalies [9]. A multiple congenital anomaly describes the condition where two or more major unrelated anomalies occur together in a baby, and the co-occurrence is presumed to be a random event [9]. A pattern of related anomalies that are known to be derived from a single primary anomaly or mechanical factor and represents a cascade of events is a sequence [9], while a syndrome is described as a pattern of multiple anomalies that is thought to be etiologically related (due to either a genetic, environmental or gene-environmental interaction) [9]. These categorizations are important from the viewpoint of counselling and prevention.

Ideally, surveillance for congenital anomalies should consider all pregnancy outcomes, namely live births, stillbirths and terminations of pregnancies. Fatal anomalies like anencephaly frequently lead to early pregnancy loss or stillbirths. Inclusion of autopsy records for stillbirths and infant deaths as a source of case ascertainment can lead to increase in the reported prevalence by up to 10% for certain internal anomalies like isolated CHDs [38]. In addition, another important factor influencing the estimation of prevalence is data on terminations of pregnancy after detection of foetal anomalies (TOPFA). The utility of ultrasound screening for foetal anomalies is well accepted. Even routine scans before 24 weeks gestation can aid the detection of major anomalies [39]. In countries where pregnancy terminations are legal, fatal anomalies when detected on ultrasound are often terminated. A single study that evaluated the impact of elective pregnancy terminations on birth defect rates indicated that inclusion of data on pregnancy terminations increased the reporting of cases for NTDs and certain syndromes by 5% or more [40]. However, obtaining data on TOPFA is challenging for various reasons in many LMICs.

An additional source of variation in birth defects prevalence is the timing when prevalence is measured. Prevalence of congenital anomalies can be measured at birth or later during childhood in countries where surveillance systems are in place. Certain internal anomalies like CHDs are usually not evident at birth. Inclusion of anomalies detected at later ages would reveal a different anomaly prevalence profile than when only those anomalies that are detected at birth are included in the surveillance. Registry data suggest that almost 60% of all major anomalies are diagnosed in the first week of life, and by the first year, this number rises to nearly 90% of all anomalies [41].

The choice of type of surveillance system (active vs. passive) and source of case ascertainment also impacts prevalence rates [32]. Surveillance for congenital anomalies can rely on either active or passive case ascertainment methods or adopt a hybrid approach. Active case ascertainment requires more resources than passive ascertainment as it involves the services of additional surveillance personnel and may be better in terms of quality and completeness of records. In contrast, passive reporting by hospital staff would be a major concern in LMICs where overburdened hospital conditions could lead to underreporting of anomalies.

In addition, involvement of specialists like neonatologists and paediatricians in the surveillance programme can positively influence prevalence estimates. Changes in diagnostic practices over time, as advanced technologies become routinely available in healthcare settings, modifications to case definitions and reporting procedures contribute to the diversity in reported prevalence rates over time.

Lastly, prevalence estimates are also impacted by prevention strategies. For example, global prevalence of spina bifida showed clear variation between geographical regions where mandatory (33.86 per 100,000 live births) as opposed to voluntary (48.35 per 100,000 live births) folic acid fortification was present [42]. Thus, existing programmes for prevention of congenital anomalies also need to be kept in mind when estimating and interpreting prevalence rates.

Birth Defects Surveillance in LMICs

It is estimated that 94% of major congenital anomalies arise in LMICs [33]. These high numbers are driven primarily by larger populations and higher fertility rates in these countries. In addition, the high prevalence of risk factors including widespread nutritional deficiencies, high incidence of maternal morbidities and infections, maternal environmental exposures in the absence of stringent regulations and overall lack of primary prevention programmes suggests a significant burden of birth defects in these countries [4, 7]. For example, the estimated number of anomaly affected births in India is more than the total of anomaly affected births occurring in several HICs [43]. Despite this, there exists a paucity of quality epidemiological data in most countries. As a result, birth defects are not prioritized as public health concerns and continue to exert a terrible toll in these countries [4]. However, as countries undergo epidemiological transition in the rates and causes of child mortality with a clear shift from infectious to non-communicable conditions including congenital anomalies [6], it becomes imperative to set up surveillance systems in such settings. Population-based surveillance systems and registries, though ideal for estimating the magnitude of congenital anomalies, are not feasible in LMICs due to their resource intensive nature, and hospital-based studies are usually the only source of data on the magnitude of congenital anomalies.

Unlike developed countries, catchment areas of hospitals in developing countries are usually undeterminable due to high patient mobility and out-of-pocket expenditure on health. In addition, the inherent biases and limited representativeness of hospital-based estimates pose serious concerns. Accurate assessment of prevalence statistics is also hampered by the limited number of qualified healthcare personnel in such settings. Despite these issues, the comparative ease of collecting hospital-based data makes it the most feasible data source for LMICs that lack rigorous birth defects surveillance systems.

In lieu of surveillance, hospital-based studies provide data on congenital anomalies (Chap. 6). Data from such hospital-based studies can be combined through systematic reviews and meta-analyses to generate best available evidence, although estimates are affected by data availability and quality. An example is the use of meta-analyses to generate an estimate of the global burden of NTDs, which are externally visible anomalies that are relatively straightforward to diagnose. Lo et al. analysed NTD data from 18 countries in 6 WHO regions and estimated that about 190,000 neonates are born each year with NTDs in LMICs alone [44]. A more extensive review and meta-analysis of the global NTD data by Zaganjor et al. using data from 75 countries identified wide variation in the presence of a registry or surveillance system for NTDs ranging from 25% in LMICs to 91% in HICs [45]. The study also highlighted the disparity in surveillance data availability across regions: minimum data from African and South-East Asian regions and majority from the American and European regions [45].

There are many challenges to birth defects surveillance in LMICs stemming from a general lack of awareness about the causes and consequences of birth defects. The limited resources and public health investment in LMICs are often focused on other health issues. Research on birth defects is non-existent or under prioritized where it does occur, leading to a dearth of local context-specific solutions. Appropriate competence and skills required for birth defects surveillance are not universally available in LMICs. Consequently, limited health system capacity, paucity of specialized personnel including neonatologists and paediatricians, dearth of advanced infrastructure and insufficiency of up-to-date diagnostic equipment like ultrasound machines and other imaging technologies to accurately diagnose birth defects may result in lack of diagnosis or even misdiagnosis. Further, the general absence of treatment and rehabilitation services coupled with the out-of-pocket nature of expenditure for available services, contributes to an underestimation of the true burden for many birth defects as many affected children never present at any health facility.

Birth defects vary widely in their manifestation, and only a limited number are discernible at birth. Many birth defects require specialist clinical expertise and technologies for accurate diagnosis. Most LMICs have limited human resource capacity, and scarcity of health professionals trained in the identification and preliminary management of anomalies is a serious concern [46]. Public health facilities frequently lack state-of-the-art infrastructure and professional ability to diagnose many birth defects. Adequate resources for data management and analysis are scarce. While the required competencies may be available in select speciality private facilities, these are usually concentrated in urban areas and out of reach of majority of the population. As a consequence of the deficient diagnostic capacity, the absence of quality data contributes to the invisibility of the problem. When available, data are usually collected from single one-time hospital studies that are not population-based. Often, the only facilities with the capacity for surveillance are tertiary referral centres, generating biased data that are of little use for planning.

Furthermore, heterogeneity in the methodologies and lack of quality result in high variation in the reported rates. In the NTD meta-analysis mentioned earlier, wide variation in the reported prevalence rates was observed for African (5.2–75.4 per 10,000 births) and South-East Asian regions (1.9–66.2 per 10,000 births) [45]. Data on vital statistics are often unreliable in LMICs [47], and this complicates mortality estimations due to congenital anomalies. Deaths may be misdiagnosed as common infectious causes, leading to serious underestimation.

The March of Dimes Report in 2006 was the first to estimate the global burden of birth defects and highlight their devastating toll in LMICs [4]. The authors provided country-specific estimates for birth defects and recognized the systematic underestimation in LMICs due to various constraints in diagnostic capabilities, unreliable health statistics and absence of birth defects surveillance systems [4]. International commitment to birth defects was mobilized through the WHA resolution that acknowledged the public health significance of birth defects [12]. It recognized the need for establishment of appropriate surveillance systems for birth defects in LMICs and indicated supportive collaborations with ICBDSR including ECLAMC and EUROCAT to aid in this endeavour [12]. Subsequently, a number of global initiatives to address birth defects have emerged in recent years. The ICBDSR has developed training programmes and online courses for health professionals in collaboration with international organizations such as the WHO and the US Centers for Disease Control and Prevention [48, 49]. Resources include a manual to facilitate the development and implementation of a congenital anomaly surveillance programme [9], an atlas that serves as a companion tool to the manual and includes photographs and illustrations [50] and a facilitator's guide [51]. The WHO South-East Asian Regional Office (SEARO) has also made significant contributions, especially in terms of helping countries develop competency in birth defects surveillance [52]. In 2014, WHO-SEARO launched the newborn-birth defects (SEAR-NBBD) database designed as an online system to support data management for integrating data on newborn health, birth defects and stillbirths [53]. Among its publications is a guide for establishing and operating a hospital-based birth defects surveillance system [54].

Birth Defects Surveillance in India

Since the mid-1970s, the pesticide endosulfan has been aerially sprayed on cashew nut plantations covering several villages in the Indian state of Kerala. Prolonged use over the years led to observations of increased occurrences of congenital anomalies among animals and children in the area. In addition to congenital anomalies like CHDs and skeletal abnormalities, exposed children also presented with low intelligence quotient (IQ) and learning disabilities as well as poor performance at school [55]. Investigation by the National Institute of Occupational Health revealed persistence of the pesticide in water samples 10 months after spraying, thus indicating continuous exposure for the local population. This coupled with the higher occurrence of congenital anomalies in the area as reported by local medical practitioners prompted the Kerala High Court to reinstate a ban on endosulfan in 2002 [55]. Unfortunately, there is yet no surveillance in place, nor were any rigorous epidemiological studies ever conducted. This example highlights the unfortunate reality of the lack of surveillance in countries like India.

With regard to absolute numbers, India ranks first in the world in terms of the largest birth cohort (25,244,000 births in 2016) [56] and contributes a quarter of global neonatal deaths. The resultant burden of neonatal mortality due to congenital anomalies is the highest worldwide [57]. Recognizing the adverse impact of birth defects, the Government of India has initiated certain services for detection and management of affected individuals. The India Newborn Action Plan has incorporated as one of its six pillars, a commitment for screening and management of birth defects including diagnosis and surgical treatment under its 'Care beyond Newborn Survival' component [58]. India also has a nation level child health programme (Rashtriya Bal Swasthya Karyakram, RBSK) that includes screening for selected birth defects including NTDs and CHDs, developmental disabilities, as well as common childhood diseases and deficiencies [59]. District Early Intervention Centres have been set up for appropriate management of detected conditions [60]. Despite these initiatives, data on the magnitude of birth defects are limited in the absence of surveillance, and lack of accurate estimates continue to impact resource allocation. There are three prominent systems that could function as a source of data on the magnitude of congenital anomalies in India; however, methodological drawbacks limit the scope of their data (Table 4.1).

The RBSK is primarily a screening service and collects limited data on birth defects related to the age, sex, place of residence and type of condition diagnosed. The data are generated from *anganwadi* (playschool) centres and government schools where the screening programmes are implemented and include children from birth to 18 years of age. The locally collected data are then transmitted to district, state and national level [59]. These data report prevalent or existing cases in the community and cannot be used to measure birth prevalence rates. Published data indicated that in the short period from April to December 2014, out of the approximately 42,000,000 children screened, 1% were detected with birth defects. Among these, 39% were CHDs, and 13% were NTDs [63]. These data do not reflect birth prevalence rates but only indicate the proportion of anomalies among screened children. These data are an underestimation as severely disabled children would not be attending *anganwadis* or schools.

One of the early registries in India is the Birth Defects Registry of India (BDRI), initiated in 2001, also a member of ICBDSR, although it is no longer an active contributor. It is a hospital-based passive registry that collates voluntary data on congenital anomalies reported across 300 out of 750 registered hospitals all over the country [61]. In addition, the SEAR-NBBD launched in 2014 is also hospital-based [52]. It is located in one of the largest referral hospitals in India and therefore is likely to report skewed data. This database does not include data on TOPFA. Recent estimates from the SEAR-NBBD documented 13,252 babies with birth defects among 1,545,258 births at the 70 reporting hospitals in the country, giving a prevalence of 0.86% [62]. In contrast, data from BDRI (last reported for 2012) indicated an NTD prevalence of 19.48 per 10,000 births including TOPFA and a CHD prevalence of 0.87

Table 4.1 Sources of data on birth defects prevalence in India

Data source		Prevalence	Characteristics
Registry	BDRI	46.85 per 10,000 births [61] NTD 19.48, CHD 0.87	Voluntary hospital-based passive surveillance
	SEAR-NBBD	0.86% (13,252/1,545,258 births) [62]	Hospital-based surveillance across 70 facilities
National programme	RBSK	1% (42,000,000 children) [63] CHD 39%, NTD 13%	National screening programme from birth to 18 years
Epidemiological studies	Hospital-based cross-sectional	<50 [64, 65] to >400 [66, 67] per 10,000 births	Hospital-based cross-sectional studies across a diverse range of hospitals, lack of uniformity in data collection
	Meta-analysis	184.48 per 10,000 births [68]	Meta-analysis of 52 hospital-based studies including 802,658 births
	PUBOS	203.51 per 10,000 births [69]	Cohort study

BDRI Birth Defects Registry of India; *NTD* neural tube defect; *CHD* congenital heart defect; *SEAR-NBBD* South-East Asia Regional Newborn and Birth Defects Database; *RBSK* Rashtriya Bal Swasthya Karyakram; *PUBOS* Pune Urban Birth Outcomes Study

per 10,000 births [61]. The limitations of these systems include lack of clarification on the number and type of hospitals reporting the data, the method of ascertainment at the registered hospitals and reliance on passive voluntary reporting that may have either overestimated or underestimated the prevalence estimates depending on the involved hospitals. The data reported in these registries represent proportions of anomalies among births occurring in the reporting hospitals and thus cannot be used for estimation of national rates.

Despite these efforts, the primary source of data on the prevalence of congenital anomalies in India remains ad hoc studies which have reported significantly varying prevalence rates ranging from below 50 per 10,000 births [64, 65] to above 400 per 10,000 births [66, 67]. The earliest study reporting the prevalence of congenital anomalies was conducted in the period 1960–1963 [70]. In this study, the authors examined 23,568 births from the single largest maternity hospital in Mumbai and reported a birth prevalence of 1.4%. Only seven cases of cardiovascular system defects were detected (0.3 per 1000 births). The overall underestimation of anomalies was recognized and attributed to lack of involvement of specialists like paediatricians and reliance solely on clinical examination to detect the defects. The authors also recognized congenital malformations as an important cause of stillbirths (10%) and perinatal mortality (9%). In a subsequent paper [71], the authors discussed the reliability of hospital records in the absence of active participation of staff and the biases introduced when large number of emergency deliveries which are typically referral cases are included in hospital-based birth prevalence estimates. These two seminal studies highlighted the issues pertaining to case ascertainment and the inherent limitations of hospital-based data, including reliability of hospital recording systems, and the issue of in or out referrals, based on the expertise available at the hospital. Subsequently, very high prevalence rates (above 400 per 10,000 births) were observed in studies that were carried out in tertiary care facilities like those hospitals attached to medical colleges and serving as referral centres [66, 67]. On the other hand, lower prevalence (below 50 per 10,000 births) was recorded among private paying hospitals catering to limited populations [64] and when authors relied solely on birth registers as a source of case ascertainment [65]. A study conducted in a referral hospital covering a wide industrial belt revealed a disproportionately high proportion (nearly 50%) of nervous system malformations [72]. Similarly, a study in the city of Lucknow showed significantly higher prevalence of NTDs in a government hospital associated with a medical college compared to another government hospital that was not associated with any teaching institute [73]. Inclusion of autopsy data when autopsies were carried out increased the detection of internal anomalies leading to higher reported prevalence rates [74].

Review of community-based studies highlighted some other issues. These studies did not include data on stillbirths and early neonatal deaths, and this would have led to an underestimation of lethal anomalies. On the other hand, influence of inclusion of anomalies diagnosed at later ages was visible in both the relative proportions of type of anomalies reported and the overall prevalence. As part of a longitudinal study of an urban birth cohort from Delhi (1969–1973), 7590 live born babies of women in the cohort were examined for congenital anomalies at birth (prevalence 25 per 1000 live births) [75]. Fifty per cent (3816) of the birth cohort was re-examined (average age five years), and 54 additional children were detected to have anomalies. This follow-up and detection of anomalies at later ages increased the original birth prevalence estimate to 32 per 1000 births. A threefold increase in the number of cardiovascular system anomalies was observed by the age of five years highlighting the later detection of certain anomalies like CHDs and the influence of extending the age of diagnosis in surveillance.

A recent systematic review and meta-analysis included data from 52 hospital-based and three community-based studies involving a total of 802,658 births [68]. Data from the meta-analysis estimated a pooled prevalence of congenital anomaly affected births in the country at 184.48 per 10,000 births (95% CI 164.74–204.21). The results clearly indicated that musculoskeletal system anomalies predominate when only live births are analysed while the prevalence of central nervous system anomalies like NTDs rises when stillbirths are also included in the analysis. The commonest reported anomalies were anencephaly and talipes equinovarus, both of which are structural anomalies that can be easily detected on physical examination. High level of heterogeneity was observed among the studies highlighting the absence of uniformity in the adopted methodologies. The review discussed key methodological issues pertaining to hospital-based surveillance of congenital anomalies and also highlighted the absence of data on TOPFA that can cause serious underestimation in the prevalence estimates. This is of particular concern as medical termination of pregnancy is legal in India [76]. Post-1990s ultrasound scans have started to become routinely available in healthcare settings in India. By 2000, basic ultrasounds scans were being performed on a routine basis in almost all major large teaching and some private hospitals surveyed [77]. This has major implications for surveillance systems and reiterates the need for inclusion of data on TOPFA in birth defects surveillance.

In order to overcome the various limitations of available hospital-based cross-sectional studies in collecting birth defects statistics, the Pune Urban Birth Outcomes study (PUBOS) used a cohort study design to estimate the birth prevalence of congenital anomalies, identify common anomalies and measure their share in the causes of neonatal mortality and childhood disability [69]. The study recruited women in early pregnancy from four public health facilities in Pune and followed them up till outcome and during the neonatal period in case of live births. Though the sample size of the cohort (2107) was relatively small, the findings have several implications for birth defects surveillance in India. Contrary to the meta-analysis, data from the cohort study revealed a greater total prevalence of congenital anomalies of 230.51 per 10,000 births (95% CI 170.99–310.11). Extrapolation at the national level would lead to 589,990 (437,674–793,445) affected births in the country. Merits of the study include the application of internationally accepted standard methodology including classification criteria [36], thus allowing for comparisons on a global level. Follow-up survey of the cohort during the neonatal period yielded additional birth defects that clearly indicated the advantages of including later ages in surveillance. In addition, the prenatal detection rate of nearly 50% for major anomalies and 100% for severe anomalies like NTDs and complex CHDs among others, reiterates the usefulness of ultrasound in prenatal detection of major congenital anomalies. One-fifth of the major congenital anomalies were detected early, and the pregnancies were terminated. This is especially relevant in an urban setting where ultrasound services are readily available and accessible (Table 4.2).

Table 4.2 Important findings from PUBOS [69]

Outcomes (n)	Major congenital anomaly affected rates
Total outcomes (1929)	230.51 per 10,000 births
Live births (1781)	168.44 per 10,000 live births
Neonatal deaths (22)	3.93 per 1000 live births
Terminations of pregnancy (including miscarriages and TOPFA, 107)	4.39 per 1000 births
Congenital anomalies typically requiring surgery	3.29 per 10,000 births

Summary and Conclusion

In conclusion, the overall lack of systematic evidence on the magnitude of birth defects at the national level argues for the establishment of birth defects surveillance in India. Given the heterogeneity of India and other LMICs, local contextual factors need to be recognized and acknowledged when setting up such a surveillance system. For any surveillance system to succeed, it needs its 'champions', along with motivated personnel and committed leaders to galvanize efforts in the right direction. Coordination through a central authority would ensure cohesion among the facilities, for example, by organizing regular contact sessions, trainings and workshops. This would also aid in timely dissemination of information that is vital to planning and implementing public health strategies. Surveillance for the sole purpose of estimating the magnitude of birth defects is meaningless without the establishment of referral linkages for the management and care of identified cases. Data generated during surveillance can serve as a foundation for further epidemiological studies. Use of such data to identify risk factors for congenital anomalies would additionally benefit other maternal health indicators due to the shared nature of these risk factors. Surveillance data would also enable multidisciplinary studies not only on the health impact of birth defects but also on their social and economic impact. In the unfortunate case of new emerging epidemics, additional surveillance objectives can be seamlessly incorporated into the existing system. The benefits of such surveillance would be immense.

References

1. Global Burden of Disease Pediatrics Collaboration (2016) Global and national burden of diseases and injuries among children and adolescents between 1990 and 2013: findings from the Global Burden of Disease 2013 Study. JAMA Pediatr 170:267–287
2. GBD 2015 DALYs and HALE Collaborators (2016) Global, regional, and national disability-adjusted life-years (DALYs) for 315 diseases and injuries and healthy life expectancy (HALE), 1990–2015: a systematic analysis for the Global Burden of Disease Study 2015. Lancet 388:1603–1658
3. Christianson A, Modell B (2004) Medical genetics in developing countries. Annu Rev Genomics Hum Genet 5:219–265. https://doi.org/10.1146/annurev.genom.5.061903.175935
4. Christianson A, Howson C, Modell B (2006) Global report on birth defects: the hidden toll of dying and disabled children. New York
5. Institute of Medicine (2003) Reducing birth defects. Meeting the challenge in the developing world. The National Academies Press, Washington
6. Oza S, Lawn J, Hogan D et al (2015) Neonatal cause-of-death estimates for the early and late neonatal periods for 194 countries: 2000–2013. Bull World Health Organ 93:19–28
7. Penchaszadeh V (2002) Preventing congenital anomalies in developing countries. Community Genet 5:61–69
8. Thacker S, Birkhead G (2008) Surveillance. In: Gregg M (ed) Field epidemiology. Oxford University Press, Oxford
9. WHO/CDC/ICBDSR (2014) Birth defects surveillance: a manual for programme managers
10. Stevenson A, Johnston H, Stewart M, Golding D (1966) Congenital malformations—a report of a study of series of consecutive births in 24 centres. Bull World Health Organ 34(Suppl):9–127

11. World Health Organization (2000) Primary health care approaches for the prevention and control of congenital and genetic disorders. Geneva
12. World Health Organization (2010) In: Sixty-third World Health Assembly, Geneva, 17–21 May 2010. Geneva
13. Grimes C, Holmer H, Maraka J et al (2016) Cost-effectiveness of club-foot treatment in low-income and middle-income countries by the Ponseti method. BMJ Glob Health 1:e000023
14. Williams J, Mai CT, Mulinare J et al (2015) Updated estimates of neural tube defects prevented by mandatory folic acid fortification—United States, 1995–2011. MMWR Morb Mortal Wkly Rep 64:1–5
15. Turnpenny P, Ellard S (2005) Emery's elements of medical genetics, 12th edn. Elsevier/Churchill Livingstone, Philadelphia
16. Graham J, Edwards M, Edwards M (1998) Teratogen update: gestational effects of maternal hyperthermia due to febrile illnesses and resultant patterns of defects in humans. Teratology 58:209–221. https://doi.org/10.1002/(SICI)1096-9926(199811)58:5%3c209::AID-TERA8%3e3.0.CO;2-Q
17. Barlow S, Kavlock R, Moore J et al (1999) Teratology society public affairs committee position paper: developmental toxicity of endocrine disruptors to humans. Teratology 60:365–375
18. Vinceti M, Malagoli C, Rothman K et al (2014) Risk of birth defects associated with maternal pregestational diabetes. Eur J Epidemiol 29:411–418. https://doi.org/10.1007/s10654-014-9913-4
19. Gregg NM (1941) Congenital cataract following German measles in the mother. Trans Ophthalmol Soc Aust 3:35–46
20. Alsdorf R, Wyszynski D (2005) Teratogenicity of sodium valproate. Expert Opin Drug Saf 4:345–353. https://doi.org/10.1517/14740338.4.2.345
21. McBride W (1961) Thalidomide and congenital abnormalities. Lancet 2:1358
22. Viteri O, Soto E, Bahado-Singh R et al (2015) Fetal anomalies and long-term effects associated with substance abuse in pregnancy: a literature review. Am J Perinatol 32:405–416
23. Feldkamp M, Carey J, Byrne J et al (2017) Etiology and clinical presentation of birth defects: population based study. BMJ 357:j2249
24. Lenz W, Pfeiffer R, Kosenow W, Hayman D (1962) Thalidomide and congenital abnormalities. Lancet 279:45–46
25. Teixeira MG, Costa M da CN, de Oliveira WK et al (2016) The epidemic of Zika virus–related microcephaly in Brazil: detection, control, etiology, and future scenarios. Am J Public Health 106:601–605. https://doi.org/10.2105/AJPH.2016.303113
26. Schuler-Faccini L, Ribeiro E, Feitosa I et al (2016) Possible association between Zika virus infection and microcephaly—Brazil, 2015. MMWR Morb Mortal Wkly Rep 65:59–62
27. Yoon P, Rasmussen S, Lynberg M et al (2001) The national birth defects prevention study. Public Health Rep 116:32–40
28. Boyd P, Haeusler M, Barisic I et al (2011) Paper 1: the EUROCAT network–organization and processes. Birth Defects Res A Clin Mol Teratol 91:S2–S15
29. Castilla EE, Orioli IM (2004) ECLAMC: the Latin American collaborative study of congenital malformations. Community Genet 7:76–94. https://doi.org/10.1159/000080776
30. Bermejo-Sánchez E, Botto LD, Feldkamp ML et al (2018) Value of sharing and networking among birth defects surveillance programs: an ICBDSR perspective. J Community Genet 9:411–415. https://doi.org/10.1007/s12687-018-0387-z
31. Mason C, Kirby R, Sever L, Langlois P (2005) Prevalence is the preferred measure of frequency of birth defects. Birth Defects Res A Clin Mol Teratol 73:690–692
32. Hobbs C, Hopkins S, Simmons C (2001) Sources of variability in birth defects prevalence rates. Teratology 64:S8–S13
33. World Health Organization (2016) Congenital anomalies. https://www.who.int/news-room/fact-sheets/detail/congenital-anomalies. Accessed 3 July 2020
34. World Health Organization (2010) International statistical classification of diseases and related health problems, 10th revision

35. British Paediatric Association, World Health Organization (1979) British Paediatric Association classification of diseases: Successor to the Cardiff Diagnostic Classification: a perinatal supplement compatible with the ninth revision of the WHO International Classification of Diseases, 1977: perinatal supplement: codes designed for use in the classification of perinatal disorders. The Association, London
36. EUROCAT (2005) EUROCAT Guide 1.3: Instruction for the registration of congenital anomalies. EUROCAT Central Registry, University of Ulster, Belfast
37. Sever LE (2012) Guidelines for conducting birth defects surveillance. Atlanta
38. Kuehl K, Loffredo C, Ferencz C (1999) Failure to diagnose congenital heart disease in infancy. Pediatrics 103:743–747
39. Whitworth M, Bricker L, Mullan C (2015) Ultrasound for fetal assessment in early pregnancy. Cochrane Database Syst Rev 14:CD007058
40. Ethen M, Canfield M (2002) Impact of including elective pregnancy terminations before 20 weeks gestation on birth defect rates. Teratology 66:S32–S35
41. Bower C, Rudy E, Callaghan A et al (2010) Age at diagnosis of birth defects. Birth Defects Res A Clin Mol Teratol 88:251–255. https://doi.org/10.1002/bdra.20658
42. Atta C, Fiest K, Frolkis A et al (2016) Global birth prevalence of spina bifida by folic acid fortification status: a systematic review and meta-analysis. Am J Public Health 106:e24–e34
43. Kar A (2014) Birth defects in India: magnitude, public health impact and prevention. JKIMSU 3:7–16
44. Lo A, Polšek D, Sidhu S (2014) Estimating the burden of neural tube defects in low- and middle-income countries. J Glob Health 4:010402
45. Zaganjor I, Sekkarie A, Tsang B et al (2016) Describing the prevalence of neural tube defects worldwide: a systematic literature review. PLoS ONE 11:e0151586
46. Farmer D, Sitkin N, Lofberg K et al (2015) Surgical interventions for congenital anomalies. In: Debas HT, Donkor P, Gawande A et al (eds) Essential surgery: disease control priorities, 3rd edn, vol 1. The International Bank for Reconstruction and Development, The World Bank, Washington
47. World Health Organization (2020) Civil registration of deaths. https://www.who.int/gho/mortality_burden_disease/registered_deaths/text/en/. Accessed 1 July 2020
48. Flores A, Valencia D, Sekkarie A et al (2015) Building capacity for birth defects surveillance in Africa: implementation of an intermediate birth defects surveillance workshop. J Glob Health Perspect 2015:1
49. ICBDSR (2020) Online self-paced course on birth defect surveillance and prevention. https://www.icbdsr.org/online-self-paced-course-on-birth-defect-surveillance-and-prevention/. Accessed 2 July 2020
50. WHO/CDC/ICBDSR (2014) Birth defects surveillance: atlas of selected congenital anomalies. Geneva
51. WHO/CDC/ICBDSR (2015) Birth defects surveillance training: facilitator's guide. Geneva
52. World Health Organization, Regional Office for South-East Asia. www.searo.who.int/en/. Accessed 14 Mar 2018
53. WHO Regional Office for South-East Asia (2020) New-born and birth defects (NBBD) surveillance initiative. https://origin.searo.who.int/entity/child_adolescent/nbbd/web/en/. Accessed 2 July 2020
54. Regional Office for South-East Asia, World Health Organization (2016) Hospital-based birth defects surveillance: a guide to establish and operate. WHO Regional Office for South-East Asia, New Delhi
55. Kumar S (2002) Pesticide ban imposed after reports of congenital abnormalities. BMJ Br Med J 325:356
56. United Nations Children's Fund (2017) The State of the World's Children 2017: Children in a digital world. UNICEF, New York
57. Liu L, Oza S, Hogan D et al (2016) Global, regional, and national causes of under-5 mortality in 2000–15: an updated systematic analysis with implications for the sustainable development goals. Lancet 388:3027–3035. https://doi.org/10.1016/S0140-6736(16)31593-8

58. Ministry of Health and Family Welfare. Government of India (2014) India newborn action plan
59. Ministry of Health and Family Welfare. Government of India (2013) Rashtriya Bal Swasthya Karyakram (RBSK). https://www.rbsk.gov.in/%0ARBSK. Accessed 29 June 2017
60. Ministry of Health and Family Welfare. Government of India (2014) Setting up district early intervention centres. Operational guidelines
61. International Clearinghouse for Birth Defects Surveillance and Research (2013) Annual Report 2013. The International Centre on Birth defects - ICBDSR Centre, Rome
62. World Health Organization. Regional Office for South-East Asia (2020) Strengthening integrated surveillance and prevention of birth defects, stillbirths and congenital Zika virus infection to accelerate reduction in newborn mortality. World Health Organization, Regional Office for South-East Asia
63. Singh A, Kumar R, Mishra C et al (2015) Moving from survival to healthy survival through child health screening and early intervention services under Rashtriya Bal Swasthya Karyakram (RBSK). Indian J Pediatr 82:1012–1018
64. Sharma P (1970) The incidence of major congenital malformations in Mysore. Indian J Pediatr 37:618–619
65. Choudhury A, Talukder G, Sharma A (1984) Neonatal congenital malformations in Calcutta. Indian Pediatr 21:399–405
66. Marwah S, Sharma S, Kaur H et al (2014) Surveillance of congenital malformations and their possible risk factors in a teaching hospital in Punjab. Int J Reprod Contracept Obstet Gynecol 3:162–167
67. Savaskar S, Mundada S, Pathan A, Gajbhiye S (2014) Study of various antenatal factors associated with congenital anomalies in neonates born at tertiary health care center. Int J Recent Trends Sci Technol 12:82–85
68. Bhide P, Kar A (2018) A national estimate of the birth prevalence of congenital anomalies in India: systematic review and meta-analysis. BMC Pediatr 18:175. https://doi.org/10.1186/s12 887-018-1149-0
69. Bhide P, Gund P, Kar A (2016) Prevalence of congenital anomalies in an Indian maternal cohort: healthcare, prevention, and surveillance implications. PLoS ONE 11:e0166408. https://doi.org/ 10.1371/journal.pone.0166408
70. Kolah P, Master P, Sanghvi L (1967) Congenital malformations and perinatal mortality in Bombay. Am J Obstet Gynecol 97:400–406
71. Master-Notani P, Kolah P, Sanghvi L (1968) Congenital malformations in the new born in Bombay. Part I. Acta Genet 18:97–108
72. Duttachoudhary A, Pal S (1997) Congenital abnormalities in Durgapur Steel Plant Hospital with special reference to neural tube defect. J Indian Med Assoc 95:135–141
73. Sharma A, Upreti M, Kamboj M et al (1994) Incidence of neural tube defects at Lucknow over a 10 year period from 1982–1991. Indian J Med Res 99:223–226
74. Bhat V, Babu L (1998) Congenital malformations at birth—a prospective study from south India. Indian J Pediatr 65:873–881
75. Ghosh S, Bhargava S, Butani R (1985) Congenital malformations in a longitudinally studied birth cohort in an urban community. Indian J Med Res 82:427–433
76. Ministry of Health and Family Welfare. Government of India. The medical termination of pregnancy act, 1971. https://mohfw.gov.in/acts-rules-and-standards-health-sector/acts/mtp-act-1971. Accessed 27 Feb 2018
77. Deka D, Malhotra N, Takkar D et al (1999) Prenatal diagnosis and assessment of fetal malformations by ultrasonography in India. Indian J Pediatr 66:737–749

Chapter 5
Rare Disease Registries: A Case Study of the Haemophilia Registry in India

Juhi Nakade

Abstract A cornerstone of public health planning is epidemiological data. Due to their low prevalence, standard epidemiological methods are unsuitable for measuring the prevalence of rare disorders. Registries not only generate data for surveillance and clinical or epidemiological research, but also help connect patients and empower advocacy for mobilizing public health services. This article provides an overview of rare disease registries, identifying that disease registries in general are resource intensive investments, which makes their functioning more feasible in high income countries. Saying this, the article describes the efforts at establishing rare disease registries in India, and their inputs in mobilizing a draft national rare disease policy in the country. The remaining section of the chapter describes the evolution of a unique registry for haemophilia, a rare, single gene disorder. The National Haemophilia Registry of India has been created by the Haemophilia Federation of India, a non-governmental organization, mentored by the World Federation of Hemophilia. The registry has evolved from lists of patients at different state-level organizations, to a nation-wide data reporting system that provides statistics on a carefully selected list of variables that can be used to study the epidemiology and trends of the disorder in India. The article will describe that while there are data quality issues with such volunteer-maintained registries, the information from the National Haemophilia Registry of India was invaluable in providing data to support public interest litigations demanding services for haemophilia in India. These litigations finally led to judicial directives to ensure that patients with haemophilia receive free of charge treatment for the condition.

Keywords Rare diseases · Registries · India · Haemophilia

Disease Registries

As early as in 1974, a registry was defined as '*a file of documents containing uniform information about individual persons, collected in a systematic and comprehensive*

J. Nakade (✉)
Birth Defects and Childhood Disability Research Centre, Pune 411020, India

way in order to serve a predetermined purpose' [1]. In 1991, another definition described a registry as '*a paper list or an electronic database containing information on the characteristics of a population affected by a given disease*' [2]. In 2012, the National Committee on Vital and Health Statistics described registries as '*an organized system for the collection, storage, retrieval, analysis and dissemination of information on individual persons who have either a particular disease, a condition (e.g. a risk factor) that predisposes to the occurrence of a health-related event or prior exposure to substances or circumstances known or suspected to cause adverse health effects*' [3]. A more complete definition has been provided by the Agency of Healthcare Research and Quality (AHRQ) as '*a patient registry is an organized system that uses observational study methods to collect uniform data (clinical or other) to evaluate specified outcomes for a population defined by a particular disease, condition or exposure and that serves one or more predetermined scientific, clinical or policy purposes*' [3].

Depending on the objectives, registries can be broadly categorized into different types [4, 5]. Public health registries are critical towards generating population-based epidemiological data. They collect data on disease occurrence for the purpose of estimating disease incidence and prevalence, for measuring variation of disease over time, and their spatial distribution. Enlisting all patients in a given population is the key to successful public health registries [6]. Population-based cancer registries are a good example. Clinical registries collate data on specific disease conditions, such as characteristics of patients, clinical course, medication response and other such data. Clinical registry data is used to describe the progression of disease and impact of interventions. Clinical registries can also be utilized to recruit patients for clinical trials and provide data on long-term outcomes which may be beyond the purview of the trial. It is important to note that unlike public health registries, case identification, rather than registration of all cases, is essential in clinical registries [6]. Product registries compile data on medical devices or pharmaceutical products such as patients using or exposed to implants and biopharmaceutical products for biologic therapy [3, 4]. Such registries provide valuable data on adverse events if any, long-term effects of therapy or implant for a chronic condition. The registry may have a one-point contact in case of implanted devices, or an extended follow-up period for chronic conditions [3, 5]. Hospital-based registries collate data from certain clinical or other centres. The focus is on a specific disease (e.g. Neurofibromatosis Registry) or a group of diseases (e.g., Registry for Neuromuscular Diseases) [7]. Rare disease registries, described further below, collect data on rare conditions.

One of the most valuable use of clinical registries is for conducting clinical trials, as they allow the opportunity to contact a group of individuals with the same clinical condition. Registry-based randomized controlled trials ensure rapid enrollment, provide reference data for making appropriate sample size calculations, provide data to control for confounding factors that would affect the overall results of a clinical trial, provide ease of data collection and follow-up, thereby reducing the overall trial cost [5, 6].

Registries can be categorized as manual, paper-based, electronic-based, or hybrid, that is using both paper and electronic documentation of data. Whatever the method of data collation, registry data collection begins with specific case definitions with respect to inclusion in the registry, and defining key domains on which the data will be collected. Specific definitions are essential for reliability and integrity of the registry data.

Registries develop based on perceived needs. Registries can be patient-driven or patient self-reported or registries can be physician-driven or professional-reported. Usually, registries are a combination of patient and professional reported information. A good example of a patient/parent-driven registry is that developed by the Cystic Fibrosis Foundation [7]. It is a not-for-profit organization that has successfully used the 'venture philanthropy' business model. This concept utilizes the techniques from venture capital and business management to achieve philanthropic goals through organizations, such as not-for-profit organizations, for a desired social impact. The Foundation involves pharmaceutical companies to generate funds and specific drug development programmes for patients suffering with cystic fibrosis. They have developed a registry of patients, who can access latest drugs that manage and increase life expectancy of patients living with cystic fibrosis. Another example of a clinician, researcher and patient-driven registry is the Cerebral Palsy Research Network [8]. This group of collaborators work towards improving treatment, and creating a pool of patients for clinical trials of drugs, devices and other non-pharmaceutical interventions aimed at improving the standards of living of patients and their families. It is essential to note that the leadership comprises of clinicians and academicians and funds are sourced from various philanthropic entities, and national and other research agencies. In contrast, the National Spina Bifida patient registry, established in 2008, is governed by the Centers for Disease Control (CDC) and funded by the Department of Health [9]. It was established to address the gaps in delivery of healthcare for these conditions. The registry data collected is primarily used to evaluate and improve existing medical services for spina bifida across various centres in the country.

Registries are not uniform in terms of drivers and funding, as explained for the three registries cited above. The registry for cystic fibrosis is patient-driven and that for cerebral palsy is professionally reported (by clinicians and academicians). The ultimate aim is to enhance the quality of life and longevity of patients, and facilitating clinical trials, for drug development with the eventual goal of improving patient care and quality of life. The sources of funding for all three organizations vary. Overall, depending on the clinical presentation, severity of the conditions, level of tertiary care required and ease with which pharmaceutical companies can initiate drug development, the utility of registry data varies from providing a pool of patients for clinical trials to simply addressing standards of health care.

The efficient functioning of a disease registry is dependent on several factors (Fig. 5.1).

1. The goals and purpose of the registry, its design and rationale need to be precise and clearly defined from the outset to successfully deliver multiple objectives. These objectives could include improvements in patient care, comparing effectiveness of treatment, monitoring current treatment or care guidelines, better

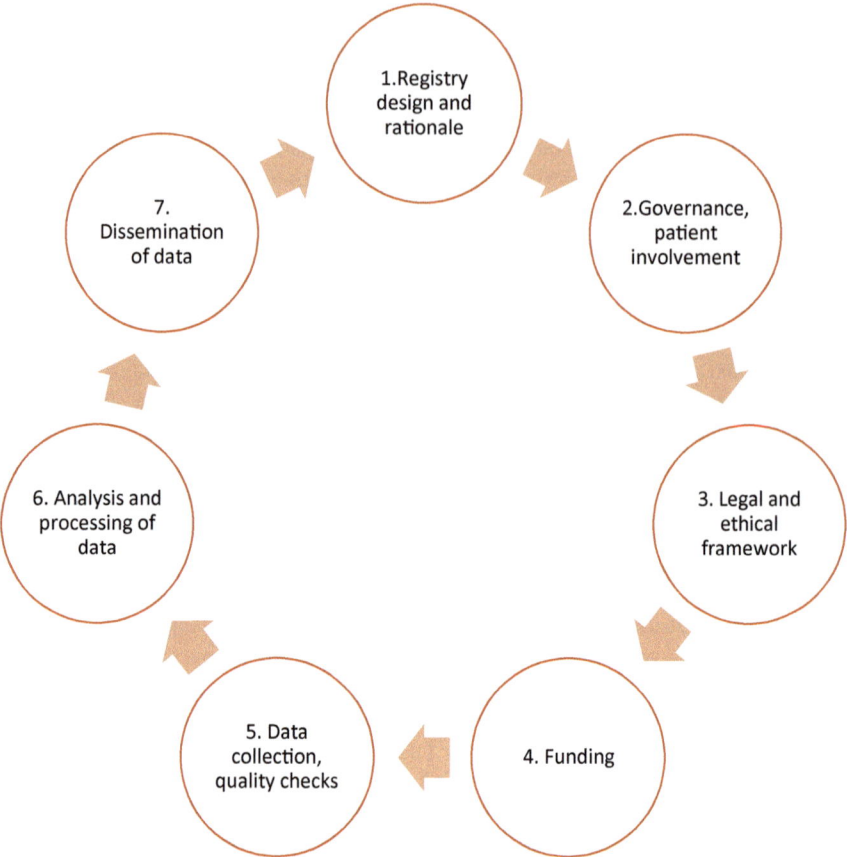

Fig. 5.1 Factors influencing successful functioning of a registry

understanding of disease epidemiology, promoting research, technological or treatment innovations, efficiency and more importantly patient decision-making in treatment or care.

2. Although registries are typically planned by professionals, patient involvement in registries whether in registry governance, designing or reporting, bolsters the relevance of the registry [10]. Patients or their representatives are as integral as clinical stakeholders in registry design, so that their needs and concerns are addressed. Committees with stakeholder representatives have the responsibility of reviewing the functioning of the registry activity, so as to ensure that the registry can yield data for the specific function that the registry has been established for. Table 5.1 lists the possible stakeholders and their roles in a disease registry.

Table 5.1 Stakeholders' expectations from a registry

Stakeholder	Role	Expectation from the registry
Patients and family/caregivers	Participatory	– Increased awareness about condition, updates on emerging knowledge – Creation of patient community – Development of newer treatments
	Advocacy	– Patient pool for interventional studies – Increasing awareness among patients, policymakers, public – Advocating for access to care, support, training and research – Raising profile of disease for funding
Clinicians	Data contribution	– Increase knowledge about the condition – Data for refining complex or undefined diagnoses – Develop and inform treatment guidelines
Researchers/academia	Principal investigators/scientific advice	– Generating knowledge for improved understanding of disease – Identifying potential therapeutics – Peer reviewed publications
Biopharma industry	Sponsorship/development	– Developing drugs for treating the disease – Conducting clinical trials to evaluate potential drugs and relevant end points – Peer-reviewed publications
Government/regulatory authorities/funding agency	Sponsorship/information receivers	– Treatment products, knowledge about the disease – Data for providing affordable health care – Designing and implementing better regulatory systems

3. Patient rights and confidentiality issues are of prime consideration and clear guidelines need to exist in order to ensure protection of patient data. It is important to get consent of patients or caregivers, if possible, appoint a registry coordinator or facilitator for follow-up, and facilitate patient access to data for transparency. Local or regional clearance from ethical committees, compliance with local or regional information legislations if any, helps transparent functioning of registry.

4.   Key to smooth functioning of a registry is funding. Registries need sufficient resources for initial planning and set-up and then long-term funding for efficient functioning of the registry to deliver its objectives. Many disease registries are funded by pharmaceutical industries as registries provide access to participants required for research. However, the role of industry in the functioning of a registry needs to be specific, especially with accessing data. Other sources of funding are hospitals participating in the registry, charities, other professional societies and, in some countries, central funding sources such as departments of health [3].

5.   Data monitoring, data analysis, as well as dissemination of data forms relevant components of the data generation cycle of registries. It is possible to develop a registry using simple filing methods, or as a web-based platform for quick and accurate data collection. Using advanced technologies, the registry can be developed as a secure platform for interactive access of data [10]. Another key element for a successful registry is data management. Registry stakeholders need to enlist essential data that will be required to fulfil the objectives of the registry and strike a balance between collecting comprehensive data and feasible data. Specified and clear definitions of variables for which data will be collected aids the data collection process. It is essential to build data sets which are as complete as possible to be useful for stakeholders. Recently, common data elements have been considered as a solution for standard data collection method for disease registries to enable comparison between registries, linking of registries or restructuring and or updating registries [3]. Registries can choose to have a flexible data set that gives the opportunity for the data set to evolve over time as the registry continues to grow. Consistency and data validation are critical in registry data collection that ensures the quality of data.

It is necessary that data handlers at the registries, whether clinicians or registry staff or volunteers, need to be educated about the importance of data quality and information management. Data quality and accuracy checks involve detecting and resolving issues with data. Data completeness is often a challenge in registries where data submission is voluntary or where long duration follow-up records on patients who may seek treatment from multiple sites are involved. Sufficient funds for data maintenance are often a challenge. The registry needs to balance the documentation process such that it is not cumbersome thereby affecting data quality. These critical issues need to be addressed during the designing phase of the registry. Data completion and quality can be maintained through allocating time, funds, regular feedback through stakeholders, data audits, issuing data quality certificates highlighting centres with high-quality data and training support. A key underlying factor influencing data quality is overall awareness on the importance of the registry data.

6.   A primary objective of disease registries, especially registries for rare conditions, is to increase the visibility of the condition by providing data on numbers. The registry data is thus an advocacy tool to seek better healthcare facilities and treatment. Registry data is an effective tool for planning services for patients. Good quality disease registry data helps support various programmes for monitoring

and evaluation of health care. Registry data, reported through publications, provide new knowledge about the condition.

7. Registry reports published timely (annually or otherwise, depending on the type of data and registry objectives) help dissemination of information. Dissemination of data and reporting help boost the data collection process itself, improving the functioning of the registry.

Rare Diseases

The World Health Organization (WHO) defines rare diseases (RD) as lifelong, debilitating health conditions that affect 1 or less individuals per 1000 population. The definition of RD varies across the world [11]. In Europe, a disease is considered rare when it impacts 5 per 10,000 (i.e. about 1 per 2000) population [12]. In the USA, RDs are those that affect less than 200,000 individuals, that is 1 in 1630 people [13]. In Japan, any condition that impacts fewer than 50,000 people is considered to be a RD. Thus, the definitions of RDs vary. This difference in definitions implies that a disease that is rare in one country (or for that matter in one part of the country due to founder effects) may not be considered rare in another area [11]. As such, there is also a confusion about the number of RDs. The Orphanet (www.orpha.net) is a 37-country network of information on RDs, with the aim of improving diagnosis, care and treatment of people with RDs. It is an expert reviewed, knowledge-base of data on RDs, malformation syndromes, morphological and biological anomalies, as well as particular clinical situations considered as 'rare' in Europe. The Orphanet enlists 2217 RDs [12]. This list includes conditions like Huntington disease, spina bifida, fragile X syndrome, Guillain–Barré syndrome, Crohn disease, cystic fibrosis, Duchenne muscular dystrophy, haemophilia and haemoglobinopathies. The latter condition is predominantly distributed in Middle-Eastern, African and South Asian countries, where the prevalence may be as high as 6% (Chap. 11). Haemoglobinopathies are found mainly among immigrants in Europe, so that while it is a RD in Europe, it may not merit this status in endemic regions of the world.

In a recent analysis, it was estimated that 85% of 5304 RDs had a point prevalence of <1 per 1,000,000. However, around 80% of the population burden of RDs were ascribed to 149 conditions with a prevalence of 1–5 per 10,000 [14]. Due to the rarity of these conditions, diagnosis is routinely delayed. Due to limited knowledge, care is also limited, and in short supply [15].

Rare Disease registries

Due to the infrequencies of the conditions, sufficient community-based data for rare conditions are difficult to obtain. RD registries are extremely valuable for diseases affecting small patient populations. By offering access to patients, registries provide the first important step in the conduct of observational descriptive studies. Though most RD registries may not have enough registered patients to provide statistically powered hypothesis testing, they nevertheless provide access to acquiring an appropriate sample size for epidemiological or clinical research. RD registries thereby provide the opportunity to address the knowledge gaps in epidemiology, treatment and care services, understand the natural history of the conditions, determine the pathogenesis that is the genetic, molecular and physiological basis of RDs, test drugs and devices and develop standard guidelines for management of the disorders. Information on the natural history of these rare disorders provides better understanding of progression, natural events and complications such as development of unusual infections, autoimmune complications, genetic/phenotypic heterogeneity and endpoints for therapeutic clinical development. The availability of biorepositories (of blood, urine, tissue or appropriate samples from patients) strengthens opportunities to evaluate drugs, medical devices and orphan products [16].

Registries provide data on quality of life and the economic consequences of RDs which may be difficult to understand using any other source. Quality of life or patient-reported outcomes are widely accepted clinical outcomes [16]. Such measures provide data to show if existing or newer disease management strategies or treatments are effectual in enhancing the standard of living of patients and families. Health economic data are a valuable addition to RD registries [16]. Most RD registries are formed with the intention of developing and improving care through newer treatments or therapies. The cost of these therapies to patients or family members is usually not well-documented. Introduction of new treatment that improves longevity could in fact introduce long-term disease management requiring financial resources. In the absence of government or insurance-based health care, out-of-pocket expenditure is likely to increase, an impact that could be incorporated in data collection.

RD registries provide a point of contact for affected patients, families and clinicians with experience in managing these conditions. Patients and families have the opportunity to connect with other families as well as clinical experts. In fact, patient registries have largely emerged due to direct or indirect suggestions or pressure/advocacy of patients and families. A RD registry thus provides a platform to engage in advocacy activities for patient communities such as advocate for support of patients' services and funds for treatment and research.

RD registries are not too different from other disease registries, but some issues are of concern. Issues of confidentiality or special considerations while enrolling of minors need to be considered. Due to the broad range of stakeholders and a variety of expectations, data ownership, data access and data communication are special issues with RD registries. RD registries need to set predefined plans for who will have access to data, access fees and durations of access. Data sharing policies must be predefined in a RD registry. Transparency is perhaps more important in RD registries. Registries need a highly motivated patient community to be engaged and participating in the registry. Regular updates on the enrollments and data collection, newsletters and information sharing through publications, full disclosure of use of data, registry funding resources are ways to achieve this transparent functioning [16].

The number of variables to be collected by the RD registry is another important consideration. Many registries may be tempted to incorporate as many data parameters as possible, leading to increasing burden on investigators and respondents, challenging completion of data collection, which might ultimately lead to the discontinuation of the registry. Patient attrition is also an issue, in case registering patients are not willing to share data on an extensive list of variables. Thus, RD registries need to balance a broad data set with the burden of data collection.

Follow-up data is a critical component of RD registries. However, collection of long-term follow-up data has some challenges such as who should provide the data (patient/specialist/clinician), retention and lost-to-follow-up patients. The regularity with which the patient approaches registry or clinician or a specialist will affect the data. The RD registry will need to invest in patient retention using pre-decided guidelines, linking to other data sources (such as linking with national birth and death registries) and policies to trace lost-to-follow-up patients. Another challenge is the fragmentation of data. The patients may not usually be lost-to-follow-up, but their sporadic interactions with registry or clinicians/specialists leads to infrequent contribution to data collection that RD registries need to consider. One key factor that contributes to high quality of data is records of mortality [17]. As such this data requires follow-up with patients' family or linking with other health information systems. In the absence of such links or active follow-up, the utility of data from registries for epidemiological research may be limited.

RD registries may encounter local regulations, and budgetary considerations may remain a perennial threat. Large pharmaceutical companies may require enough motivation to introduce new drugs/devices where regulatory processes are rigorous or lead to pricing issues. The number of patients using this product/device is small, and the product does not support cost of time and effort [16]. Lack of interest from pharmaceutical companies, often major financial investors in the registry, threatens the long-term survival of the registry. Lastly governance of a RD registry is a complex issue due to the variety of stakeholders involved in it and their expectations from a registry. Responsibilities of funding, role of financial sponsors, setting of agenda and roles and responsibilities of stakeholders, ownership of registry and the data, protection of patient's privacy, outreach for participation, creating collaborations and publications are some of the critical governance issues with respect to a RD registry.

This discussion on disease and RD and registries demonstrates their utility for conditions that are rare, where routine epidemiological methods are inapplicable. Disease registries are high cost investments, as they require professional inputs for design of the data to be collected with reference to the needed objectives. Case definitions and case ascertainment methods must be identified. Staff collecting and reporting data need to be trained. Due to such high financial investment, disease registries are mostly located in industrialized countries. There are some examples of disease registries in lower middle income countries (LMICs). In India, expertise for establishing and maintaining a disease registry is available. For example, the National Cancer Registry Programme aims to create a system of regional community-based cancer registries [18]. The objective of this registry is to collate data and increase the outreach of the registry especially in rural areas.

RDs and registries in India

The utility of rare disease registries is evident from not only the increasing number of publications on these conditions, but also the increase in the number of journals in this field. In India, around 450 RDs have been reported, primarily from tertiary care hospitals [19]. Table 5.2 shows some of the more familiar RDs reported in India. The data in the table indicates that these conditions are rare, data is scattered and actual incidence in newborn babies in India or prevalence in the Indian population is unclear. Treatment for these conditions is costly and requires repeated visits to selected tertiary care centres, resulting in high treatment and management costs. However, it is evident that patient organizations help support families and patients with RDs. These organizations may or may not have a dedicated registry but are advocates for improved and subsidized treatment in India. It is necessary to note that there are multiple organizations or parent/patient groups in India which are currently advocates for a single or multiple rare disease or conditions. They may have associations with international organizations and may have regional chapters/organizations that help in increasing the outreach of the organization.

Several RD organizations have originated from common experiences of delayed diagnosis due to inadequate awareness about these conditions among the medical community, and lack of financial resources for diagnosis and management [74]. Nearly all globally used therapies and diagnostics are available in India, but are located in urban centres. Financial costs were the key factor limiting access to treatment, as there are no government policies or insurance schemes to support patients with these disorders. The RD organizations in India play a significant role in reaching out to patients who may have no other support to turn to. They have worked towards increasing public awareness about these conditions, supporting caregivers and parents on management of the condition, increasing awareness among parents about genetic counselling, and improving knowledge among providers about these disorders.

The biggest achievement of these organizations, and of individual patients and their families is in mobilizing a National Policy for Rare Diseases [19]. Regrettably, the policy is still in the draft stage, and has not been implemented. The genesis of the policy was in response to writ petitions submitted by patients in 2013 and in 2016 [75]. Several patients who were eligible for health services under the Employees State Insurance Scheme (ESIS) were denied reimbursement, as these insurance policies had an upper ceiling on costs. Based on court directives to formulate a policy for care for patients with RD, a draft policy was submitted to the court by the Ministry of Health and Family Welfare in 2017.

The dilemma of allocating funding for RD in a resource-limited public health system was deliberated by the Ministry of Health and Family Welfare. As a first step

Table 5.2 RD organizations in India

Name of RD addressed by organization	Name of organizations in India	Incidence	Availability of registry in India
Umbrella organization	Indian Organization for Rare Diseases (I-ORD) [20]		Registry not available
Umbrella organization	Organizations for Rare Diseases India (ORDI) [21]		Patient registry available
Down syndrome	Down Syndrome Federation of India (DSFI) [22]	1 in 700 in USA [23] Incidence 1 in 826 live births 1 per 1250 reported in India [24]	In the process of establishing a registry
Duchene muscular dystrophy and other muscular dystrophy	Dystrophy Annihilation Research Trust (DART) [25]	1 in 3600 live male births [26] 1 in 3500 to 1 in 5000 boys [27]	Not established but intend to contribute to a national registry
Spinal muscular atrophy	The Cure SMA Foundation [28] Molecular Diagnostics, Counselling, Care and Research Centre (MDCRC) [29]	1 in 6000–10,000 [30]	Registry available with MDCRC
Haemangiomas		4–5% of infants [31]	No
Systemic lupus erythematous	Lupus Trust India [32]	0.3/100,000 to 23.2/100,000 [33] 3.2/100,000 in India [34]	Registry available
Primary Immunodeficiency Diseases in children (PID)	Indian Patients Society for Primary Immunodeficiency (IPSPI) [35]	True incidence or prevalence undocumented however calculated prevalence 1 in 2000 children as reported in USA 1 in 1200 general population (USA) [36]	Registry available [37]
Pompe disease	Pompe Foundation India [38]	1 in 5000–8000 live births in USA, Europe and Australia [39]	Registry available

(continued)

Table 5.2 (continued)

Name of RD addressed by organization	Name of organizations in India	Incidence	Availability of registry in India
Cystic fibrosis		1 in 2500 Caucasians Ethnic variations reported Prevalence in India undocumented Incidence in migrant Indian population: 1 in 40,000 (in USA) 1 in 10,000–12,000 (in UK) [40] Estimated number of children born in India could be 10,908 presuming incidence of 1 in 2500 live births [41]	Not available
Thalassemia	Thalassemics India [42] Thalassemia and Sickle Cell Society (TSCS) [43]	Prevalence of beta thalassemia is 3–4% in India with estimated 8000–10,000 new births each year [44]	Registry available
Sickle cell anaemia	Thalassemia and Sickle Cell Society [43]	Estimated 42,016 babies with sickle cell anaemia born in 2010 [45]	Registry available
Haemophilia	Hemophilia Federation of India (HFI) [46]	1 in 5000 live male births [47] 0.9 per 100,000 population in India [48]	Registry available
Fragile X syndrome	Fragile X Society of India (FXS-I) [49]	1 in 5000 men 1 in 4000–6000 women [50] Indian prevalence unknown [51]	Registry available
Rett syndrome	Indian Rett Syndrome Foundation (IRSF) [52]	1 in 10,000 to 1 in 22,000 [53]	Registry available
Cerebral palsy	Indian Institute of Cerebral Palsy (IICP) [54]	Prevalence ranging from 1.5 to 4 per 1000 live births [55] 2.95 per 1000 births (pooled prevalence in India) [56]	Registry available

(continued)

Table 5.2 (continued)

Name of RD addressed by organization	Name of organizations in India	Incidence	Availability of registry in India
Lysosomal storage disorder	Lysosomal Storage Disorders Support Society (LSDSS) [57]	1 in 7000 to 8000 in India [58]	No but intend to contribute to a national registry
Inborn errors of metabolism	Metabolic Errors and Rare Disease Organization of India (MERD India) [59]	1 in 2497 in India [60]	Not a formal registry, information of parents recorded
Multiple sclerosis	Multiple Sclerosis Society of India [61]	Globally 5–20 per 100,000 1.33 per 100,000 [62] 7–10 per 100,000 [63]	Registry available
Sjögren's syndrome	Sjögren's India [64]	0.1–0.4% worldwide [65] 3 in 100 to 1 in 1000 [66]	Not available
GNE myopathy	World Without GNE Myopathy (WWGM) [67]	Worldwide prevalence approximately 4–21 per 1,000,000 [68]	Not available
Retinoblastoma	Iksha Foundation [69]	1 in 20,000 live births [70]	Associated with national retinoblastoma study, that advocates data driven management guidelines
Motor neurone disease	Asha Ek Hope Foundation [71]	Median prevalence—4.48 per 100,000 Standardized incidence rate 1.68 per 100,000 [72] 1.5 to 2.7/100,000/year prevalence of 3 to 5/100,000 [73]	Registry available

towards reducing treatment costs, a committee was to be constituted with representatives of several ministries. Ministries were to examine specific aspects, such as indigenous drug development (Department of Pharmaceuticals), removal of import duty on treatment products (Department of Revenue), encouraging corporate houses to finance treatment (Ministry of Corporate Affairs), revision of the insurance act (Department of Financial Services) and explore the possibility of raising the financial ceiling for RD patients under the ESIS (Ministry of Labour) [75]. Unfortunately, an initial fund of one billion rupees for management of RDs has not been established. The policy has not been approved. It is noteworthy that the policy does not have an elucidation for rare diseases, but enlists major RDs, excluding haemophilia, thalassemia and sickle cell anaemia, for which treatment is already being provided.

The Haemophilia Registry in India

One of the least visible registries is that of haemophilia, a rare single gene disorder, that is responsible for morbidity, disability and premature mortality. This registry has been one of the strong points in advocacy for obtaining treatment for haemophilia in India. The remaining section of this article describes the haemophilia registry in India, its remarkable evolution as a volunteer-based reporting system, mentored by the World Federation of Hemophilia (WFH).

Haemophilia

Haemophilia is an inherited medical condition, caused by a single gene mutation either in the coagulation factor VIII gene (haemophilia A) or the coagulation factor IX gene (haemophilia B). The deficient synthesis of clotting factor VIII (in case of haemophilia A) or clotting factor IX (in case of haemophilia B) predisposes patients to spontaneous or trauma-related bleeding episodes, which vary according to the severity of the disorder [76, 77]. Haemathrosis is an important clinical manifestation in patients with poorly treated severe haemophilia, along with soft and deep tissue haematomas and haematuria. Intracranial haemorrhage was and remains an important cause of mortality [78]. Untreated haemophilia results in mortality within the first decade of life. The complications of haemophilia arise from joint damage caused by repeated haemorrhages, leading to progressive and severely crippling disability. Transfusion transmitted infections (TTIs), such as HIV, hepatitis B or C, and development of allo-antibodies (inhibitors) are other complications that gravely impair the quality of life of patients and caregivers [79].

Replacement therapy with clotting factor concentrate (CFC, anti-haemophilic factor) is the mainstay for treatment of haemophilia. Historically whole blood transfusion was the first method of replacement therapy. The treatment of haemophilia progressed to the use of plasma concentrates or fresh frozen plasma (FFP) which became the preferred therapy for bleeding episodes. In 1964, the discovery of FVIII-rich cryoprecipitate from slow thawing of FFP, by Judith Pool, was a turning point for management of haemophilia. The arrival of some pharmaceutical companies in early 1970s led to production of relatively large amounts of CFCs in lyophilized formulations which were of intermediate purity. The 1970s also saw the discovery of desmopressin that turned out to be a risk-free and inexpensive treatment alternative for patients with mild haemophilia [80, 81].

Earlier, the main concern of treatment with plasma products was the risk of TTIs, as these products did not undergo viral inactivation [82]. The treatment of haemophilia has, however, progressed from plasma derived concentrates to recombinant concentrates. Although the first generation of recombinant products had an exemplary record of efficacy and safety, manufacturers have invested in further improvements that have led to products with high concentrations of FVIII or FIX and with a greater half-life [81]. A serious complication of replacement therapy is the development of allo-antibodies called inhibitors [83]. These are high affinity IgG molecules to factor VIII which neutralize CFC, so that the patient becomes non-responsive to treatment [84].

Bleeding episodes are managed either on-demand, or through prophylaxis. Prophylaxis drastically reduces the annual number of bleeding episodes, as compared to on-demand therapy [85]. Prophylaxis has become the favoured treatment option for management of haemophilia in the US and several European nations. Prophylaxis has markedly enhanced the quality of life of patients and has reduced the occurrence of joint complications among patients [86–93]. On-demand treatment remains the main form of treatment for patients in developing countries, where CFC has to be purchased through out-of-pocket expenditure [48, 94].

In India, a study conducted in 2010 identified the impact of prohibitive costs of CFC on treatment of haemophilia. In this prospective study of 113 bleeding events, one bleeding episode was untreated, seven were treated with commonly available pain killers, 23 were treated only with ice, 38 (34%) were treated with first aid (i.e. cold compression and rest) and CFC was used for 28/113 (25%) of bleeding episodes. Among those using CFC, only 32% of the estimated required CFC was used, due to inability to afford the cost of the treatment product. The consequence of poor treatment was evident as a total of 127 school/work days were lost due to immobilization during the study period, with an average of 7–8 days of school/work day lost per patient due to bleeding [95]. In a follow-up study, the estimated expenditure on treatment on families was determined. Using stringent assumptions that a single infusion would control the bleeding, the study identified that if all bleeding episodes were addressed therapeutically, families would be compelled to spend between 21 and 314 times their monthly earnings. This expenditure would have had catastrophic consequences on 70% of families [96].

The impact of suboptimal treatment was evident from the prevalence of disability among patients with haemophilia. Another Indian study identified that only 9 out of 148 patients with haemophilia randomly selected from across the country were disability-free. The risk factors notably linked with disability were patients age, socio-economic status, number of persons in the family, family history of haemophilia, frequency of physiotherapy exercises, home use of coagulation factor concentrate and type of blood product(s) used, that is clotting factor concentrate or cryoprecipitate [97]. The study reiterated the need to ensure coagulation factor concentrate in adequate amounts to prevent disability.

Evolution of Haemophilia Services in India

The evolution of haemophilia care, globally, has occurred due to advances in treatment and resource allocations through health budgets [98]. Where haemophilia services are funded through either insurance, or through federally funded programmes, standard diagnosis, treatment and management guidelines are available [99–102]. Till 2013, haemophilia services were delivered exclusively through the efforts of the Haemophilia Federation of India (HFI) [46]. The HFI is a nongovernmental organization (NGO) that was established in 1983, with the goal of facilitating access to treatment for patients, and support for families. Although at present, state governments provide free of cost treatment for patients, this was only achieved by the advocacy of the HFI.

Figure 5.2 shows the organizational structure of the HFI, which has member organizations (chapters) in all states of the country. Currently, there are 86 affiliated and non-affiliated chapters in India (Fig. 5.2) [46]. The number of chapters in any particular state is variable and depends on the number of patients. Larger states, and states with more numbers of diagnosed patients have larger numbers of organizations within a state.

Figure 5.3 enlists the primary functions of the NGO. Till 2013, the HFI was the only source of CFC for patients with bleeding disorders in the country. CFC was purchased and imported by the HFI and distributed to 'chapters' across the country (Fig. 5.2). The CFC was sold to patients at subsidized costs. A proportion of the CFC was donated on humanitarian grounds to the most economically vulnerable patients. Each chapter is associated with a medical facility, where at least one specialist is trained in the management of haemophilia through a capacity building programme, mentored by the World Federation of Hemophilia (WFH) (described below). The HFI has helped patients test for inhibitors and has provided appropriate treatment to patients with inhibitors through corporate partnerships. Globally, as management of haemophilia has benefited from prophylaxis, the HFI has initiated prophylaxis at some centres [104].

In addition to providing diagnostic and care facilities for people with haemophilia, another important function of the NGO has been to bring together patients and families and provide psychosocial support, especially after the diagnosis of the disorder [105]. Another role has been to establish prenatal testing, carrier detection and genetic counselling facilities for families, by engaging with academic institutions that can conduct prenatal testing [106, 107]. A crucial activity of the HFI is advocacy for the rights of haemophilia patients in India. The HFI has extensively engaged in intense advocacy that has resulted in state government supported haemophilia care in all states in India [104]. The HFI advocacy has led to the inclusion of haemophilia under the Rights of Persons with Disability Act 2016 [108]. One of the main functions of the HFI has been in haemophilia surveillance, through a volunteer mediated registry. This aspect is further described below.

Fig. 5.2 State-wise distribution of haemophilia member organizations across India [103]. The state of Jammu & Kashmir was reorganized into union territories of Jammu and Kashmir and Ladakh in 2019

World Federation of Hemophilia : Global mentorship

The WFH, a not-for-profit organization, was established in 1963 [109]. The objective of this programme is to ensure haemophilia care worldwide. The organization was established as there was little appreciation of the extent of the problem faced by patients with this condition. The rarity of the condition was compounded by poor awareness among clinical practitioners and lack of confirmed diagnosis of haemophilia due to unavailability of trained laboratory professionals. Treatment,

Fig. 5.3 Main functions of
the haemophilia NGO

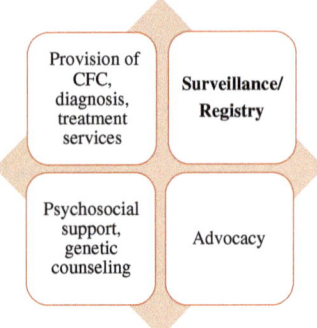

if available, was not easily accessible to patients, especially for the economically vulnerable populations from LMICs. The WFH has assisted with the formation of national associations for people with haemophilia and their families across the world, so that currently there are 125 National Member Organizations (NMOs) globally.

The WFH promotes capacity building of NMOs through workshops and training programmes that address a range of topics including fundraising, volunteer recruitment, media relations, advocacy and patient outreach, relationship building with pharmaceutical companies and selection and purchasing treatment products. It developed the WFH Twinning Programs in 1994 that helped pair treatment centres in developed countries with those in developing countries, helping in training and capacity building activities among medical and paramedical personnel associated with haemophilia care [110]. Further, it developed a successful project called Operation Access. Through this project WFH was instrumental in creating a treatment coalition with companies who donated treatment products, and the haemophilia organization played a major role in education and advocacy. The WFH was the catalyst and advisor throughout this process, targeting the development and implementation of country level national haemophilia programme. The WFH Global Alliance for Progress (GAP) initiative [111, 112] aims at addressing the global disparity in care among developed and developing countries [112]. The WFH has now expanded its scope, and its 'Treatment for All' strategy to include other inherited bleeding disorders and women with bleeding disorders.

Annual Global Surveys and the National Hemophilia Registry of India

With access to CFC being of critical importance, one of the vital activities of the HFI is the maintenance of records of patients. The primary purpose of patient registration was to monitor the use of CFC. Linked to clinical information, these records serendipitously included key variables that could be used to describe the epidemiology of haemophilia in the country. The data included demographic and clinical data on patients (place of residence, age at registration, date of birth, data on coagulation deficiency, severity status, occupation of patient and parent, annual income, reported age at first bleed, date of diagnosis and family history). The potential epidemiological utility of these data was illustrated through a description of the epidemiology of haemophilia in India's second largest state of Maharashtra between 1989 and 2005 [107, 113, 114]. The unique aspect of this registry was that the data was maintained by volunteers of the haemophilia chapters, and there was no specific funding for this activity. Although the limited involvement of professionals did lead to sources of error, the data illustrated the potential promise of a registry, maintained through primarily voluntary efforts [115].

Early on, the WFH had realized the importance and utility of data to support its research and advocacy activities. The Annual Global Survey (AGS) of the WFH, established in 1998 with the first compiled report published in 1999, has evolved from a restricted set of data, to a robust source of data on the demographic and clinical description of patients with haemophilia and bleeding disorders across the world. The survey now provides information on access to care along with treatment product use, data on infectious complications and haemophilia care across the world. The rich data from the AGS is used to mobilize policymakers to address the goal of comprehensive care for haemophilia across the world [116].

Data collection for the AGS is initiated by the WFH, which sends out questionnaires to the participating organizations, including the HFI. Participation in AGS is voluntary, and the data is self-reported. The WFH carries out checks for inconsistencies and seeks clarifications before finalizing the data. Of the 125 countries that report data, there are differences in data quality. Registries from high-income countries provide fairly consistent information, validating its use in program planning, whereas the quality and availability of data from LMICs may be limited.

The database of the WFH has now been upgraded to a World Bleeding Disorders Registry (WBDR) [117]. The initial data variables have been expanded with addition of functional scales for haemophilia (Haemophilia Joint Health Score, Functional Independence Score in Haemophilia, Joint Disease, Range of motion, WFH score), along with quality of life scales. The WBDR has a Data Quality Accreditation (DQA) program, with a data quality team providing training and feedback on quality of data to all participating haemophilia treatment centers (HTCs) [118].

The clinical and epidemiological data on burden of haemophilia and other bleeding disorders available through the AGS provides healthcare providers and policymakers an overview of trends of haemophilia and its treatment. In 2018, India reported 20,778 people with haemophilia, 676 people with Von Willebrand's Disorder and 457 people with other bleeding disorders [119]. Between 2016 and 2018, there was a 13% increase in patients with haemophilia, 27% increase in patients with von Willebrand's Disorder and 28% increase in people diagnosed with other rare bleeding disorders [119, 120]. There were 769 people with haemophilia A with inhibitors. Among patients with haemophilia A, 38% were aged between 19 and 44 years, whereas only 2% were aged between 0 and 4 years. A quarter of patients with haemophilia A were between 5 and 18 years of age. The lower numbers of patients in the younger age groups, especially 0–4 years could arise from lack of diagnosis and registration, or genetic counseling [106]. The utilization of CFC had improved slightly from 0.155 in 2017 to 0.23 in 2018, but was low as compared to developed countries [119–121].

Registry data for mobilizing access to care

By 2000, with increasing diagnosis it was not possible to keep up supply of CFC for patients across India. The HFI first approached all state authorities to provide haemophilia treatment to patients. Some small states with small number of patients readily agreed to provide treatment [104]. However, for other states, haemophilia chapters in each state filed public interest litigation in the High Court of the state, demanding CFC as the right to life for people with haemophilia. As each court demanded lists of patients, the relevance of the National Hemophilia Registry was understood. The National Hemophilia Registry,was computerized through a completely voluntary effort. Duplicate entries were removed. The data was the basis for judicial directives to state governments to provide free clotting factor concentrate for patients across India [104].

This case study on the haemophilia registry in India identifies that registries are an important tool to understand and document the epidemiology of RDs and conditions. The haemophilia case study shows that even in LMICs, partnering with global organizations, and with dedication and effort, a RD registry can be established, maintained and used for mobilization of care for patients. This model of global–national partnership can be used by other RD organizations for advocating for care for patients.

References

1. Brooke EM, World Health Organization (1974) The current and future use of registers in health information systems. World Health Organization
2. Viviani L, Zolin A, Mehta A, Olesen HV (2014) The European Cystic Fibrosis Society Patient Registry: valuable lessons learned on how to sustain a disease registry. Orphanet J Rare Dis 9(1):81
3. Gliklich RE, Dreyer NA, Leavy MB (eds) (2014) Registries for evaluating patient outcomes: a user's guide, No. 13. Government Printing Office
4. Santoro M, Coi A, Di Paola ML et al (2015) Rare disease registries classification and characterization: a data mining approach. Public Health Genomics 18(2):13–122
5. Jansen-van der Weide MC, Gaasterland CM, Roes KC et al (2018) Rare disease registries: potential applications towards impact on development of new drug treatments. Orphanet J Rare Dis 13(1):1–11
6. Kodra Y, Weinbach J, Posada-De-La-Paz M et al (2018) Recommendations for improving the quality of rare disease registries. Int J Environ Res Public Health 15(8):1644
7. Cystic Fibrosis Foundation. About the Cystic Fibrosis Foundation. https://www.cff.org/About-Us/About-the-Cystic-Fibrosis-Foundation/Our-History/. Accessed 22 July 2020
8. Cerebral Palsy Research Network. Cerebral Palsy Research Network fact sheet. https://cprn.org/wp-content/uploads/2019/12/cprn-fact-sheet-2019.pdf/. Accessed 22 July 2020
9. Spina Bifida Patient Registry. About national spina bifida patient registry. https://www.cdc.gov/ncbddd/spinabifida/nsbprregistry.html. Accessed 22 July 2020

10. Mandavia R, Knight A, Phillips J et al (2017) What are the essential features of a successful surgical registry? A systematic review. BMJ Open 7(9)
11. Ferreira CR (2019) The burden of rare diseases. Am J Med Genet Part A 179(6):885–892
12. Orphanet. An online database of RDs and orphan drugs. https://www.orpha.net/consor/cgi-in/Disease_Search_List.php?lng=EN. Accessed 22 July 2020
13. Boat TF, Field MJ (eds) (2011) RDs and orphan products: accelerating research and development. National Academies Press
14. Wakap SN, Lambert DM, Olry A et al (2020) Estimating cumulative point prevalence of RDs: analysis of the Orphanet database. Eur J Med Genet 28(2):165–173
15. Valdez R, Ouyang L, Bolen J (2016) Public health and RDs: oxymoron no more. Prev Chronic Dis 13
16. Gliklich RE, Dreyer NA, Leavy MB (eds) (2014) Rare disease registries. In: Registries for evaluating patient outcomes: a user's guide, No. 13. Government Printing Office
17. Coi A, Santoro M, Villaverde-Hueso A et al (2016) The quality of rare disease registries: evaluation and characterization. Public Health Genomics 19(2):108–115
18. Development of an atlas of cancer in India. https://www.ncdirindia.org/ncrp/ca/about.aspx. Accessed 22 July 2020
19. Ministry of Health and Family Welfare, Government of India (2020) National policy for treatment of RDs. https://main.mohfw.gov.in/sites/default/files/Rare%20Diseases%20Policy%20FINAL.pdf. Accessed 22 July 2020
20. Indian Organization for Rare Diseases (2005) https://www.i-ord.org/. Accessed 22 July 2020
21. Organization for Rare Diseases India (2014) https://ordindia.in/. Accessed 22 July 2020
22. Down Syndrome Federation India (2016) https://www.downsyndrome.in/. Accessed 22 July 2020
23. Centers for Disease Control and Prevention. Data and statistics on Down syndrome. https://www.cdc.gov/ncbddd/birthdefects/downsyndrome/data.html. Accessed 22 July 2020
24. Jayalakshamma MM, Amudha S, Tilak P et al (2010) Cytogenetic analysis in Down syndrome. Int J Hum Genet 10(1–3):95–99
25. Dystrophy Annihilation Research Trust (2003) https://dartindia.in/. Accessed 22 July 2020
26. Singh RJ, Manjunath M, Preethish-Kumar V et al (2018) Natural history of a cohort of Duchenne muscular dystrophy children seen between 1998 and 2014: an observational study from South India. Neurol India 66(1):77
27. Nalini A, Polavarapu K, Preethish-Kumar V (2017) Muscular dystrophies: an Indian scenario. Neurol India 65(5):969
28. Cure SMA Foundation of India (2014) https://curesmaindia.org/. Accessed 22 July 2020
29. Molecular Diagnostics, Counseling, Care and Research Centre. In: Registry. https://www.mdcrcindia.org/ourservices/registry/. Accessed 22 July 2020
30. Farrar MA, Kiernan MC (2015) The genetics of spinal muscular atrophy: progress and challenges. Neurotherapeutics 12(2):290–302
31. Vivar KL, Mancini AJ (2018) Infantile hemangiomas: an update on pathogenesis, associations, and management. Indian J Paediatr Dermatol 19(4):293
32. Lupus Trust India. https://www.lupustrustindia.org/about. Accessed 22 July 2020
33. Rees F, Doherty M, Grainge MJ et al (2017) The worldwide incidence and prevalence of systemic lupus erythematosus: a systematic review of epidemiological studies. Rheumatology 56(11):1945–1961
34. Malaviya AN, Singh RR, Singh YN et al (1993) Prevalence of systemic lupus erythematosus in India. Lupus 2(2):115–118
35. Indian Patients Society for Primary Immunodeficiency (2017) https://www.ipspiindia.org/. Accessed 22 July 2020
36. Boyle JM, Buckley RH (2007) Population prevalence of diagnosed primary immunodeficiency diseases in the United States. J Clin Immunol 27(5):497–502
37. Gupta S, Madkaikar M, Singh S et al (2012) Primary immunodeficiencies in India: a perspective. Ann N Y Acad Sci 1250(1):73–79
38. Pompe Foundation. https://pompeindia.org/. Accessed 22 July 2020

39. Muranjan M, Karande S (2018) Enzyme replacement therapy in India: lessons and insights. J Postgrad Med 64(4):195
40. Sarkar A (2002) Cystic fibrosis: Indian experience. Indian Pediatr 39:813–818
41. Mandal A, Kabra SK, Lodha R (2015) Cystic fibrosis in India: past, present and future. J Pulm Med Respir Res 1
42. Thalassemics India (2013) https://www.thalassemicsindia.org/index.php. Accessed 22 July 2020
43. Thalassemia and Sickle Cell Society (1998) https://tscsindia.org/. Accessed 22 July 2020
44. Mohanty D, Colah RB, Gorakshakar AC et al (2013) Prevalence of β-thalassemia and other haemoglobinopathies in six cities in India: a multicentre study. J Community Genet 4(1):33–42
45. Hockham C, Bhatt S, Colah R et al (2018) The spatial epidemiology of sickle-cell anaemia in India. Sci Rep 8(1):1–10
46. Hemophilia Federation of India. https://www.hemophilia.in/. Accessed 22 July 2020
47. Haldane JB (1935) The rate of spontaneous mutation of a human gene. J Genet 31(3):317
48. Kar A, Phadnis S, Dharmarajan S et al (2014) Epidemiology & social costs of haemophilia in India. Indian J Med Res 140(1):19
49. Fragile X Society—India. https://www.fragilex.in/. Accessed 22 July 2020
50. Saldarriaga W, Tassone F, González-Teshima LY et al (2014) Fragile X syndrome. Colomb Med 45(4):190–198
51. Sachdeva A, Jain P, Gunasekaran V et al (2019) Consensus statement of the Indian Academy of Pediatrics on diagnosis and management of fragile X syndrome in India. Indian Pediatr 56(3):221–228
52. Indian RETT Syndrome Foundation (2010) https://www.rettsyndrome.in/. Accessed 22 July 2020
53. Kumar S, Alexander M, Gnanamuthu C (2004) Recent experience with Rett syndrome at a tertiary care center. Neurol India 52(4):494
54. Indian Institute of Cerebral Palsy (2012) https://www.iicpindia.org/. Accessed 22 July 2020
55. Centers for Disease Control. Data and statistics for cerebral palsy. https://www.cdc.gov/ncbddd/cp/data.html. Accessed 22 July 2020
56. Chauhan A, Singh M, Jaiswal N et al (2019) Prevalence of cerebral palsy in Indian children: a systematic review and meta-analysis. Indian J Pediatr 86(12):1124–1130
57. Lysosomal Storage Disorders Support Society (2012) https://www.lsdss.org/. Accessed 22 July 2020
58. Phadke SR (2015) Lysosomal storage disorders: present and future. Indian Pediatr 52(12):1025–1026
59. MERD India Foundation (2011) https://www.merdindia.com/. Accessed 22 July 2020
60. Lodh M, Kerketta A (2013) Inborn errors of metabolism in a tertiary care hospital of Eastern India. Indian Pediatr 50(12):1155–1156
61. Multiple Sclerosis Society of India. https://www.mssocietyindia.org/. Accessed 22 July 2020
62. Bhatia R, Bali P, Chaudhari RM (2015) Epidemiology and genetic aspects of multiple sclerosis in India. Ann Indian Acad Neurol 18(Suppl 1):S6
63. Singhal BS, Advani H (2015) Multiple sclerosis in India: an overview. Ann Indian Acad Neurol 18(Suppl 1):S2
64. Sjögren's India (2016) https://www.sjogrensindia.org/. Accessed 22 July 2020
65. Qin B, Wang J, Yang Z et al (2015) Epidemiology of primary Sjögren's syndrome: a systematic review and meta-analysis. Ann Rheum Dis 74:1983–1989
66. Kishore M, Panat SR, Aggarwal A et al (2014) Sjögren's syndrome: a review. SRM J Res Dent Sci 5(1):31
67. World Without GNE Myopathy. https://gne-myopathy.org/. Accessed 22 July 2020
68. Celeste FV, Vilboux T, Ciccone C et al (2014) Mutation update for GNE gene variants associated with GNE myopathy. Hum Mutat 35(8):915–926
69. Iksha Foundation. https://ikshafoundation.org/. Accessed 22 July 2020
70. Seth R, Singh A, Guru V et al (2017) Long-term follow-up of retinoblastoma survivors: experience from India. South Asian J Cancer 6(4):176

71. Asha Ek Hope Foundation of MND/ALS, India. https://www.ashaekhope.com/. Accessed 22 July 2020
72. Logroscino G, Piccininni M, Marin B et al (2018) Global, regional, and national burden of motor neuron diseases 1990–2016: a systematic analysis for the Global Burden of Disease Study 2016. Lancet Neurol 17(12):1083–1097
73. Al-Chalabi A, Hardiman O (2013) The epidemiology of ALS: a conspiracy of genes, environment and time. Nat Rev Neurol 9(11):617
74. Choudhury MC, Saberwal G (2019) The role of patient organizations in the rare disease ecosystem in India: an interview based study. Orphanet J Rare Dis 14(1):117
75. Bhuyan A (2017) Government submits rare disease policy to Delhi HC, recommends Rs 100 crore for genetic diseases. https://thewire.in/health/rare-disease-policy. Accessed 22 July 2020
76. Gomez K, Chowdary P (2014) Haemophilia B: molecular basis. In: Christine A, Lee CA, Berntorp EE, Hoots KW (eds) Textbook of Haemophilia, 3rd edn. Wiley-Blackwell, pp 97–102
77. van den Berg HM, Fischer K (2010) Phenotypic–genotypic relationship. In: Christine A, Lee CA, Berntorp EE, Hoots KW (eds) Textbook of Haemophilia, 3rd edn. Wiley-Blackwell, pp 33–37
78. Hoots KW (2010) Emergency management of haemophilia. In: Christine A, Lee CA, Berntorp EE, Hoots KW (eds) Textbook of Haemophilia, 3rd edn. Wiley-Blackwell, pp 394–400
79. Phadnis S, Kar A (2017) The impact of a haemophilia education intervention on the knowledge and health related quality of life of parents of Indian children with haemophilia. Haemophilia 23(1):82–88
80. Coppola A, Di Capua M, Di Minno MND et al (2010) Treatment of haemophilia: a review of current advances and ongoing issues. J Blood Med 1:183
81. Mannucci PM (2008) Back to the future: a recent history of haemophilia treatment. Haemophilia 14:10–18
82. Evatt BL, Austin H, Leon G et al (1999) Haemophilia therapy: assessing the cumulative risk of HIV exposure by cryoprecipitate. Haemophilia 5(5):295
83. Saint-Remy JMR, Jacquemin MG (2014) Inhibitors to factor VIII: immunology. In: Christine A, Lee CA, Berntorp EE, Hoots KW (eds) Textbook of Haemophilia, 3rd edn. Wiley-Blackwell, pp 41–47
84. Iorio A, Halimeh S, Holzhauer S et al (2010) Rate of inhibitor development in previously untreated haemophilia A patients treated with plasma-derived or recombinant factor VIII concentrates: a systematic review. J Thromb Haemost 8(6):1256–1265
85. Kavakli K, Yang R, Rusen L et al (2015) Prophylaxis vs. on-demand treatment with BAY 81-8973, a full-length plasma protein-free recombinant factor VIII product: results from a randomized trial (LEOPOLD II). J Thromb Haemost 13(3):360–369
86. Royal S, Schramm W, Berntorp E et al (2002) Quality-of-life differences between prophylactic and on-demand factor replacement therapy in European haemophilia patients. Haemophilia 8(1):44–50
87. Coppola A, Di Capua M, De Simone C (2008) Primary prophylaxis in children with haemophilia. Blood Transfus 6(Suppl 2):S4
88. Duncan N, Shapiro A, Ye X et al (2012) Treatment patterns, health-related quality of life and adherence to prophylaxis among haemophilia A patients in the United States. Haemophilia 18(5):760–765
89. Poon JL, Zhou ZY, Doctor JN et al (2012) Quality of life in haemophilia A: haemophilia utilization group study Va (HUGS-Va). Haemophilia 18(5):699–707
90. García-Dasí M, Aznar JA, Jiménez-Yuste V et al (2015) Adherence to prophylaxis and quality of life in children and adolescents with severe haemophilia A. Haemophilia 21(4):458–464
91. Gringeri A, Leissinger C, Cortesi PA et al (2013) Health-related quality of life in patients with haemophilia and inhibitors on prophylaxis with anti-inhibitor complex concentrate: results from the Pro-FEIBA study. Haemophilia 19(5):736–743
92. Mondorf W, Kalnins W, Klamroth R (2013) Patient-reported outcomes of 182 adults with severe haemophilia in Germany comparing prophylactic vs. on-demand replacement therapy. Haemophilia 19(4):558–563

93. Schwarz R, Ljung R, Tedgård U (2015) Various regimens for prophylactic treatment of patients with haemophilia. Eur J Haematol 94:11–16
94. Srivastava A (2003) Factor replacement therapy in haemophilia—are there models for developing countries? Haemophilia 9(4):391–396
95. Dharmarajan S, Gund P, Phadnis S et al (2012) Treatment decisions and usage of clotting factor concentrate by a cohort of Indian haemophilia patients. Haemophilia 18(1):e27–e29
96. Dharmarajan S, Phadnis S, Gund P et al (2014) Out-of-pocket and catastrophic expenditure on treatment of haemophilia by Indian families. Haemophilia 20(3):382–387
97. Kar A, Mirkazemi R, Singh P et al (2007) Disability in Indian patients with haemophilia. Haemophilia 13(4):398–404
98. Ludlam CA, Lee CA, Berntorp EE et al (2005) Comprehensive care and delivery of care: the developed world. In: Christine A, Lee CA, Berntorp EE, Hoots KW (eds) Textbook of Haemophilia, 3rd edn. Wiley-Blackwell, pp 350–365
99. Hay CR, Brown S, Collins PW et al (2006) The diagnosis and management of factor VIII and IX inhibitors: a guideline from the United Kingdom Haemophilia Centre Doctors Organisation. Br J Haematol 133(6):591–605
100. United Kingdom Haemophilia Centre Doctors' Organisation (UKHCDO) (2003) Guidelines on the selection and use of therapeutic products to treat haemophilia and other hereditary bleeding disorders. Haemophilia 9(1):1–23
101. Keeling D, Tait C, Makris M (2008) Guideline on the selection and use of therapeutic products to treat haemophilia and other hereditary bleeding disorders: a United Kingdom Haemophilia Center Doctors' Organisation (UKHCDO) guideline approved by the British Committee for Standards in Haematology. Haemophilia 14(4):671–684
102. Wilde JT, Mutimer D, Dolan G et al (2011) UKHCDO guidelines on the management of HCV in patients with hereditary bleeding disorders. Haemophilia 17(5):e877–e883
103. Map of India. https://mapchart.net/india.html. Accessed 22 July 2020
104. Ghosh K (2019) Evolution of hemophilia care in India. Indian J Hematol Blood Transfus 35(4):716–721
105. Ghosh K, Shetty S, Sahu D (2010) Haemophilia care in India: innovations and integrations by various chapters of Haemophilia Federation of India (HFI). Haemophilia 16(1):61–65
106. Nakade J, Potnis-Lele M, Kar A (2013) Impact of genetic counselling? The potential utility of haemophilia surveillance data in developing countries. Haemophilia 19(6):e388–e390
107. Potnis-Lele M, Kar A (2003) Modification of family size in families reporting history of haemophilia from Maharashtra, India. Int J Epidemiol 32(2):316–320
108. Government of India, Ministry of Law and Justice Legislative Department (2016) The Rights of Persons with Disabilities Act 2016. https://legislative.gov.in/sites/default/files/A2016-49_1.pdf. Accessed 22 July 2020
109. World Federation of Hemophilia. https://www.wfh.org/en/home. Accessed 22 July 2020
110. World Federation of Hemophilia. Twinning programs. https://www.wfh.org/en/our-work-global/twinning-program. Accessed 22 July 2020
111. World Federation of Hemophilia. Regional and national programs. https://www.wfh.org/en/regional-and-national-programs. Accessed 22 July 2020
112. Skinner MW (2012) WFH: closing the global gap—achieving optimal care. Haemophilia 18:1–12
113. Kar A, Potnis-Lele M (2001) Descriptive epidemiology of haemophilia in Maharashtra, India. Haemophilia 7(6):561–567
114. Kar A, Potnis-Lele M (2004) Haemophilia data collection in developing countries: example of the haemophilia database of Maharashtra. Haemophilia 10(3):301–304
115. Kar A (2010) Factors influencing haemophilia prevalence estimates from the volunteer-supervised Indian registry in Maharashtra. Haemophilia 16(6):952–954
116. World Federation of Hemophilia. Annual Global Surveys. https://www1.wfh.org/publications/files/pdf-1731.pdf. Accessed 22 July 2020
117. World Federation of Hemophilia. World Bleeding Disorders Registry. https://www.wfh.org/en/our-work-research-data/world-bleeding-disorders-registry. Accessed 22 July 2020

118. Coffin D, Herr C, O'Hara J et al (2018) World bleeding disorders registry: the pilot study. Haemophilia 24(3):e113
119. Report on the Annual Global Survey (2018) World Federation of Hemophilia. https://www1.wfh.org/publications/files/pdf-1731.pdf. Accessed 22 July 2020
120. Report on the Annual Global Survey (2016) World Federation of Hemophilia. https://www1.wfh.org/publications/files/pdf-1690.pdf. Accessed 22 July 2020
121. Report on Annual Global Survey (2017) World Federation of Hemophilia. https://www1.wfh.org/publications/files/pdf-1714.pdf. Accessed 22 July 2020

Chapter 6
Magnitude of Congenital Anomalies in India

Anita Kar and Dhammasagar Ujagare

Abstract Epidemiological data on birth defects is usually derived from surveillance systems. Systematic surveillance for birth defects is yet to be established in India. Data on the magnitude and trends of birth defects have to be derived from other sources, primarily research studies. This review provides an overview of sources of data on congenital anomalies in India, and the magnitude, mortality and disability associated with these conditions in the country. Available data indicate a birth prevalence of 18.44–23.05 per 1000 births, which translates into 472,177 and 581,899 affected births occurring annually in the country. Estimates of congenital anomaly neonatal mortality range from 54,000 to 99,000. Epidemiological data identify that as congenital anomaly mortality is averted, the numbers of disability survivors increase. Data from limited studies suggest that 70% of birth defects are non-fatal. The Modell Global Database of Congenital Disorders estimates that there would be over 150,000 survivors with congenital disability at five years of age per birth cohort. The review identifies the need for a nation-wide birth defects surveillance system for monitoring these large numbers in India. The Rashtriya Bal Swasthya Karyakram, a community-based screening and early identification programme for common childhood diseases, nutritional deficiencies, birth defects and developmental delays and disabilities, is a potential opportunity for collecting nation-wide data on birth defects. Till surveillance is established, robust primary studies are needed, that can inform the true prevalence of congenital anomaly affected births, mortality and survivors with disability. Strategically established population-based surveillance can be utilized to monitor the types and prevalence of congenital anomalies in India.

Keywords Birth defects · Congenital anomalies · India · Prevalence · Disability · Mortality · Child health

Epidemiological data describe the person, place and time of occurrence of a disease or a health condition. The data provide information on the magnitude and impact of

A. Kar (✉)
Birth Defects and Childhood Disability Research Centre, Pune 411020, India

D. Ujagare
ICMR—National AIDS Research Institute, Pune 411026, India

© Springer Nature Singapore Pte Ltd. 2021
A. Kar (ed.), *Birth Defects in India*,
https://doi.org/10.1007/978-981-16-1554-2_6

143

the condition in terms of mortality and disability, morbidity patterns and trends, and risk and causative factors for the disease or health condition. Such data inform public health action by identifying healthcare needs, the required services and the financial investment for implementing these services. Epidemiological data can be obtained through several sources. Typically, health information systems provide information on the numbers affected, and the consequences in terms of morbidity, mortality and disability. For example, HIV surveillance determines the number of people affected. Healthcare usage by patients with HIV can be determined from hospital in-patient or out-patient data. Vital statistics provide data on cause-specific mortality, for example, the number of AIDS-related deaths. Routinely available data from public health information systems and vital records provide information for understanding the epidemiology of the health condition, and the impact of implemented interventions.

The challenge of describing the epidemiology of diseases/health conditions arise when data collection systems are unavailable, or the data lack in quality and time-liness, or are incomplete. The latter is especially relevant in India, where 70% of healthcare providers belong to the private health sector. These providers do not report health data to the public health information system, resulting in a large gap in the understanding of the epidemiology of prevalent conditions. In such instances, surveys and research studies form the major source of data.

Obtaining data on birth defects is especially challenging, as these are conditions with low prevalence. Visible structural defects (congenital anomalies, congenital malformations) like spina bifida, orofacial clefts or limb defects are relatively easy to recognize at birth. Some common chromosomal anomalies or single gene disorders may be suspected in very sick newborns, or those with abnormal physical appearance. The signs and symptoms of some common conditions, especially congenital heart defects may not be immediately apparent at birth, and may be detected later during childhood.

Data on congenital anomalies are reported as birth prevalence. Birth prevalence of congenital anomalies is the number of congenital anomaly affected newborns per 1000 population. This measure is distinct from population prevalence, which reports the number of individuals with congenital anomalies resident in a population. Due to the higher mortality associated with severe birth defects, birth prevalence of congenital anomalies is higher than population prevalence. However, if prevalence is measured at later ages (for example, among children at 15 years of age), then population prevalence may be higher than the birth prevalence due to cumulative cohorts of surviving children. Measurement of birth prevalence has the advantage that denominator data, that is the number of births occurring in the population, are routinely collected by health information systems. Thus, the relative ease of recognition of visible structural defects, as well as the availability of denominator data on numbers of births makes birth prevalence the most appropriate indicator to report the prevalence of congenital anomalies.

In industrialized countries, and some Latin American countries, birth defects surveillance systems report data on congenital anomalies [1–8]. Birth defects surveil-lance systems were established after the thalidomide incident, as described in Chap. 3. The understanding that drugs could cross the placental barrier and cause foetal malformations, led to the establishment of birth defects surveillance, with the specific purpose of monitoring teratogenic exposures. Birth defect surveillance is a resource intensive activity that requires careful planning and funding [9]. The essential steps of establishing a surveillance are summarized in Box 6.1 and Chap. 4. The key point to note is that in order to produce quality data, the surveillance system has to be supported by sufficient resources.

Box 6.1 Key considerations while establishing a birth defects surveillance system (from Ref. [9])

Logistical issues : Establishing goals and objectives (why is this data needed, how will the data be utilized) Funding, staffing, partners, Data privacy and confidentiality issues, Data dissemination plan
Population coverage : National, sub-national, selected districts, villages, towns ?
Surveillance method : Population-based (birth defects will be reported from selected population; which populations ? how will they be selected?) or Hospital-based (hospitals will report birth defects data; which hospitals ? how will a random sample of hospitals be selected so that the data does not lead to over-or under-estimation of prevalence?)
Case ascertainment : how will the data be collected ? Active (staff will collect the data), passive (data will be voluntarily reported) or hybrid
Case finding : sources from which cases will be identified (delivery room records, neonatal intensive care unit records, pediatric department records etc)
Case inclusion : Will all or selected congenital anomalies be included in the surveillance ?
Description of anomalies : case abstraction form to provide sufficient details for coding
Age of inclusion : will data be collected among neonates prior to discharge, neonatal period (0-27 days), post-neonatal (28-364 days) or older children ?
Pregnancy outcomes : will the data be reported for livebirths only, livebirths and fetal deaths, or livebirth, fetal deaths and pregnancy terminations
Data collection on other variables (risk factors eg maternal age, religion, ethnicity, obstetric history, occupation, rural/urban residence, paternal characteristics etc)
Data collection method and tools : Paper based or electronic
Diagnosing and Coding defects
Data quality checks
Data analysis (prevalence, types, risk characteristics) and reporting

The birth prevalence of congenital anomalies is typically 2–3%, that is, it ranges between 20 and 30 per 1000 births. However, this prevalence is influenced by the way the data is reported, that is, the number of conditions included, methods of coding and categorization used, and the age of inclusion, that is, the age till which the registry collects and reports data. For example, the South Australian Birth Defects Register which includes data on congenital anomalies, chromosomal conditions, cerebral palsy and haematological/immunological conditions in children till five years of age and includes data on pregnancy terminations and stillbirths, reports a prevalence of around 6% [3]. It is, however, important to remember that birth defect rates are usually constant and show rare and limited fluctuations unless there is a teratogenic exposure in the population. Box 6.2 illustrates the utility of investing in birth defects surveillance, as the data provides a rich source of information on the epidemiology and required healthcare interventions.

Box 6.2 Utility of birth defects surveillance (data from South Australia birth defects registry [3])

• Proportion of total births with birth defects (1986-2013) was 6.0%	
• No fluctuation in annual rates during the period 1986-2013	
• Most commonly reported birth defects were	
	~ urogenital abnormalities (16.7 per 1,000 births)
	~ musculoskeletal (16.2 per 1,000 births)
	~ cardiovascular (12.2 per 1,000 births)
	~ gastrointestinal (6.4 per 1,000 births)
	~ chromosomal (4.5 per 1,000 births)
	~ nervous system (4.2 per 1,000 births)
	~ respiratory (1.8 per 1,000 births)
	~ metabolic (1.4 per 1,000 births)
	~ haematological /immune conditions (0.5 per 1,000 births).
• Significant downward trends in the prevalence of anencephaly, spina bifida, congenital heart defects, renal agenesis /dysgenesis and developmental dysplasia of the hip	
• No changes in the trends of cleft lip/palate	
• upward trends in hypospadias	
• significant increase in trends of Down syndrome, even after adjusting for maternal age	
• Over 50% of cardiovascular, urogenital, haematological/immune and metabolic defects were notified after discharge from the birth hospital	
• ethnic difference in birth defects, proportion of affected births in Caucasian and aboriginal mothers was nearly similar (5.7 % and 5.1%), but lower among births to Asian mothers (4.2%)	
• Birth defects were higher in males (by 37 %) than females (due to the reporting of conditions like undescended testicles and hypospadias).	
• higher among multiple births (twins or triplets) than singleton pregnancies	
• 15% of spontaneous stillbirths, and 31% of neonatal deaths were associated with birth defects.	

Data are reported for children till five years of age. The registry uses the British Paediatric Association (BPA) Classification of Diseases, 1979, for coding of congenital anomalies.

Where such surveillance systems are not in place, such as in India, other sources of data provide information on congenital anomalies. This article reviews the sources of data and describes the magnitude of congenital anomaly mortality and disability in India, beginning with an overview of child health in the country, which underlines the need for determining the trends of congenital anomalies in India.

Child Health in India

India reported over 24 million births in 2016. In 2017, the Neonatal Mortality Rate (NMR) was 23 per thousand live births. The Infant Mortality Rate (IMR) was 33, while the under-5 mortality rate was 37. There are large variations in indicators across states of the country. IMR, for example, ranged from 10 in the state of Kerala to 47 in Madhya Pradesh, while mortality rates among children younger than five years of age ranged from 12 in Kerala to 55 in Madhya Pradesh [10].

Indian national data reflected major decline in child deaths. Deaths among children below five years of age reduced by 37% since 2012, from 1.4 million to 882,000. Infant deaths reduced by 34% from 1.09 million to 721,000 between 2015 and 2017. Neonatal deaths reduced by 29%, from 779,000 to 549,000. Although child mortality from common infectious causes such as pneumonia, diarrhoea, and measles, and neonatal mortality due to sepsis, birth asphyxia or trauma declined significantly, neonatal deaths due to prematurity and low birth weight remained a major concern, and significant focus of public health action [11].

Rigorous data on congenital anomalies are not collected in India, but estimated data indicated that in 2017, congenital anomalies were not insignificant contributors to child mortality. The data suggested that 35% of neonatal deaths were associated with preterm births, 24% with intra-partum events, 14% were due to sepsis and meningitis, while 11% were associated with congenital anomalies [12]. Trends indicated that the proportion of deaths due to prematurity and congenital anomalies increased, as other causes of child deaths reduced [12]. Such data signal the urgency of monitoring data on birth defects in the country.

Sources of Data

In lieu of birth defects surveillance in India, data on congenital anomalies have to be obtained from different sources of data (Fig. 6.1).

1. Prevalence (magnitude)
   Congenital anomaly *birth prevalence* data are available from different hospital-based cross-sectional studies, and a single cohort study. *Population prevalence* data are available from limited community-based cross-sectional studies. *Estimated data on the magnitude* of congenital anomalies are available from the Modell Global Database of Congenital Disorders (MGDb). Routine data on

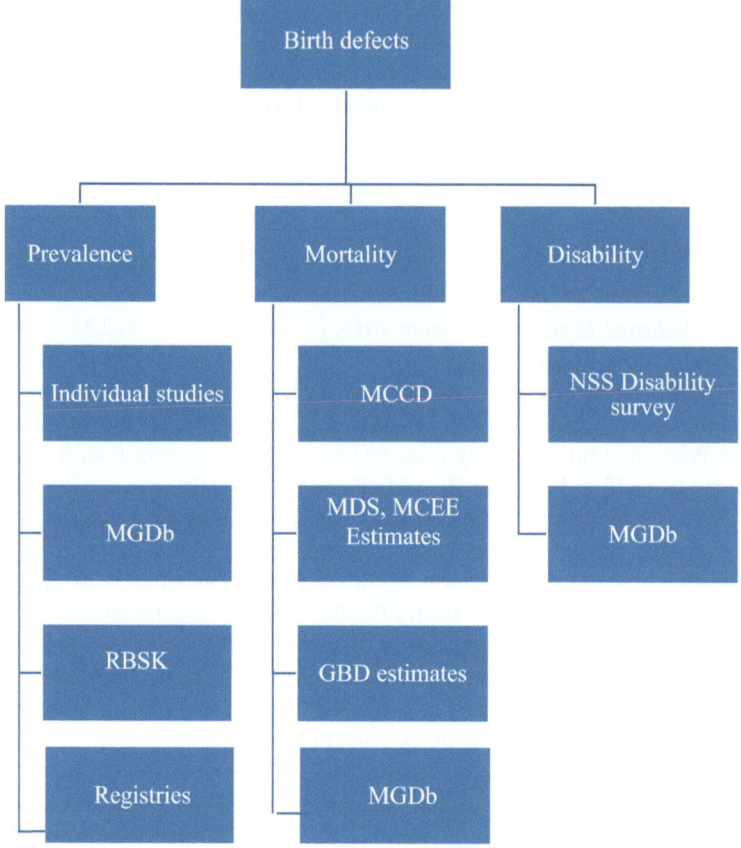

Fig. 6.1 Sources of data to describe the epidemiology of congenital anomalies in India. MGDb = Modell Global Database of Congenital Disorders, RBSK = Rashtriya Bal Swasthya Karyakram, SEARO-NBDD = South East Asia Regional Office of the World Health Organization—Neonatal-Birth Defects Database, MCCD = Medical Certification of Cause of Death, MDS = Million Death Study, MCEE = Maternal and Child Epidemiology Estimation Group, GBD = Global Burden of Disease, MGDb = Modell Global Database of Congenital Disorders, NSSO = National Sample Survey

prevalence of birth defects are available from public health information systems. Data on birth defects among beneficiaries of the Rashtriya Bal Swasthya Karyakram (RBSK) child screening service are infrequently reported. There is and have been attempts to establish birth defects surveillance, but the data are not systematic and based on voluntary reporting.

2.   Congenital anomaly mortality

These data are available from cause-specific child mortality estimates from the Million Death Study (MDS), estimates from the Maternal and Child Epidemiology Estimation Group (MCEE), from of the MGDb and the Global Burden of Disease (India) data.

3. Data on congenital disability at 5 years of age are available from a study, the MGDb, and from a national disability survey that enquired about the time of onset of disability. Onset of disability since birth provides data on the magnitude of functional impairments caused by birth defects (and developmental disabilities).

Congenital Anomaly Prevalence

Individual Research Studies

In India, estimates of congenital anomaly prevalence are available from cross-sectional studies, and a single cohort study. Cross-sectional studies are standard epidemiological methods for measuring prevalence. However, unlike other diseases or health conditions, population-based cross-sectional studies are challenged by the low prevalence of congenital anomalies. Reporting birth prevalence is the most appropriate method, as the surveys can be set in maternity hospitals. The number of congenital anomalies among all births occurring at the hospital will provide data on the proportion of congenital anomaly affected births. In India, hospital-based prevalence will not provide true estimates of congenital anomalies. Selection bias will occur if a major referral hospital is selected. Data will be over-estimates, due to referral of complicated cases. Hence, hospital-based cross-sectional studies, although feasible, will only report proportion of birth defects among all births occurring at the selected hospitals.

Population-based surveys of congenital anomalies require very large populations, which make these projects expensive. Data may be collected through door-to-door survey of households in a defined population. Although expensive, this approach is likely to yield the most accurate population prevalence rates. The major limitation is recall or willingness to share data on birth defects, and the availability of accurate clinical records. Certain key considerations for conducting studies to measure the prevalence of congenital anomalies are summarized in Box 6.3.

Box 6.3 Cross-sectional studies for measurement of congenital anomalies in India and other mixed healthcare settings[*]

Characteristics of cross-sectional studies

- Observational study, that is, the investigator does not alter exposure variables.

- Outcome and exposure variables are measured at the same time. For example, a study to measure the prevalence of spina bifida (outcome variable) can simultaneously measure data on exposure variables such preconception folate supplementation, age of mother, socio-economic status, access to antenatal care, etc.
- Can be used to measure association between outcome and exposures, but cannot be used to establish causality.
- Useful method to know the magnitude of congenital anomalies when surveillance systems are not in place.
- Necessary to keep in mind that the prevalence of a disease/disorder is dependent on incidence (new cases) and duration of survival. Congenital anomalies are rare conditions, and survival is likely to be low in LMICs. As such, prevalence will be low.
- Low prevalence indicates the necessity to survey large populations, making the research expensive.

Important points to keep in mind while planning the study

1. **Population**
   Population where the study is being conducted should be the population at risk and should have the same characteristics as the general population. Otherwise, it will not be possible to generalize the estimates to the general population. For example, women delivering at referral facilities for complicated cases would not represent the general population of women delivering at all birthing facilities.
2. **Study setting**: Hospital based or population based?

   - Hospital based: Will yield data on *proportion of affected births among all births occurring at the hospital(s)*. Selection bias has to be avoided by ensuring representativeness of study hospitals in the sample. For example, the selection of a general hospital versus a hospital providing care for complicated cases will influence prevalence, as the numbers of cases of congenital anomalies are likely to be higher in the second type of facility. Prevalence measurements will be over-estimates.
   - Population based: Challenging as large populations will have to be surveyed. Timing (newborn, infants or older ages) will be an important determinant of prevalence. For older ages, it is necessary to ensure proper recall of events. Under-estimates are likely due to reduced survival, out-migration for treatment, lack of availability of clinical records for surviving children, recall/willingness to disclose details of termination of pregnancy or death or disability, sufficient knowledge about the birth defect in case of a stillbirth, availability and accessibility to antenatal care (ANC), type of ANC service, etc.

3. **Measurement of outcome**

    Measurement of outcome (congenital anomaly) has to keep in mind measurement bias arising from misclassification (misdiagnosed or undiagnosed cases). It is necessary to follow globally used descriptions being used by international registries. The WHO/CDC/ICBDSR manual [9] is an excellent source to understand coding and classification. An online course is available on the ICBDSR website. Steps to consider are:

    - Selection of type of malformations to be studied: all/selected congenital anomalies?
    - Case definitions.

4. **Defining exposure variables**

    Depending on the objectives of the study, these may be characteristics such as teratogenic exposures, habits and health status, family history and folic acid use.

5. **Sampling**

    - For hospital-based studies, it is necessary to draw a random selection of hospitals.
    - Denominator will be all women delivering at selected hospitals.
    - For population-based studies, all birth occurring to women resident in a geographic area are included.
    - Age needs to be defined, that is, will the data be collected for measuring the prevalence of congenital anomalies among neonates, infants and children under five years of age or older.
    - Denominator will be all pregnant women *resident in the area* for the duration of the survey.
    - Data will be incorrect if data on deliveries of resident women occurring at facilities outside the study area (e.g. maternal residence of the woman) are excluded, and women from outside the area, delivering at one of the hospitals is included.

6. **Data reporting**

    1. Live birth prevalence of congenital anomalies = live birth cases/total live births × 10,000.
    2. Birth prevalence of congenital anomalies = live birth cases + foetal death (stillbirths) cases/total live births + foetal deaths (stillbirths) × 10,000 [9].
    3. Total prevalence of congenital anomalies = live birth cases + foetal death (stillbirths) cases + ETOPFA cases/total live births + total foetal deaths (stillbirths) + total ETOPFA × 10,000 [9].

4. Data may be reported as *point prevalence* (prevalence as measured at any given period of time) or as *period prevalence* (prevalence as measured over a period of time).

*Mixed healthcare settings are typically those where there are government and private health facilities. There is no compulsory data reporting from private health facilities, which are used by the majority of the population. As a significant proportion of health care is financed through out of pocket expenditure, there is population mobility. In context of these population characteristics, it is critically important to ensure appropriate study designs for estimating prevalence.

Several studies have reported the magnitude of congenital anomalies in India. A systematic review and meta-analysis pooled the data from these studies to derive a national estimate [13]. The systematic review conducted in 2015 identified 52 hospital-based and 3 population-based studies. The quality of studies varied. Hospitals were selected in a non-random manner, with most being large referral hospitals where high risk women were likely to be referred. None of the studies used (or mentioned) case definitions and case ascertainment methods. ICD codes were rarely used/reported. Physical examination for visible birth defects was the method of detection of malformations in all studies. None of the studies included data on pregnancy terminations.

The pooled national prevalence among 802,658 births estimated a congenital anomaly birth prevalence of 184.48 per 10,000 births (95% CI 164.74–204.21) [13]. The pooled live birth prevalence from hospital-based studies was 203.33 per 10,000 live births (95% CI 171.32–235.34) for 44,392 live births. Population-based studies reported a higher pooled prevalence of 261.05 per 10,000 live births (95% CI 199.13–322.96) among 10,193 live births. Table 6.1 shows the system-wise prevalence of anomalies, and the differences due to the inclusion of stillbirths among hospital-based studies (community-based studies reported data on live births only). Central nervous system anomalies were most frequently reported in both hospital and community-based studies, followed by anomalies of the musculoskeletal system (75.85 per 10,000 births, (95% CI 58.80–92.90) and 65.64 per 10,000 births (95% CI 52.97–78.31), respectively). Cardiovascular system anomalies, the most prevalent type of congenital anomaly, had the lowest birth prevalence across both hospital and community settings [13].

Table 6.2 shows the prevalence of selected congenital anomalies. Anencephaly was the most commonly reported anomaly with a birth prevalence of 21.1 per 10,000 births (95% CI 16.91–25.29). Talipes equinovarus was the next most prevalent congenital anomaly (birth prevalence 17.9 per 10,000 births, 95% CI 15.09–20.71), followed by orofacial clefts (birth prevalence 14.94 per 10,000 births, 95% CI 12.64–17.24) and hypospadias (birth prevalence 12.20 per 10,000 births, 95% CI 9.79–14.60). The analysis found that among live births, the pooled prevalence of clubfoot was highest (35.08 per 10,000 live births, 95% CI 16.88–53.29).

Table 6.1 System-wise prevalence of congenital anomalies from hospital-based and population-based studies [13]

System	Birth prevalence per 10,000 births ($n = 14$ studies)	Live birth prevalence per 10,000 live births ($n = 3$ hospital-based studies)	Live birth prevalence per 10,000 live births ($n = 3$ community-based studies)
Central nervous system	75.85 (95% CI 58.80–92.90)	28.93 (95% CI 13.64–44.22)	26.19 (95% CI 15.55–36.83)
Musculoskeletal system	65.64 (95% CI 52.97–78.31)	79.38 (95% CI 32.32–126.44)	65.88 (95% CI 23.13–108.63)
Cardiovascular system	27.06 (95% CI 20.03–34.09)	23.04 (95% CI 4.69–41.39)	9.32 (95% CI −0.81 to 19.45)
Gastrointestinal system	50.19 (95% CI 42.50–57.87)	37.72 (95% CI 26.41–49.03)	–
Genitourinary system	39.08 (95% CI 27.86–50.30)	28.41 (95% CI 16.18–40.65)	37.42 (95% CI 13.14–61.70)

Table 6.2 Prevalence of selected congenital anomalies from hospital-based and population-based studies [13]

Anomaly	Birth prevalence per 10,000 births ($n = 25$ hospital studies)	Live birth prevalence per 10,000 live births ($n = 5$ hospital studies)
Anencephaly	21.10 (95% CI 16.91–25.29)	17.11 (95% CI 13.59–20.63)
Exomphalos/omphalocele	4.65 (95% CI 3.23–6.07)	1.60 (95% CI 0.46–2.74)
Gastrochisis	7.00 (95% CI −4.56 to 18.56)	1.60 (95% CI 1.60–1.60)
Hypospadias	12.20 (95% CI 9.79–14.60)	5.39 (95% CI 3.19–7.59)
Orofacial clefts	14.94 (95% CI 12.64–17.24)	15.69 (95% CI 11.74–19.63)
Spina bifida	5.85 (95% CI 4.48–7.21)	8.45 (95% CI 3.08–13.81)
Talipes	17.90 (95% CI 15.09–20.71)	35.08 (95% CI 16.88–53.29)

PUBOs Study Estimates

None of the hospital or community-based studies included data on pregnancy terminations. In order to circumvent, the methodological issues associated with a cross-
sectional study, the Pune Urban Birth Outcome study (PUBOs) recruited a cohort of
2107 women at 9 ± 3 weeks of gestation and followed them up till outcome (stillbirths, pregnancy termination, live births, neonatal deaths) [14]. The advantage of a
cohort design was that all pregnancy outcomes could be measured. There was a 9%
loss to follow-up, so that data was available for 1910 women. The characteristics of
the cohort were similar to that of the general population, and the pregnancy indicators
matched those that were reported for the general population. The total prevalence of
congenital anomalies was 230.51 (95% CI 170.99–310.11) (Table 6.3). The congen-

Table 6.3 Measures of congenital anomaly affected events from the PUBOs cohort (data from [14])

	Congenital anomaly affected rates
Total outcomes	230.51 per 10,000 births
Live births	168.44 per 10,000 live births
Neonatal deaths	3.93 per 1000 live births
Early neonatal deaths	3.36 per 1000 live births
Stillbirths	2.19 per 1000 births
Perinatal deaths	5.5 per 1000 births
Terminations of pregnancy after detection of foetal anomalies	4.39 per 1000 births

ital anomaly neonatal mortality rate was 3.93 per 1000 live births. There were 4.39 pregnancy terminations for foetal malformation per 1000 births [14].

Table 6.4 shows the rates per 10,000 for specific congenital anomalies identified during the neonatal period. Due to the small size of the study, it was not possible to compare the data with those reported by large registries. However, the prevalence of 230.51 per 10,000 births indicated over 530,000 congenital anomaly affected births per year in India. There are likely to be 151,488 cases of congenital heart defects, 75,000 with talipes equinovarus, over 60,000 neural tube defects, 88,000 with congenital urogenital disorders, over 25,000 hypospadias, and over 10,000 cases of orofacial clefts (Table 6.4).

The data from the PUBOs cohort indicated that while one in five births were low birth weight, one in nine were preterm births and one in 20 pregnancies resulted in a miscarriage, one in 44 births were affected with a major congenital anomaly, which was similar to the number of stillbirths. In terms of type of anomaly, the cohort data suggested that one in 152 births would be affected with a congenital heart defect, one in 304 births would be affected with clubfoot or a renal anomaly, while one in 364 births would present with a neural tube defect.

Table 6.4 Rates of some congenital anomalies and estimated numbers of affected births in India (data from [14])

Rates congenital birth defects	Rate per 10,000 diagnosed within the neonatal period	Number of births per year
Major congenital anomalies	230.51	530,208
Congenital heart defects	65.86	151,488
Talipes equinovarus	32.93	75,744
Neural tube defects	27.44	63,116
Cleft palate	5.49	12,628
Hypospadias	10.98	25,256
Urinary system anomalies	38.42	88,372

Estimates from the Modell Database

The Modell Global Database of Congenital Disorders (MGDb) is a database of modelled estimates of congenital anomaly prevalence, mortality and disability [15]. The work was initiated in 1980 by Prof. Bernadette Modell. The first estimates were reported in the highly cited March of Dimes Report of 2006 [16]. Updated estimates for all countries are freely available (https://discovery.ucl.ac.uk/1532179/). Table 6.5 shows that the estimated baseline birth prevalence (total affected stillbirths and live births per 1000 births) for India was 20.83 per 1000, similar to that reported by the meta-analysis of congenital anomaly magnitude in India. In absolute numbers, this would indicate 540,421 affected births. Data are available for neural tube defects, orofacial clefts, congenital heart defects and other lethal and sub-lethal congenital anomalies. However, these estimates are based on existing primary data. The MGDb estimates 14,087 pregnancy terminations for foetal malformations. Most importantly, the MGDb is the only source of data on survivors with disability, a crucial public health indicator for birth defects. The MGDb estimated that for each birth cohort, there would be 156,218 survivors with disability at the age of five years (Table 6.5).

Rashtriya Bal Swasthya Karyakram Data

Population-based data on the magnitude of birth defects are available from the Rashtriya Bal Swasthya Karyakram (RBSK) [17], further explained in Chap. 12. Children are screened in the community and referred to District Early Intervention Centres for confirmatory diagnosis and treatment for nine common birth defects (neural tube defects, Down syndrome, cleft lip/palate, clubfoot, developmental dysplasia of hip, congenital cataract, congenital deafness, congenital heart defects, retinopathy of prematurity, congenital hypothyroidism, sickle cell anaemia and thalassemia) and selected developmental disabilities. In an early publication, the prevalence of congenital anomalies among children between the age of 2 and 18 years was reported to be 1% of 42 million children screened by this programme. Among 135,000 congenital anomaly affected births, 39% were congenital heart defects and 13% were cases of neural tube defects [18]. The latest available data mentions that in 2016, of 187 million screened children, 346,000 were diagnosed with a birth defect [19] (Table 6.6). These data suggest that children with birth defects make up less than 0.2% of cases.

The RBSK data has several limitations. There is a possibility that screening children above the age of two years may miss sick children or children with disabilities, as they may not attend schools or play centres. The RBSK does not screen children attending private schools. Nevertheless, the RBSK is a promising option for surveillance for birth defects in India.

Table 6.5 Estimates of the magnitude of congenital anomalies in India from the Modell Database

Without care annual numbers	Total	NTD	OFC	CHD	Other lethal	Other sub-lethal
Total affected births including stillbirths and live births (baseline births per 1000)	540,421 (20.83)	4,103,781 (4)	28,616 (1.1)	86,399 (3.33)	187,088 (7.21)	134,537 (5.19)
Foetal deaths (per 1000)	37,887 (1.46)	26,066 (1.0)	4713 (0.18)	828 (0.03)	5374 (0.21)	906 (0.06)
Under-5 deaths attributable to the disorder group (per 1000)	314,602 (12.13)	68,553 (2.64)	22,435 (0.86)	72,071 (2.78)	145,259 (5.60)	6284 (0.24)
Survivors at 5 years living with disability	169,244 (6.52)	2226 (0.09)	4897 (0.19)	10,913 (0.42)	30,652 (1.18)	120,556 (4.65)
Estimated actual annual numbers 2010–14						
Reduction due to pregnancy termination	14,087 (0.54)	8977 (0.35)	90 (0.0)	343 (0.01)	3985 (0.15)	692 (0.03)
Actual foetal deaths	34,574 (1.33)	27,209 (1.05)	430 (0.02)	821 (0.03)	5214 (0.2)	900 (0.03)
Actual under-5 deaths attributable to the disorder	278,060 (10.72)	60,581 (2.33)	19,423 (0.75)	63,165 (2.43)	129,076 (4.97)	5815 (0.22)
Actual survivors at 5 years living with disability	156,218 (6.02)	4838 (0.19)	4644 (0.18)	11,428 (0.44)	30,552 (1.18)	104,756 (6.02)
Actual survivors at 5 years effectively cured	38,775 (1.49)	0 0	3096 (0.12)	7839 (0.30)	12,245 (0.47)	15,596 (0.06)

Table 6.6 Birth defects and disabilities among RBSK screened cases [19]

	(Millions)	(00,000)			
	Total number of children screened	Childhood diseases	Deficiencies	Birth defects	Developmental delays including disabilities
2014–15	106	72	26	2.3	14
2015–16	187	83	26	3.4	19

Data from Birth Defects Surveillance Systems in India

There are two attempts at establishing birth defects surveillance in India. The Birth Defects Registry of India (BDRI) was a private initiative by the Fetal Care Research Foundation, located in Chennai, India [20]. Between 2000 and 2015, nearly 278 hospitals voluntarily reported data on birth defects. These data were reported to the ICBDSR for a brief period. Due to the voluntary nature of these activities, and the passive method of surveillance, the data were not systematic. The WHO-SEARO has established a Newborn-Birth Defects Database (SEAR-NBDD) [21]. The network includes 170 hospitals from seven countries of the region. The purpose of this network is to develop capacity for birth defects surveillance, with the purpose of providing data on the epidemiology of birth defects in these countries. However, this registry is also limited by the passive method of surveillance, and the lack of specific funding to support the activity.

Congenital Anomaly Mortality

Data on congenital anomaly neonatal and child mortality are available. They identify that congenital anomalies cause considerable numbers of neonatal and child deaths in India.

Data from the Medical Certification of Cause of Death

Data on vital statistics in India are available from the Registrar General of India. Medical certification of cause of death is available for only 22% of total registered deaths across all ages occurring in the country. Most of these medically certified deaths are reported from hospitals located in urban areas. The most recent data from the Medical Certification of Cause of Death (MCCD) is available for 2017, which covered 1,411,060 deaths from across the country [22]. A total of 105,605 infant deaths were recorded, among which 8829 (8.4%) were certified as being

Table 6.7 Causes of infant mortality from the medical certification of cause of death report 2017

	Category	Sub-category	Number	Proportion
1	Certain conditions originating in perinatal period		81,014	76.7
		Hypoxia, birth asphyxia and other respiratory conditions	32,321	30.6
		Slow foetal growth, foetal malnutrition and immaturity	30,183	28.6
2	Certain infectious and parasitic diseases		5877	5.6
		Septicaemia	4539	4.3
3	Congenital malformations, deformations and chromosomal abnormalities		5595	5.3
		Congenital malformations of the circulatory system	3369	3.2
4	Diseases of respiratory system		4078	3.9
		Pneumonia	1720	1.6
5	Diseases of the circulatory system		1801	1.7
		All forms of heart diseases including pulmonary circulation	1492	1.4

caused by 'congenital malformations, deformations, and chromosomal abnormalities'. According to the MCCD data, congenital malformation, deformation and chromosomal abnormalities caused 5.3% of deaths. In contrast, 77% of infant mortality was due to 'certain conditions originating in the perinatal period', followed by infectious and parasitic diseases (5.6%) (Table 6.7).

Congenital malformations of the circulatory system (4802, 54%) were the major type of congenital anomalies, followed by 'other' congenital malformations (40% of mortality). Spina bifida caused 464 (5.3%) of deaths, with more deaths (327) occurring in the age group above one year than among infants (137). The MCCD recorded only 51 (0.7%) deaths due to cleft lip/palate. Majority of congenital anomaly deaths occurred in the first year of life (5595, 63%). The MCDD data are obviously not systematically collected, and therefore, cannot be used to determine congenital anomaly mortality in India.

National Child Mortality Estimates

Data on congenital anomaly mortality are available from studies that have estimated cause-specific child mortality in India.

Cause-Specific Mortality Estimates

Data on congenital anomaly mortality is available from the Million Death Study (MDS), a study that estimated cause-specific mortality trends from among 1.3 million households in India [23]. These households were randomly selected from the Sample Registration System (SRS), an ongoing surveillance system that was established in 1971 by the Registrar General of India to collect vital statistics data. The SRS includes randomly selected villages, and urban blocks, from which data on approximately 140,000 births and 460,000 deaths are recorded annually. The MDS used the WHO verbal autopsy questionnaire to collect information on cause-specific mortality. The field data were assigned underlying cause of death by trained physicians using ICD (International Statistical Classification of Diseases and Related Health Problems) coding guidelines.

Between 2000 and 2015, the MDS reported 696,000 neonatal deaths and 505,000 deaths among children between 1 and 59 months [23]. The congenital anomaly neonatal mortality rate per 1000 live births was 1.1 and the rate for death among children between 1 and 59 months was 1.0 per 1000 live births. In absolute numbers, the MDS data indicated 28,000 neonatal 26,000 congenital anomaly deaths in the 1–59 month age group, respectively, for 2015. The MDS reported that between 2000 and 2015, average annual neonatal congenital anomaly mortality had decreased by 5.1%, while child (1–59 months) congenital anomaly mortality had reduced by 3.5%.

The Maternal and Child Epidemiology Estimation Group (MCEE) reported cause-specific under-5 mortality in India [24]. The methodology adopted by this group differs from that used by the MDS. The study estimated 71,802 (Uncertainty Interval (UI) 56,681–92,640) congenital anomaly neonatal deaths, and 28,035 (UI 21,820–33,777) deaths in the age group of 1–59 months. According to these estimates, in 2015, congenital anomalies were the fourth major cause of neonatal death, with prematurity accounting for 44%, intra-partum complications for 19%, sepsis/meningitis for 14% and congenital anomalies for 10% of mortality. Congenital anomalies were the fourth-largest cause of child deaths in the 1–59 month age group (31% mortality due to pneumonia, 21% due to diarrhoea, 7% due to injuries and 6% due to congenital anomalies).

The study reported a relationship between under 5 mortality rate and congenital anomalies [24]. In regions of the country that had achieved the rate of <25 deaths per 1000 live births, the leading causes were preterm birth complications (26.4%), followed by congenital anomalies (17.1%). In contrast, in regions with high under 5 mortality (>65 deaths per 1000 live births), preterm birth complications (27.4%), pneumonia (18.7%) and diarrhoea (11.2%) were the leading causes of mortality. The two leading causes of child death across the country were preterm birth complications and pneumonia, but in the southern regions (with the lowest under-5 mortality rates), preterm birth complications and congenital anomalies were the leading causes of mortality.

Global Burden of Disease Study

The India State-Level Disease Burden Initiative as part of the Global Burden of Diseases, Injuries and Risk Factors Study (GBD) 2017 has reported national and sub-national trends of cause-specific under-5 mortality [25]. This study reported that congenital anomalies were the fourth leading cause of neonatal deaths, causing 8.6% of neonatal deaths, after mortality caused by preterm birth 27.7%, encephalopathy due to birth asphyxia and trauma 14.5%, and lower respiratory tract infections, 11%. The study reported a decline in the death rate for all major causes of child deaths, but the smallest declines were observed for congenital anomalies. For example, the decline was 82 and 69% for measles and diarrhoeal diseases, while it was the least (15%) for congenital anomalies. Although the study observed a strong inverse correlation between infectious diseases like measles and socio-demographic index (SDI), no such relationship was observed for congenital birth defects.

GBDI Congenital Birth Defect Data

A more focussed analysis on congenital anomaly mortality was conducted using the data available from the Global Burden of Disease (India) (GBDI) database [26]. The GBD uses the term congenital birth defects to include a number of congenital anomalies (neural tube defects, congenital heart anomalies, orofacial clefts, congenital musculoskeletal and limb anomalies, urogenital congenital anomalies, digestive congenital anomalies, other congenital birth defects), and chromosomal abnormalities (i.e. Down syndrome, Turner syndrome, Klinefelter syndrome, other chromosomal abnormalities). Condition-specific mortality data are available from the GBD India Visualization Hub (IHME GBD India Compare 2017).

The GBD modelled data estimated a total of 501,764 congenital birth defect deaths globally among children below five years of age in 2017, with over 70% of these deaths occurring in low and low-middle SDI countries (Fig. 6.2a). Estimated data indicated that birth defects caused over 82,436 deaths among children below five years of age in India in 2017. Within the neonatal period, birth defect mortality was highest in the early and late neonatal periods (7842 and 846 per 100,000 population, respectively) (Fig. 6.2b).

In comparison with other causes of child mortality which showed considerable decline between 1990 and 2017, trend analysis indicated a relatively smaller decline in birth defect mortality between 1990 and 2017 (Fig. 6.3a). In the early neonatal period, for example, neonatal encephalopathy deaths reduced by 52%, preterm birth complications reduced by 45%, and neonatal sepsis reduced by 30%, while there was only 11% reduction in birth defect mortality between 1990 and 2017. The proportion of birth defect mortality in children below five years of age increased from 4.0% in 1990 to 7.9% in 2017 (Fig. 6.3b). The proportionate increase in congenital anomaly mortality was caused by reduction in other major causes of mortality in India [26].

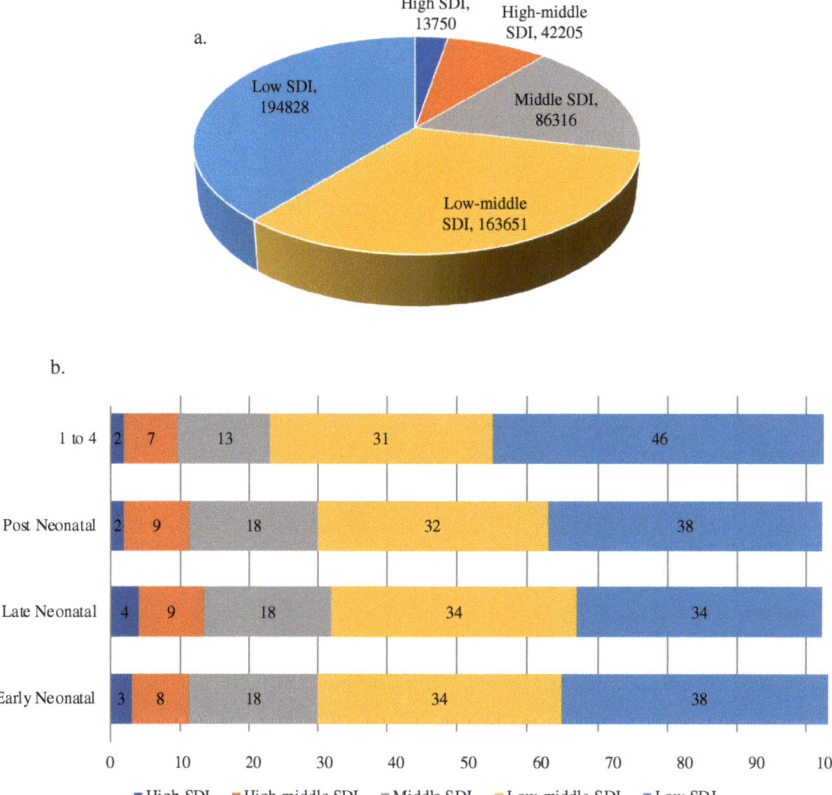

Fig. 6.2 Birth defect mortality. a Birth defect mortality by SDI regions showing that the largest numbers of deaths occur in lower SDI regions. **b** Proportion of birth defect mortality by age (early neonatal, late neonatal, post-neonatal and 1–4 years). The magnitude of the problem in lower SDI regions is apparent

The modelled estimates of the GBDI suggest that congenital anomalies are emerging as significant causes of mortality in the more developed states of India. Among the states, birth defects were the second largest cause of mortality in Kerala. In 17 out of 31 states/regions of the country, birth defects were the third leading cause of neonatal mortality [26]. Although the GBD data are modelled estimates, and they do not include data on pregnancy terminations, they provide a best-available source of data to understand the epidemiology of selected birth defects in India.

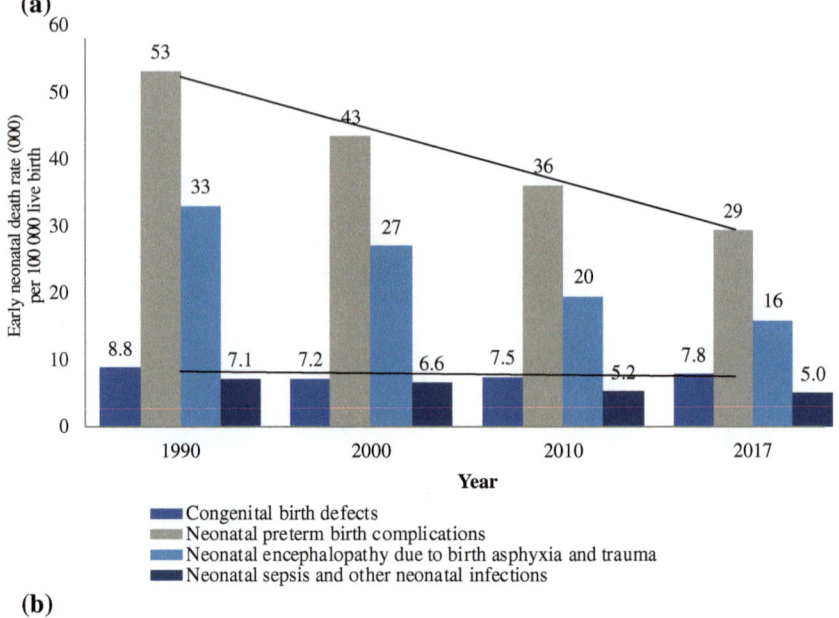

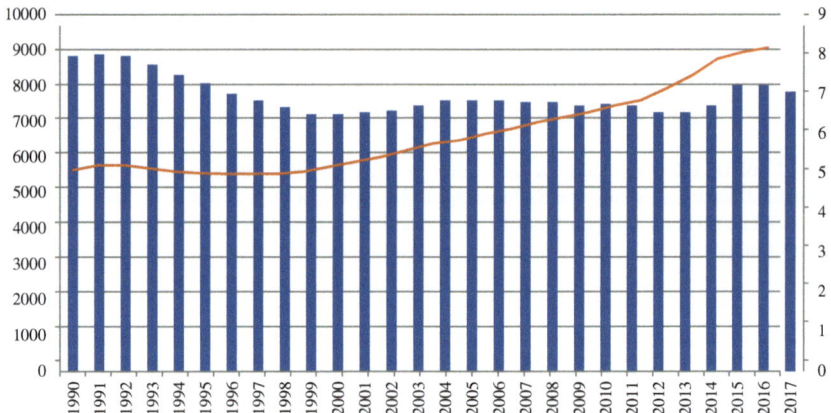

Fig. 6.3 Decline in cause-specific neonatal mortality in India. a The decline in neonatal preterm birth complications, sepsis and encephalopathy has been more rapid than that of birth defect deaths. **b** shows the small decline in birth defect mortality rate between 1990 and 2017 (solid bars). The proportion of birth defect mortality among all-cause mortality shows an increase between 1990 and 2017 (line)

Mortality Data from MGDb

In addition to providing baseline estimates of the prevalence of congenital anomalies, the MGDb provides data on foetal deaths, and mortality in children under five years of age [15]. The data are estimated for two scenarios, if interventions were available, and if interventions were unavailable. The effect of two interventions, folic acid fortification and pregnancy termination are used for estimation. For India, the MGDb computed 37,887 foetal and 314,602 deaths in children less than five years of age without either of these interventions (Table 6.5).

Disability

One of the significant findings from the PUBOs cohort was that 70% of congenital anomaly affected neonates were live born (congenital anomaly live born rate 168.44 per 10,000 births) [14]. Congenital anomaly live births included neonates with congenital heart defects, congenital talipes equinovarus, congenital hydronephrosis, hypospadias, undescended testicles and ear anomalies. The data reflect the magnitude of children surviving with disabilities.

The MGDb is the only source that estimates survivors with disability [15]. The data predicts over 150,000 survivors with congenital disability at five years of age per birth cohort (Table 6.8). One of the most important issues highlighted in the MGDb is that as congenital anomaly deaths are averted, the numbers of children surviving with disabilities increase. The MGDb offers three indicators, decrease in under-5 mortality, increase in disability, and the ratio of increase in disability to decrease in under- 5 mortality. The MGDb estimates that existing interventions would result in

Table 6.8 Survival with disability

	Without intervention		Actual estimates after intervention	
	Disability rate at 5 years/1000	Numbers of disability survivors at 5 years	Disability rate at 5 years/1000	Numbers of disability survivors at 5 years
Total congenital malformations	6.52	169,244	6.02	156,218
NTD	0.09	2226	0.19	4838
OFC	0.19	4897	0.18	4644
CHD	0.42	10,913	0.44	11,428
Other potentially lethal	1.18	30,652	1.18	30,552
Other potentially sub-lethal	4.65	120,556	4.04	104,756

36,543 less congenital anomaly deaths among children below five years of age in India, but this would increase the numbers of children with disability by 25,749.

Another source of data on congenital disability is from the National Sample Survey, a nation-wide disability survey [27]. The survey covered 1172.86 million individuals across the country, of which 2.2% were persons with disability (25.802 million). The proportion of individuals with disability since birth was 30% among all persons with disabilities (approximately 7.74 million). The survey data are not yet available, but the results of an earlier survey, conducted in 2002, also reported a 30% prevalence of disability since birth (Chap. 8).

Summary

A number of observations can be made from this review. Firstly, estimates on the magnitude of congenital anomalies in India are challenged by the lack of reliable data. The potential of the RBSK surveillance remains under-utilized, as data are unavailable for children less than 2 years of age. If made functional, the RBSK would be a useful system to monitor birth defects across the country. The RBSK collects data from a diversity of settings, both urban and rural. As RBSK teams monitor areas frequently, the RBSK has the potential for quickly detecting a teratogenic exposure.

In lieu of any such surveillance systems, data on the magnitude of congenital anomalies in India are available from studies and estimates. Individual studies suffer from poor methodology, with little consistency between the findings of different studies. Cardiovascular defects, the most common anomalies had the lowest prevalence in reported studies, but visible anomalies like anencephaly, talipes, orofacial clefts and hypospadias were commonly reported. These studies, however, remain the main source of data on the magnitude of congenital anomalies in India, yielding a national birth prevalence of 184.48 per 10,000 births. These numbers suggest that annually congenital anomalies affect 472,177 births in India. The MGDb estimated the total birth prevalence (live births and stillbirths) of congenital anomalies to be 20.83 per 1000, which would indicate 540,421 affected births. The total birth prevalence (pregnancy termination, live birth and stillbirths) reported from the PUBOs cohort was higher at 230.51 per 10,000 births (95% CI 170.99–310.11). In absolute numbers, these data indicated 581,899 (431,647–782,841) annual congenital anomaly affected births in India.

The data on congenital anomaly child mortality reported by the MDS, MCEE and the India State-Level Disease Burden Initiative differ in their methodology. But the data indicate that anywhere between 54,000 (MDS), over 99,000 (MCEE) and 82,436 (GBD) deaths are likely to have been caused by congenital anomalies in India, among children below five years of age. The Global Burden of Disease data indicate that the decline in the numbers of congenital anomaly deaths has been smaller than the decline in other common causes of neonatal deaths. Such data suggest that congenital anomalies will account for increasing numbers of deaths, as is being observed in the southern parts of India.

A major knowledge gap is in the magnitude of disability survivors. PUBOs reported congenital anomaly live birth prevalence of 168.44 per 10,000 live births. These data on survival till the end of the neonatal period, when extrapolated would suggest 415,630 disability survivors (till the end of the neonatal period) annually. The MGDb estimates that per birth cohort, there would be 156,218 children surviving with disabilities at the age of 5 years. Cumulative cohorts could explain the nearly 8 million persons with congenital disabilities in India.

Based on the challenges of interpreting hospital-based surveillance data from India, population-based birth defects surveillance, among carefully selected communities remains a possibility that needs to be explored. India has several high risk situations and areas. Industrial catastrophes like the Bhopal gas tragedy or reports of children with severe birth defects in areas where banned pesticides are being used are examples of potential sites for long-term surveillance of communities. The other alternative is to support ongoing academic study sites, where cohorts are being followed up for different types of maternal and child health studies [28]. Such sites could not only provide data on congenital anomalies, but also data on risk factors and outcomes (such as morbidity, mortality and hospitalizations) [29]. India already has in place a maternal health surveillance system, the Mother and Child Tracking System. Although there are questions on the data quality [30, 31], this system can be used to have an ongoing surveillance on birth defects. Overall, till systems are put in place for collecting data on birth defects, rigorously performed studies still remain the main source of birth defects data in India.

References

1. Mai CT, Isenburg JL, Canfield MA et al (2019) National population-based estimates for major birth defects, 2010–2014. Birth Defects Res 111(18):1420–1435
2. Lanzoni M, Morris J, Garne E et al (2017) European monitoring of congenital anomalies. JRC-EUROCAT report on statistical monitoring of congenital anomalies (2006–2015). EUR 29010 EN. Publications Office of the European Union, Luxembourg. ISBN 978-92-79-77305-1. https://doi.org/10.2760/157556, PUBSY No. JRC109868. https://publications.jrc.ec.europa.eu/repository/bitstream/JRC109868/kjna29010enn.pdf. Accessed 3 Sept 2018
3. Gibson CS, Scott H, Rice R, Scheil W (2017) Birth defects in South Australia 2013. SA Birth Defects Register, Women's and Children's Health Network, Adelaide. Available at https://www.wch.sa.gov.au/services/az/other/phru/documents/2013_sabdr_annual_report.pdf. Accessed 3 Sept 2018
4. Correa-Villaseñor A, Cragan J, Kucik J, O'Leary L, Siffel C, Williams L (2003) The metropolitan Atlanta congenital defects program: 35 years of birth defects surveillance at the centers for disease control and prevention. Birth Defects Res Part A Clin Mol Teratol 67(9):617–624
5. Mai CT, Kirby RS, Correa A, Rosenberg D, Petros M, Fagen MC (2016) Public health practice of population-based birth defects surveillance programs in the United States. J Public Health Manag Pract 22(3):E1–E8
6. Botto LD, Robert-Gnansia E, Siffel C, Harris J, Borman B, Mastroiacovo P (2006) Fostering international collaboration in birth defects research and prevention: a perspective from the International Clearinghouse for Birth Defects Surveillance and Research. Am J Public Health 96(5):774–780

7. Irgens LM (2000) The medical birth registry of Norway. Epidemiological research and surveillance throughout 30 years. Acta Obstet Gynecol Scand Spec Issue Rev 79(6):435–439
8. Groisman B, Bidondo MP, Gili JA, Barbero P, Liascovich R (2013) Strategies to achieve sustainability and quality in birth defects registries: the experience of the National Registry of Congenital Anomalies of Argentina. J Registry Manag 40(1):29–31
9. Birth defects surveillance: a manual for programme managers, second edition. Geneva: World Health Organization; 2020. Licence: CC BY-NC-SA 3.0 IGO
10. Office of the Registrar General & Census Commissioner, India. SRS statistical report 2017. In: Estimates of mortality indicators, Chap 4. Available at https://censusindia.gov.in/vital_sta tistics/SRS_Report_2017/11.%20Chap%204-Estimates%20of%20Mortality%20Indicators-2017.pdf. Accessed April 2021
11. UNICEF, WHO, World Bank, UN-DESA Population Division. Levels and trends in child mortality report 2019. Estimates developed by the UN Inter-Agency Group for Child Mortality Estimation. Available at https://www.unicef.org/media/60561/file/UN-IGME-child-mortality-report-2019.pdf. Accessed Sept 2018
12. Hug L, Alexander M, You D, UN Inter-agency Group for Child Mortality Estimation (2019) National, regional, and global levels and trends in neonatal mortality between 1990 and 2017, with scenario-based projections to 2030: a systematic analysis. Lancet Glob Health 7(6):e710–e720
13. Bhide P, Kar A (2018) A national estimate of the birth prevalence of congenital anomalies in India: systematic review and meta-analysis. BMC Pediatr 18(1):175
14. Bhide P, Gund P, Kar A (2016) Prevalence of congenital anomalies in an Indian maternal cohort: healthcare, prevention, and surveillance implications. PLoS ONE 11(11):e0166408
15. Modell B, Darlison MW, Moorthie S, Blencowe H, Petrou M, Lawn J (2016) Epidemiological methods in community genetics and the Modell Global Database of Congenital Disorders (MGDb). Available at https://discovery.ucl.ac.uk/id/eprint/1532179/17/Epidemiological%20Methods%20in%20Community%20Genetics%20and%20the%20Modell%20Global%20D atabase%202017-04.pdf. Accessed 5 Sept 2018
16. Christianson A, Howson CP, Modell B (2006) March of Dimes global report on birth defects: the hidden toll of dying and disabled children. March of Dimes Birth Defects Foundation, White Plains
17. Government of India, Ministry of Health and Family Welfare, National Health Mission (2013) Rashtriya Bal Swasthya Karyakram. A child health screening and early inter-vention services under NRHM. Ministry of Health and Family Welfare. Available at https://nhm.gov.in/images/pdf/programmes/RBSK/Operational_Guidelines/Operational%20Guidelines_RBSK.pdf. Accessed 6 Aug 2019
18. Singh AK, Kumar R, Mishra CK, Khera A, Srivastava A (2015) Moving from survival to healthy survival through child health screening and early intervention services under Rashtriya Bal Swasthya Karyakram (RBSK). Indian J Pediatr 82(11):1012–1018
19. Government of India, Ministry of Health and Family Welfare (2019) Answers data of Rajya Sabha questions for session 240/year wise physical status Rashtriya Bal Swasthya Karyakram (RBSK) during 2014–15 and 2015–16. Available at https://data.gov.in/node/3978901/dow nload. Accessed 18 Nov 2019
20. Birth Defects Registry of India. Available at https://fcrf.org.in/bdri_abus.asp. Accessed 11 Oct 2017
21. SEAR NBBD newborn and birth defects surveillance initiative. https://origin.searo.who.int/entity/child_adolescent/nbbd/web/en/#:~:text=In%202014%2C%20WHO%2DSEARO%20c reated,for%20submission%20to%20the%20system. Accessed 6 Aug 2019
22. Office of the Registrar General India. Report on medical certification of cause of death 2017 Available at https://censusindia.gov.in/2011-Documents/mccd_Report1/MCCD_Report-2017. pdf. Accessed Dec 2018
23. Fadel SA, Rasaily R, Awasthi S, Begum R, Black RE, Gelband H et al (2017) Changes in cause-specific neonatal and 1–59-month child mortality in India from 2000 to 2015: a nationally representative survey. Lancet 390(10106):1972–1980

24. Liu L, Chu Y, Oza S, Hogan D, Perin J, Bassani DG et al (2019) National, regional, and state-level all-cause and cause-specific under-5 mortality in India in 2000–15: a systematic analysis with implications for the Sustainable Development Goals. Lancet Glob Health 7(6):e721–e734
25. Dandona R, Kumar GA, Henry NJ et al (2020) Subnational mapping of under-5 and neonatal mortality trends in India: the Global Burden of Disease Study 2000–17. Lancet 395(10237):1640–1658
26. Ujagare D, Kar A (2021) Birth defect mortality in India 1990–2017: estimates from the Global Burden of Disease data. J Community Genet 12(1):81–90. https://doi.org/10.1007/s12687-020-00487-z. Epub 2020 Oct 15. PMID: 33063164; PMCID: PMC7846616
27. Ministry of Statistics and Programme Implementation. Persons with disabilities in India. NSS report no. 583 (76/26/1), July–Dec 2018. Available at https://www.mospi.gov.in/sites/default/files/publication_reports/Report_583_Final_0.pdf. Accessed 5 Sept 2018
28. Fall CHD (2018) Nutrition in fetal life and childhood and its linkage with adult non-communicable disease: lessons from birth cohort studies in India. Proc Indian Natl Sci Acad 84(4):881–889
29. Botto LD, Mastroiacovo P (2018) Triple surveillance: a proposal for an integrated strategy to support and accelerate birth defect prevention. Ann N Y Acad Sci 1414(1):126–136
30. Nagarajan TJP, Goel S (2016) Is mother and child tracking system (MCTS) on the right track? An experience from a northern state of India. Indian J Public Health 60(1):34–39
31. Gera R, Muthusamy N, Bahulekar A, Sharma A, Singh P, Sekhar A, Singh V (2015) An in-depth assessment of India's mother and child tracking system (MCTS) in Rajasthan and Uttar Pradesh. BMC Health Serv Res 15(1):315

Chapter 7
Magnitude of Developmental Disabilities in India

Humaira Ansari

Abstract Developmental disabilities are a group of conditions responsible for physical, social, emotional, behavioral, cognitive, and motor impairments in children. This article reviews available data on the magnitude of developmental disabilities in India. Estimation of these conditions is challenged by the lack of validated, culturally adapted screening tools in low- and middle-income countries. There is a paucity of good quality studies, so that prevalence estimates vary between studies. Nevertheless, data from selected studies identify a notable magnitude of developmental disabilities in India. Comparison between an Indian study with data from a US surveillance for autism and selected developmental disabilities identified that the magnitude of autism and attention-deficit/hyperactivity disorders were similar, learning disabilities were lower in prevalence, but other developmental disabilities such as intellectual disabilities, epilepsy, and hearing and vision impairment were markedly higher in India. The Global Burden of Disease study estimates that developmental disabilities may affect more than 11 million children under the age of five years in India. The article identifies the need for well-designed studies using validated screening tools so that the data could yield better estimates of the magnitude of the problem in India. The high numbers suggested by existing studies indicate the need to urgently expand services for prevention, care and rehabilitation of children with developmental disabilities.

Keywords Prevalence · Developmental disabilities · India · Low and middle income countries

Childhood Development

Early childhood, that is the first five years of life, form the critical period of growth and development of children. Growth is increase in physical size. Development refers to functioning and capability, which increase with the development of motor, cognitive, emotional, and social functions. Development is a continuous process, correlated with the change and maturation of the central nervous system [35]. Although development

H. Ansari (✉)
Symbiosis International (Deemed) University, Pune, Maharashtra, India

© Springer Nature Singapore Pte Ltd. 2021   169
A. Kar (ed.), *Birth Defects in India*,
https://doi.org/10.1007/978-981-16-1554-2_7

has been divided into specific domains of gross motor, fine motor, language, cognition and social/emotional growth, substantial overlaps exist. Studies have established specific ages when certain milestones are achieved, and there is a range of variation observed among children [67]. Failure to achieve a set of skills by a specific age is indicative of a developmental delay. Crossing the window of achievement without achievement of milestones is strongly indicative of the need for interventions.

Table 13.1 (Chap. 13) summarizes the developmental milestones. Early development is marked by primitive reflexes, such as the Moro (startle) reflex, palmar grasp reflex and rooting and suck reflex. As the child grows, these reflexes are integrated. Development is reflected in improvement of gross motor and fine motor skills, cognitive and language development and personal social interaction. The child holds its neck by three months, sits with support by six months and without support by eight to nine months, stands with support by eight months, walks with support by 10 months. Crawling is seen at 11 months and walking without support by 12 months. The child runs by 18 months and climbs stairs by 24 months.

Fine motor skills are those involving small muscle groups. By four months, the child grasps objects placed in the hand, by five months reaches out to an object, by seven months palmar grasp and by nine months pincer grasp develops. Language development is evident at one month of age, when the newborn turns its head toward a sound, starts cooing by three months, and starts producing monosyllables by six months, bi-syllables by nine months and two words with meaning by 12 months. By 18 months, the child speaks ten words and can communicate with simple sentences by two years of age. Personal social development includes smiling by two months, recognizing mother by three months, smiling at mirror image by six months and waving goodbye by nine months. Parallel play develops by 18–24 months. These milestones are seen in all normally developing infants.

Developmental Delays and Disabilities

Development is a multifactorial process. Environmental factors such as nutrition and stimulation, disease and psychological factors interact with the genetic predisposition of the child to determine the developmental pattern. Studies have identified many factors that affect development. These include genetic factors (hereditary conditions, consanguineous marriages), maternal complications (prolonged labor, eclampsia/pre-eclampsia), nutritional deficiencies, poverty, infection, illness and injury (including febrile illness, injury or trauma, damage to the central nervous system), prematurity, low birth weight, exposure to environmental toxins such as smoke and psychosocial stress [4, 56]. Motor development is determined by family patterns, prolonged illness, or pathophysiological conditions such as cerebral palsy and intellectual disability. Delay in language development is commonly linked to hearing loss. Cognitive development, that is the intellectual maturation of the child, is associated with nurturing care and strong relationships. Emotional and behavioural developments are more individual-specific.

Delayed milestones, that is not achieving skills within a specific time frame, or persistence of primitive reflexes is termed as developmental delay. Modifiable risk factors like poor nutrition, poverty, and infections are more common in the low- and low-middle income countries (LMICs). The higher prevalence of these factors can be attributed to the high prevalence of developmental delays in these countries [14, 44]. These risk factors form targets of public health programmes, in order to improve growth and development in the early years. Specific medical conditions may be a cause for delayed or disordered development. Developmental disability refers to a childhood intellectual, physical or behavioral impairment or combination of these impairments that cause substantial functional limitations in major life activities.

Developmental Disabilities

A developmental disability (neuro-developmental disability, neuro-developmental disorders, NDD) arises when age-specific skills are not achieved within a specific time frame, affecting the functioning and skill performance of the child. Developmental disabilities are a group of conditions where the child has an impairment in physical, learning, language, sensory, motor, cognitive, social, emotional skills and behavior [12]. The most common developmental disabilities are epilepsy or seizures, sensory impairments (hearing or vision loss), cerebral palsy, attention-deficit/hyperactivity disorders (ADHDs), autism spectrum disorders (ASDs), and intellectual disability (ID). Developmental disabilities may affect several functions. For example, in children with epilepsy, 22% also report ASD, 33% have ADHD, and 30–50% have behavioral and emotional problems [4]. This raises issues not only on the disabling nature of these conditions, but also regarding measurement of the magnitude of the conditions, as there is a risk of over-counting.

Epilepsy is a neurological condition, characterized by two unprovoked seizures more than 24 hours apart [62]. Epileptic seizures are caused by abnormal signaling of neurons, causing involuntary movements, loss of awareness, sensations, behaviors and emotions. Intractable epilepsy (i.e., epilepsy that cannot be controlled with medications) accounts for 30–40% of all epilepsy. A European study on the quality of life (QoL) of people with epilepsy identified that achieving better seizure control and reducing the side-effects of medications were related to improved quality of life [3]. An Indian study identified that the QoL was impaired in all patients, but more so among women, older patients, those with simple partial seizures and those with recent seizures [61]. Hearing loss present at birth (congenital hearing loss) is caused by genetic factors, and by other factors such as prematurity and low birth weight. Congenital infections, especially congenital cytomegalovirus infections, are associated with congenital hearing loss [37]. Childhood blindness and vision impairment constitutes only 4% of blindness [36], but both these congenital sensory organ impairments (hearing loss and blindness) can severely affect the QoL, education, and employability of individuals. Cerebral palsy is a heterogeneous group of non-progressive neuro-motor disorders that affects balance and movement. It is one of the most common causes of childhood locomotor disability [55]. ASDs are characterized by impairments in social interactions, repetitive behaviors, and restricted interests. Pervasive developmental disorders (PDDs) are disorders that include a broad range of social communication deficits. Appropriate interventions can improve the behavior and language achievement, but most people with ASD are dependent on caregivers throughout life [40]. ADHD is another common disabling neuro-developmental disorder that is marked by inattentiveness, hyperactivity and impulsiveness [18]. Intellectual disabilities are one of the largest group of disabling conditions, characterized by below-average intellectual function and limitations in adaptive functioning [46]. Developmental disabilities impact activities of daily living, causing participation restriction and affecting the educational and employment potential of the individual. Within families, developmental disabilities affect family functioning and QoL of the child, parents and siblings. Chapter 2 provides a further overview of the disabling nature of these conditions.

Table 7.1 shows the exposures/factors associated with some common developmental disabilities [15, 18, 25, 33, 37, 38, 40, 59]. These include genetic factors, gene-environmental factors, environmental factors, maternal health status and health service factors. The prevalence of these factors are higher in LMICs, which might account for the higher prevalence of developmental disabilities in these countries. The factors associated with developmental disabilities form the targets of maternal health services during the prenatal and perinatal periods. Increasing institutional deliveries in LMICs, for example, can reduce some of the adverse complications for developmental disabilities like epilepsy and cerebral palsy.

Tools for Measuring Developmental Disabilities

Developmental disabilities are diagnosed through a step-wise process, with screening followed by diagnosis [7]. Screening tools examine early child development and can detect a developmental delay and disability. A screening test is meant to identify a child with a developmental delay, but further evaluation is required to confirm the presence or absence of a developmental difficulty. Screening tests are therefore followed by specific diagnostic tests. For example, after screening for hearing loss, further evaluation and diagnosis is done using audiometry. For children with intellectual impairment, the Vineland Social Maturity Scale (VSMS) is a diagnostic instrument to determine social maturity.

Over 100 screening tools have been developed, which have been reviewed comprehensively and are available as the World Bank's Toolkit for Measuring Early Child Development (ECD) in low-income and middle-income countries [19]. An ECD measurement inventory which summarizes and lists a total of 147 tools for children up to 8 years of age is also available [16]. The tools may be for population-level, or individual-level screening. The tools may or may not screen for the nine developmental domains listed in the World Bank Toolkit (cognitive, language, motor, socioemotional/temperament, attention/executive function,

Table 7.1 Risk factors for common developmental disabilities

Condition	Risk/associated factors
Epilepsy	Structural etiology (stroke, trauma, infection, congenital), genetic etiology (familial syndromes, mutation), infectious etiology (malaria, tuberculosis, HIV, toxoplasmosis, congenital Zika, cytomegalovirus, etc.), metabolic etiology (disorders such aminoacidopathies, uremia, etc.) immune etiology (autoimmune mediated central nervous system inflammation), unknown etiologies
Cerebral palsy	Birth complications (neonatal encephalopathy, birth asphyxia, trauma), breech position, preterm birth, mechanical ventilation, post-natal administration of steroids for lung maturation, systemic inflation in premature born infants, low birth weight, fetal hypothyroxinemia, genetic factors, multiple births, disadvantaged populations, prepregnancy obesity, maternal pre-eclampsia, fetal growth restriction, maternal infections, cerebral malformations, perinatal stroke, kernicterus
Congenital hearing loss	Admission to a neonatal intensive care unit, low gestational age and birth weight, medical interventions (assisted ventilation, venous access and aminoglycoside use), genetic factors, autosomal recessive genetic factors, congenital infections, primarily cytomegalovirus infection, socioeconomic factors, access to prevention services such as rubella immunization
Congenital vision impairment (VI) and blindness	Congenital anomalies (uveal coloboma, anophthalmos, microphthalmos), infantile glaucoma, retinal dystrophies, Leber's congenital amaurosis, congenital cataract, retinoblastoma, ophthalmia neonatorum, retinopathy of prematurity, optic nerve lesions, cerebral visual impairment
ASD	Genetic factors (approximately 40–90% heritability). Environmental risk factors (neonatal hypoxia, maternal obesity, gestational diabetes mellitus, short interval between pregnancies, older sibling with ASD, paternal age >50, maternal age >40, valproate use during pregnancy) Not associated with vaccination, prolonged labor, cesarean section or assisted vaginal delivery, use of assisted reproductive technologies and premature rupture of membranes

(continued)

Table 7.1 (continued)

Condition	Risk/associated factors
ADHD	Gene-environmental genetic (70–80% heritability) Male sex, ethnicity and low socioeconomic status, prenatal and perinatal factors, including maternal smoking and alcohol use, low birth weight, premature birth and exposure to environmental toxins, like organophosphate pesticides, zinc, lead, and polychlorinated biphenyls
Intellectual disability	Genetic (chromosomal abnormalities, single-gene disorders, inherited conditions), non-genetic (advanced maternal age, maternal black race, low maternal education, third or more parity, maternal alcohol use or tobacco use, maternal diabetes, hypertension, epilepsy and asthma, preterm birth, male sex and low birth weight)

personal-social/adaptive, academic/pre-academic, approaches to learning, disability screener) [19]. Table 7.2 enlists some of the widely used screening and diagnostic tools.

There are several reviews on these developmental tools [21, 43]. A recent review of ECD instruments identified the limited numbers of population level screening tools (five) [5]. The review noted that the tools did not cover all domains, or did not rate high on accuracy and feasibility. Cognitive, language, and motor domains were measured frequently, with gaps across other domains. Vision, hearing, and disability screeners were missing or absent in all population-level tools. Most of the widely used screening tools (Table 7.2) have been developed in high-income countries, but several culturally adapted and validated tools have been developed in LMICs. These include a number of tools from India (Table 7.3). The reliability and validity of the tools are important, as they would influence prevalence estimates.

Among the tools developed in India, both the Baroda Developmental Screening Test and the Trivandrum Developmental Screening Chart (which are derived from Bayley Scales of Infant Development), as well as the ICMR Psychosocial Developmental Screening Test have not been re-validated since their inception [47, 48]. The INCLEN Neurodevelopmental Screening Test is a recently developed and validated tool [29] that has been used to report data on the prevalence of developmental disabilities from five settings across the country [2].

Sources of Data

Data on developmental disabilities can be obtained from surveillance systems, national surveys and from ad hoc studies. Estimates are also available from the Global Burden of Disease (GBD) analyses.

Table 7.2 Routinely used screening tools

Tool	Domains and time required	Age group
Bayley Scales of Infant Development (BSID-I, 1st edition; BSID-II, 2nd edition; BSID-III, 3rd edition)	Used to assess development across domains such as cognition, language, motor skills, socioemotional and personal/social adaptive skills. Useful in diagnosing and planning interventions for developmental delay. Requires specialist training and 30–90 min for administration	One month–3.5 years
British Ability Scales (BAS)	Includes domains such as cognition, language and pre-academic and academic screening. The main purpose is to develop and support interventions. Requires specialist training and 30–45 min for administration	Three years–17.9 years
Denver Developmental Materials II (formerly DDST)	It includes cognition, language, motor, and personal-social/adaptive domains. It is used for screening and not diagnosis. Requires specialist training and time required is 10–20 min	One month–6 years
Stanford Binet Intelligence Scale	Used for studying cognitive skills, language, and attention/executive functioning. It includes 15 sub-tests. Requires specialist training. Time required for each subset is 5 min	Two years–85 years
Ages and Stages Questionnaire (ASQ)	Used to screen for domains such as: communication, gross motor, fine motor, problem solving, and personal-social skill. Requires minimal training and 10–20 min for administration	One month–5.5 years
Vineland Adaptive Behavior Scales II	Includes domains such as language, motor skills, socioemotional skills, personal/social and pre-academic and academic skills. Requires moderate training for use. Time required for administration is 20–90 min	Birth–90 years

Table 7.3 Tools validated for use in India

Name of tool	Age range	Description (domains measured)	Sensitivity	Specificity	Time required	Special training required	Country used	Cost
Trivandrum screening chart (TDST)	0–2 years, 3–6 years	Mental, motor, hearing, and vision	66.7%	78.8%	5–10 min	No	India	Free
Development Assessment Tool for Anganwadi's (DATA) DATA II	1.6–3 years 3–4 years	Gross and fine motor, language, social skills	Not available	Not available	Not available	Limited training	India	Free/low cost
Lucknow development screen (LDSC)	6 months–2 years	Gross and fine motor, language, social skills	95.9%	73.1%	10–15 min	Limited training	India	Free/low cost
Baroda Development Screening Test (BDST)	0–30 months	Gross and fine motor, language, social skills	95%	65%	10–15 min	Moderate training	India	Free/low cost
Rashtriya Bal Swasthya Karyakram (RBSK) screening tool	0–6 years	Gross and fine motor, language, social skills	Not available	Not available	30 min	Minimal	India	Free
Guide for monitoring child development (GMCD)	0–3.5 years	Communication, gross and fine motor, socioemotional, self-help skills	88%	93%	7–10 min	Minimal	Turkey, India, South Africa	Free/low cost

(continued)

Table 7.3 (continued)

Name of tool	Age range	Description (domains measured)	Sensitivity	Specificity	Time required	Special training required	Country used	Cost
Development Assessment Scale for Indian Infants (DASII)	1–30 months	Gross and fine motor, cognitive, personal, social	Not available	Not available	60 min	Yes, specialist	India	Copyright
Disability Screening Schedule (DSS)	0–83 months	Physical, motor, sensory, cognitive	89%	98%	5 min	Minimal	India	Free/low cost
Screening test battery for assessment of psychosocial development (STBAPD)	0–6 years	Gross and fine motor, language, vision, hearing, concept development, self-help skills, social skills	Not available	Not available	30 min	Yes	India	Inexpensive
Language Evaluation Scale Trivandrum (LEST)	0–6 years	Language	66.7%	94.8%	10 min	Minimal	India	Expensive
WHO Ten Question Screen	2–9 years	Cognitive disability, movement disability, seizures, vision, hearing impairment	100%	Not available	10–15 min	Minimal	Multiple	Free

(continued)

Table 7.3 (continued)

Name of tool	Age range	Description (domains measured)	Sensitivity	Specificity	Time required	Special training required	Country used	Cost
Developmental Milestones Checklist (DMC, DMC-II)	1 month–8 years	Language, motor, personal-social	Not available	Not available	10–20 min	Minimal	Multiple (Cambodia, Kenya, Burkina Faso)	Free/low cost
ICMR Psychosocial Development Screening Test	0–6 years	Cognition, language, motor, social, personal	Not available	Not available	10–20 min	Minimal	India	Free
Caregiver-Reported Early Child Development Index (CREDI)	0–3 years	Cognition, language, motor, social, attention	Not available	Not available	5–15 min	Minimal	Multiple	Free
Intergrowth 21st Neurodevelopment Assessment (INTER-NDA)	1.8–2.2 years	Cognition, language, motor, social	66.7%	98.6%	35–45 min	Moderate	Brazil, Kenya, India, Italy, UK	Free
Profile of Socio-Emotional Development (PSED)	5 months–3 years	Social	Not available	Not available	15–20 min	Moderate	Multiple	Free
12 month screener	0–12 months	Cognition, language, motor, social	79%	85%	Not available	Yes	India, Pakistan, Zambia	Free/low cost

(continued)

Table 7.3 (continued)

Name of tool	Age range	Description (domains measured)	Sensitivity	Specificity	Time required	Special training required	Country used	Cost
East Asia Pacific Early Child Developmental Scales	3–5 years	Cognition, language, motor, social, academics, learning	Not available	Not available	45–60 min	Moderate	East-Asia Pacific	Free
Engle Scale and Survey	24–59 months	Cognition, language, motor	Not available	Not available	30–40 min	Not available	East-Asia Pacific	Free
Rapid Pre-Screening Denver Questionnaire	0–6 years	Gross motor, fine motor, language, personal-social	100%	7.8%	20 min	Yes	India	Not available
Woodside Screening Technique	6 weeks–24 months	Social, hearing, language, vision, gross and fine motor	88%	83%	Not available	Yes	India	Free/low cost
INCLEN Neurodevelopmental Screening Test	2–9 years	Vision, speech, hearing, language, cognition, gross, and fine motor	75%	87%	Not available	Yes	India	Free/low cost

Developmental Disability Surveillance

Several surveillance programs for specific developmental disabilities have been established. For example, there are 27 surveillance programs for cerebral palsy, located in Europe, Australia, and North America, that provide data on these conditions [26]. The Autism and Developmental Disabilities Surveillance System in the USA collects data on the prevalence and trends of intellectual disability, cerebral palsy, hearing loss, vision impairment, and epilepsy. The data are collected from the health and special education records of 8-year-old children who live in one of 11 surveillance sites across the USA. This surveillance program emerged from the Metropolitan Atlanta Developmental Disabilities Study (MADDS) that was established in 1984. The goal of the Autism and Developmental Disabilities Monitoring (ADDM) Network is to provide prevalence data among 8-year-old children, describe the characteristics of affected children, and identify risk factors for these conditions. The surveillance system forms the basis for further research, such as characterizing these conditions, identifying risk factors for these conditions, and improving diagnostic tools.

The utility of a surveillance system for developmental disabilities is evident from the data reported by the ADDM Network. In 2016, the prevalence of ASD among 8-year-old children was 18.5%, with ASD being four times more likely to affect boys than girls. There was no difference in ASD prevalence between black and white children, but autism prevalence among Hispanic children was lower. One-third of children with ASD also had intellectual disability. Nearly 84% of children had been diagnosed by the age of 4 years [41]. Due to methodological issues, ASD prevalence varied between 1.5 and 3.1% between different reporting sites. The utility of such surveillance data is that it can identify the needed services and support for children and adults with ASD. The data forms the background knowledge for conducting further research to understand the etiology of autism and other developmental disabilities.

Data on developmental delays and disabilities in India are collected by the Rashtriya Bal Swasthya Karyakram [60], a screening and early intervention programme. Under this programme, children are screened in community settings, for common childhood diseases, nutritional deficiencies, birth defects and developmental delays and disabilities [60]. As developmental delays are included with disabilities, the numbers are not specific for developmental disabilities. For example, in 2015–16, this programmes screened 187 million children across the country, of whom 83 million had a diagnosis of common childhood diseases, 26 had nutritional deficiencies, 3 million had birth defects, and 19 million had developmental delays and disabilities [28]. The RBSK data is, however, infrequently reported, and not representative as it is restricted to users of the RBSK service. Furthermore, the data is an over-estimate, as it includes both developmental delays and disabilities.

National Surveys

The magnitude and trends of developmental disabilities have been reported through several national level surveys in different countries. In the USA, for example, the National Health Interview Survey (NHIS) is a source of data for reporting the prevalence and trends of specific developmental disabilities (ADHD, cerebral palsy, ASD, ID, seizures, hearing loss, blindness, learning disorders (LDs), stuttering or stammering, and other developmental delay) among children aged 3–17 years [6, 71]. Data are collected from randomly sampled households through personal interviews conducted by trained interviewers. Data are collected on selected demographic and broad health measures, following which one adult and one child is randomly selected and interviewed using a more detailed health questionnaire. The most recent data included 88,530 children aged 3–17 years [71].

The study reported that the prevalence of any developmental disability in the USA between 2009 and 2017 was 16.93%. It was 9% for ADHD, 1.74% for ASD, 0.16% for visual impairment, 0.31% for cerebral palsy, 0.63% for hearing loss, 7.7% for learning disabilities, 1.1% for intellectual disabilities, 0.77% for seizure disorders, 2% stuttering/stammering and 4% for other developmental delay. For some conditions such as cerebral palsy, there were small differences in prevalence by age group, but some conditions like learning disability and ADHD had higher prevalence in the school years, when they are likely to be recognized.

The data indicated that the prevalence of children diagnosed with any developmental disability had increased in the USA from 5.76% in 2014, to 6.99% in 2016. The prevalence of children with ADHD, ASD, and ID had increased (ADHD increased from 8.47 to 9.54%, an increase of 12.6%; ASD increased from 1.12 to 2.49%, an increase of 122.3%, and ID increased from 0.93 to 1.17%; an increase of 25.8%). The diagnosis of autism was higher at older (8–12 years) than younger ages (3–7 years). As noted earlier, the prevalence of developmental disabilities was higher in boys than girls, and among white children.

In India, there is no nation-wide survey equivalent to the NHIS for measurement of the prevalence of developmental disabilities. The Census of India collects data on disability prevalence, through a single question that records data on impairment of vision, hearing speech, movement, cognition, and multiple disabilities. Two national disability surveys have been conducted by the National Sample Survey Organization (NSSO) [53], one in 2002 and another recently in 2018. (The findings from these surveys have been described in Chap. 8.) In 2002 survey, data on household, sociodemographic, and disability characteristics were collected from a random sample of 396,943 individuals from across the country.

Prevalence data reported by the Census 2011 and the NSSO 2002 for children and young adults less than 20 years of age identified that the disability prevalence was 2.2% and 1.8% of the total Indian population respectively. The proportion of children with disabilities below five years of age was estimated to be between 0.5 and 1%, that is 0.54–1.29 million. The National Sample Survey 2002 reported that among the 4.63 million children under 18 years of age, 58% were reported to have been born with disability (2.70 million children, prevalence 64 per 10,000), while the remaining 42% (1.93 million children, prevalence 49 per 10,000) had acquired disability (Chap. 8). The data indicated that 88% of speech disability, 85% of multiple disability, 78% of cognitive disability, 63% of visual disability were reported to have been present since birth.

Ad Hoc Studies

Most of the data on the prevalence of developmental disabilities are available from independent studies. Systematic reviews and meta-analysis have been conducted for most of the common developmental disabilities, so that global prevalence estimates are available. However, majority of studies are available from industrialized countries. Nearly, all systematic reviews identify the paucity and poor quality of studies from LMICs. Maulik and Darmstadt [44] identified that the poor quality of research was responsible for a significant knowledge gap and frequently questionable data.

For example, from a systematic review and meta-analysis of 51 studies of selected NDDs (epilepsy, hearing and vision impairment, ADHD, cerebral palsy, ASD, behavioral disorders, motor impairment, and other neurological impairments), the authors estimated that the prevalence of NDDs was 7.6 per 1000. Majority of studies were on epilepsy. The prevalence was highest for behavioral problems, i.e. 362 per 1000, followed by mental disorders 232 per 1000, ADHD was 61 per 1000, epilepsy 8 and ASD 0.6 per 1000, respectively. Most of the studies were from the Asia Pacific region. The highest pooled prevalence was from Latin America. There were very wide variations in prevalence, which led to conclusions such as epilepsy being more common in Asia and Africa, whereas ADHD and hearing impairment were common in South America. Conditions like ASD appeared to have a very low prevalence in LMICs. The paucity of good quality studies in LMICs identified the challenge of estimating the true prevalence of developmental disabilities in these countries [4].

Prevalence of Developmental Disabilities in India

A multicentric study by Arora et al. is perhaps the strongest study to report the prevalence of developmental disabilities from India [2]. The study recruited 3977 children from five sites across the country, using cluster sampling method. The children were in the 2–<6 and 6–9-year age groups. The neuro-developmental disabilities that were included in the study were vision impairment, epilepsy, cerebral palsy, hearing impairment, speech and language disorders, autism spectrum disorders, and intellectual disability. Children aged between 6 and 9 years were additionally assessed for ADHD and learning disorders.

All children were assessed using a validated version of the Diagnostic and Statistical Manual of Mental Disorders, Fourth Edition (Diagnostic and Statistical Manual),

Text Revision (DSM-IV-TR) guidelines. Culturally relevant tools, that were feasible for community-based use were developed and validated for epilepsy [58], cerebral palsy [29], ASD [34] and ADHD [48]. Specific diagnostic instruments were used for confirming diagnosis [2].

The prevalence of neuro-developmental disorders was 12% (95% CI 11.0–13.0%) (475 out of 3964). Among children with neuro-developmental disorders, nearly 22% had more than two disorders. ASD, cerebral palsy, and epilepsy were most frequently associated with comorbidities. Hearing impairment, intellectual disability, speech and language disorders, epilepsy, and learning disorders (LD) were the most common types of developmental disabilities identified across all sites. Site-specific variations in prevalence were observed. The survey did not find difference in prevalence of developmental disabilities among boys and girls, urban and rural residence and by religion. The study identified several modifiable risk factors associated with developmental disabilities. These were home delivery, history of perinatal asphyxia and neonatal illness, post-natal brain infections, stunting, low birth weight/prematurity. The population attributable fraction was nearly 37% for these factors (Table 7.4).

A large body of work on developmental disabilities has been contributed by Nair and colleagues from the Child Development Centre, Kerala [11, 51]. Several culturally adapted, valid tools that could be used by community health workers have been developed. Using two such validated tools, the Trivandrum Developmental Screening Chart (TDSC) 0–3 and Language Evaluation Scale Trivandrum (LEST) 0–3, a survey of 32,664 children less than three years of age across the state of Kerala was conducted [49, 50]. Screened children were referred to pediatricians for re-evaluation. In this age group, the prevalence of developmental disability was 2.5%. Among 1110 children who were clinically evaluated, 69% had developmental delay, 14% had speech delay, 6% had global delay, 5% had gross motor delay, and 4% had hearing impairment.

Table 7.4 Prevalence of developmental disabilities (%) [2]

Description	Age group (years)	
	2-<6	6–9
Any neuro-developmental disorder	9.2	13.6
>1 neuro-developmental disorder	2.3	2.6
Visual impairment	0.7	0.6
Epilepsy	1.1	2.2
Neuro-motor impairment (NMI)-cerebral palsy	2.1	1.3
Hearing impairment	3.3	2.6
Speech/language	1.6	1.6
Autism spectrum disorder	1	1.4
Intellectual disability	3.1	5.2
Attention-deficit/hyperactivity disorder		1
Learning disabilities		1.6

Selected Conditions

Epilepsy

A recent systematic review and meta-analysis of 222 studies examined the prevalence and incidence of epilepsy globally. The point prevalence of active epilepsy was 6.38 per 1000 persons (95% CI 5.57–7.30), the lifetime prevalence was 7.60 per 1000 persons (95% CI 6.17–9.38). The incidence rate was 61.44 per 100,000 person-years (95% CI 50.75–74.38). The prevalence of epilepsy did not differ significantly by age group, sex, or study quality. Epilepsy prevalence was higher in low- to middle-income countries [20].

A systematic review of epilepsy prevalence in Europe identified that population-based epidemiological studies on epilepsy were available mainly from the UK and the Nordic, Baltic, and western Mediterranean countries. The study estimated 0.9 million cases (prevalence 4.5–5.0 per 1000) among children and adolescents, 1.9 million in ages 20–64 years (prevalence 6 per 1000), and 0.6 million in ages 65 years and older (prevalence 7 per 1000). The study reported that 20–30% of patients would have more than one seizure per month. The estimated number of new cases per year among European children and adolescents was 130,000 (incidence rate 70 per 100,000), 96,000 in adults 20–64 years (incidence rate 30 per 100,000), and 85,000 in the elderly 65 years and older (incidence 100 per 100,000) [22].

Several studies to estimate the prevalence of epilepsy in India have been conducted, but data quality are affected by issues of case definitions, sample size, data collection tools, research setting (i.e., urban versus rural), and the inclusion of acute symptomatic seizures (which is not epilepsy) [1]. A systematic review and meta-analysis of 20 studies estimated a prevalence of 5.34 per 1000 (4.25–6.41 per 1000) [64]. The estimated rural rate was 5.5 per 1000, and the urban rate was 5.1 per 1000. These estimates were similar to prevalence reports of other studies (Table 7.5). Incidence rates of epilepsy have varied between 0.2 and 0.6 per 1000 population. Although earlier studies had reported a higher prevalence of epilepsy among males, this gender difference has narrowed due to better care seeking among women [1].

Cerebral Palsy

Analysis of 49 global cerebral palsy prevalence studies conducted among children born between 1985 and 2011 reported a pooled prevalence of 2.11 per 1000 live births (95% CI 1.98–2.25) [55]. The pooled prevalence of cerebral palsy was associated with birth timing and weight, being highest in children weighing 1000–1499 g at birth (59.18 per 1000 live births, 95% CI 53.06–66.01), and among children born before 28 weeks of gestation (111.80 per 1000 live births, 95% CI 69.53–179.78). The study reported that the overall prevalence of cerebral palsy appeared to have remained constant, despite the increased survival of preterm and low or very low

Table 7.5 Selected Indian epilepsy prevalence studies

Studies	Prevalence	Region	Sample	Tool used
Mani et al. [42]	5.4 per 1000 (lifetime prevalence); 4.63/1000 active epilepsy prevalence	Rural population, Yelundar, Karnataka	64,963 individuals, house to house survey	Modified ICEBERG (International Community Based Epilepsy Research Group) screening instrument
Radhakrishnan et al. [57]	4.9 per 1000	Urban population, belonging to 10 panchayats of Thrissur, Palakkad and Malappuram districts, Kerala	238,102 population	Modified WHO screening questionnaire
Bangalore Urban Rural Neuro-Epidemiological Survey (BURNs) [27]	Overall—8.82 per 1000 5.8 per 1000 for urban 11.9 per 1000 for rural	Bangalore, Karnataka	102,572	Modified WHO protocol
Das and Diswas [13]	5.7 per 1000 for the	Urban population, Kolkata	52,377	NIMHANS screening questionnaire

birth weight infants. Himpens et al. [30] reported a prevalence of 1.13 per 1000 live births (95% CI 0.93–0.14) per 1000 in term born infants, and an increased cerebral palsy prevalence among infants born at 22–26 weeks of gestation (146 per 1000 live births , 95% CI 125–170). Hirtz et al. [31] reported the prevalence of cerebral palsy at 2.4 per 1000 live births. The prevalence in preterm births and in children with low birth weight (11.2 per 1000 live births) and very low birth weight (63.5 per 1000 live births) was higher than term born infants. Winter et al. [70] estimated the prevalence of cerebral palsy from the MADDS data. For the period between 1975 and 1991, the reported prevalence of cerebral palsy was 2.0 per 1000, showing a modest increase from 1.7 per 1000 in 1971, primarily among infants of normal birth weight. No change in prevalence was seen among low birth weight or very low birth weight infants. The prevalence was higher in boys, African-American children with normal birth weight and in white children with low birth weight. Spastic cerebral palsy was the most common subtype of cerebral palsy identified [70].

The challenge of estimating the prevalence of cerebral palsy in resource limited settings was reported in a systematic review of 20 studies. These studies were published between 1990 and 2009. The authors of the systematic review reported lack of appropriate study designs, case classifications and definitions. Most studies were hospital, rather than population-based. Such methodological issues resulted in skewed prevalence rates of 31–160 per 1000. Pooled prevalence rates for India (2–2.8/1000) were however similar to data from Western countries [23].

An identical paucity of quality studies was observed by Chauhan et al. [8] in a systematic review and meta-analysis of studies on cerebral palsy in India. Globally, the prevalence of cerebral palsy ranges from 1.5 to 4 per 1000 births. The systematic review extracted eight community based studies of cerebral palsy prevalence in children aged 1–18 years in India. The studies were located in either rural or urban areas, or in both geographical locations, and included children of different ages . The studies used several different screening and diagnostic tools (INCLEN Diagnostic Tool for Neuro-Motor Impairments, Trivandrum Developmental Screening Chart (TDSC), Denver Developmental Screening Test (DDST), pre-tested Performa for Disabled Children, Lucknow Neurodevelopmental Screen (LNDS) and WHO questionnaire). The overall pooled prevalence was 2.95 (95% CI 2.03–3.88), with lower prevalence in rural areas (1.83; 95% CI 0.41–3.25) than urban areas (2.29; 95% CI 1.43–3.16).

Intellectual Disabilities

A systematic review of 52 studies conducted between 1980 and 2009 reported that the prevalence of IDs was 10.37 per 1000 [45]. The prevalence was higher among low- and middle-income countries and among children rather than adults. The estimates varied by country, income group, age group of the study population and the study design adopted. The authors noted the importance of using appropriate tools for measuring prevalence, as using psychological assessment tools yielded higher estimates when compared to those using standard diagnostic systems or disability assessment instruments. A more recent systematic review and meta-analysis of twenty two studies conducted between 2010 and 2015 identified the prevalence of ID between 0.05 and 1.55%. The authors reiterated that different methodological approaches, age groups and different case definitions were the key reason for differences in prevalence data [46].

The prevalence of intellectual disabilities in Indian studies is very heterogeneous. Studies have reported that the prevalence of intellectual disability varies by age, gender, urban versus rural residence, but study quality issues are associated with these findings [39]. Using national disability data published by the National Sample Survey 2002, the disability prevalence was 10.5 per 1000 population. ID prevalence was higher in urban than rural areas.

Autism Spectrum Disorders

Prevalence estimates of ASD are also challenged by methodological issues. Williams et al. [69] estimated that the prevalence of typical autism was 7.1 per 10,000 (95%

CI 1.6–30.6). The prevalence estimates varied by the diagnostic criteria used (ICD-10 or DSM-IV or others), age of children and study location. Elsabbagh et al. [17] reported the prevalence of ASD from a review of studies that spanned over a period of 50 years. The studies varied in terms of diagnostic category, criteria, age at prevalence evaluation, and geographical setting. These factors led to a large variation in prevalence, ranging from 0.19/1000 (for autistic disorder) to 11.6/1000 for PDD. Tsai [66] updated these data, finding nearly no difference in prevalence estimates (1.32/1000 for AD and 6.19/1000 for PDD/ASD). These estimates were further updated after 2014 [10]. This study concluded that there appeared to be increasing prevalence within regions, but methodological differences in case detection and study designs could have influenced the data. A systematic review of studies published from South Asia between 1962 and 2016 showed that the prevalence ranged from 0.09% in India to 1.07% in Sri Lanka. Three percent prevalence was reported from Dhaka. Prevalence studies from Pakistan, Nepal, Bhutan, Maldives, and Afghanistan were either unavailable, or not eligible for inclusion in the review [32].

A systematic review and meta-analysis of studies on ASD in India identified 195 records, of which four studies were included for determining the prevalence of ASD. However, the prevalence data were limited by study quality, especially the diagnostic tools used, and the sample size of the studies [9].

Attention-Deficit/Hyperactive Disorders

ADHD is difficult to diagnose, which influences prevalence estimates. A systematic review of 39 eligible studies conducted between 1992 and 2006 reported wide variation in prevalence from 2.2 to 17.8% [63]. The review identified a higher prevalence in boys as compared to girls, reduction in prevalence of ADHD by age, and lower prevalence among Asian children as compared to non-Hispanic, white children. The type of study tool, and the type of respondent that is parents or teachers, influenced prevalence data. A meta-analysis of 86 studies conducted between 1994 and 2010, all of which used DSM-IV reported that the pooled prevalence of ADHD ranged between 5.9 and 7.1% [68]. Another systematic review conducted in 2015 examined the pooled prevalence by DSM criteria, and by other factors such as informants, sampling frames, measurements, full versus part DSM criteria and regions on the prevalence of ADHD. There were 175 eligible studies which yielded a pooled estimate of 7.2% (95% CI 6.7–7.8). The study did not find any difference in prevalence between DSM editions used in data collection. The analysis also identified a 2% higher prevalence in the US as compared to studies done in Europe [65].

In India, a tool that can be used at the community level by clinicians has been developed and used to report data on ADHD from a systematically drawn sample. The prevalence of ADHD was 1% [2, 47].

Data from Global Burden of Disease Study

The Global Burden of Disease 2016, presented modeled estimates of the magnitude of epilepsy, intellectual disability, hearing loss, vision loss, ASD, and ADHD in children less than 5 years from 195 countries. The GBD estimated 53 million children with any of the six developmental disabilities, as compared to 52.9 million in 1990. Nearly, 94% (around 50 million) children lived in LMICs, while just 5% (2.7 million) were resident in high-income countries. The male-to-female proportions depended on the type of developmental disability, but was slightly higher (54%) among males. The most prevalent developmental disability was vision loss (26.4 million), followed by hearing loss which affected nearly 15 million children. Although the absolute numbers of children with hearing loss increased, the prevalence decreased between 1990 and 2016. ADHD was the least prevalent of all disabilities (890,229 cases). The years lived with disability (YLD) was the highest for intellectual disability, followed by epilepsy, hearing loss, vision loss, ASD, and ADHD. The GBD 2016 data estimated that the prevalence of developmental disabilities had increased in sub-Saharan Africa, North Africa, and Middle East. The highest prevalence of developmental disabilities was in South Asia, whereas the lowest prevalence was in North America [24].

India

The modeled estimates indicated that there were 11.5 million (11,560,118 (10,518,238–12,554,824)) cases of developmental disabilities in India in 2016, which was a small reduction from 1990 (10,524–10,308 cases per 100,000 population). India had the highest number of cases globally. By type, there were 800,000 cases of epilepsy, which constituted 42% of the estimated 1,979,233 cases occurring globally, over 800,000 cases of ASD which constituted 36% of 2,366,873 cases worldwide. There were an estimated three million cases of intellectual disability, which was 47% of 6,830,618 cases estimated worldwide. There were three and a half million individuals with hearing loss, and five million with vision loss, constituting 40% of 8,872,948 and 37% of 13,427,729 global cases. The proportion of ADHD cases (16%) was relatively lower (67,000 out of 429,470 cases). In both numbers of cases and YLDs, India was ranked first for all these conditions, with the exception of ADHD where it was ranked second after China. However, in terms of rates per 100,000 population, India was not among the top ten ranked nations, with the exception of intellectual disability (8th rank) and hearing loss (10th rank), globally. The highest YLDs were found in India for all disabilities except ADHD (Table 7.6).

Table 7.6 Prevalence of developmental disabilities in India in 2016 (with 95% uncertainty interval)

	Prevalence		YLD	
	Number	Rate	Number	Rate
Epilepsy	823,482.41 (657,410.91–1,058,323.00)	734.33 (586.24–943.75)	306,008.29 (217,777.39–439,166.06)	272.88 (194.20–391.62)
Intellectual disability	3,190,464.53 (2,523,762.73–3,872,888.30)	2845.06 (2250.53–3453.60)	395,583.05 (286,635.52–544,194.07)	352.76 (255.60–485.28)
Hearing loss	3,533,324.00 (3,096,058.16–4,026,636.56)	3150.80 (2760.87–3590.70)	232,015.36 (159,169.76–325,299.93)	206.90 (141.94–290.08)
Vision loss	5,097,650.84 (4,575,076.44–5,690,437.75)	4545.77 (4079.77–5074.38)	233,730.90 (154,052.91–360,515.41)	208.43 (137.37–321.49)
Autism spectrum disorder	850,811.93 (696,988.94–1,037,775.28)	758.70 (621.53–925.42)	114,372.85 (73,029.39–165,266.55)	101.99 (65.12–147.37)
Attention deficit hyperactivity disorder	66,937.19 (60,271.03–75,229.70)	59.69 (53.75–67.09)	802.50 (463.09–1273.98)	0.72 (0.41–1.14)

Conclusion

In conclusion, this review identifies a substantial number of children with developmental disabilities in the country. Table 7.7 compares the findings from the multicentric Indian study [2] with the period prevalence estimates reported from the USA [71]. The comparison shows that the prevalence of AHD and ASD were more or less similar, the prevalence of learning disability was markedly lower, but for all other developmental disabilities (vision impairment, cerebral palsy, hearing impairment and intellectual disability), the prevalence was notably higher in India. The GBD estimates that India is likely to harbor the highest number of children with developmental disabilities, and any specific type of developmental disability. The healthcare implications can be understood, as the GBD estimates over 3 million children with intellectual disability, 3.5 million, and 5 million children with hearing and vision loss in the country. The magnitude of developmental disabilities may contribute to the magnitude of childhood disabilities, captured in the Census 2011 data [54].

The high prevalence of developmental disabilities may be influenced by the higher prevalence of several well-documented maternal, environmental and healthcare system-related risk factors, identifying the need for a specific package of maternal health and health service-related interventions for these conditions. Developmental disabilities trends show negligible reduction over time [24] while at the same time, other common causes of neonatal and child mortality have reduced. Such trends imply that health services of LMICs including India are likely to be overwhelmed by the magnitude of these conditions. The functioning and quality of life of children with several types of developmental disabilities can be improved by early intervention (Chap. 13). At present, there is no developmental screening for children in India, although children discharged from neonatal intensive care unit

Table 7.7 Comparison of data on the prevalence of developmental disabilities

	India		The United States	GBD (India) (rate per 100,000)
	2-<6 (%)	6–9 (%)	<8 years (%)	<5
Attention-deficit/hyperactivity disorders		1	9	59.69
Autism spectrum disorders	1	1.6	1.74	758.7
Visual impairment	0.7		0.16	4545.7
Cerebral palsy	2.1	1.3	0.31	–
Hearing impairment	3.3	2.6	0.63	3150.8
Learning disorders	1.6		7.7	–
Intellectual disability	3.1	5.2	1.1	2845.0
Epilepsy	1.1	2.2	0.77	734.33

are supposed to be followed up by community health workers for a period of one year under the RBSK programme. Increasing parental awareness about developmental milestones, and appropriate child development and nurturing may increase the detection of developmental disabilities. Translation of knowledge into action has already been demonstrated through the extensive community based work done by the Child Development Centre, Kerala [52].

Estimation of prevalence of developmental disabilities brings to the fore the issue of screening tools. One of the major challenges of tools developed in industrialized countries is that they may not be culturally appropriate, and they may loose their psychometric properties after translation [2]. Another reality that is more pronounced in India is that there are 22 official, and a total of 121 languages in the country [54]. Several tools have been developed in India, but with the major focus on ensuring that they can be used by healthcare workers, many of them lack optimal psychometric properties. Developing context specific tools will help in early identification, referral and intervention which in turn will help to reduce the magnitude of developmental disabilities.

References

1. Amudhan S, Gururaj G, Satishchandra P (2015) Epilepsy in India I: epidemiology and public health. Am Indian Acad Neurol 18:263–277
2. Arora NK, Nair MKC, Gulati S et al (2018) Neurodevelopmental disorders in children aged 2–9 years: population-based burden estimates across five regions in India. PLoS Med 15:1–19
3. Baker GA, Jacoby A, Buck D et al (1997) Quality of life of people with epilepsy: a European study. Epilepsia 38:353–362
4. Bitta M, Kariuki SM, Abubakar A, Newton CRJC (2018) Burden of neurodevelopmental disorders in low and middle-income countries: a systematic review and meta-analysis. Welcome Open Res 2:1–28
5. Boggs D, Milner KM, Chandna J et al (2019) Rating early child development outcome measurement tools for routine health programme use. Arch Dis Child 104:S22–S33
6. Boyle CA, Boulet S, Schieve LA et al (2011) Trends in the prevalence of developmental disabilities in US children, 1997–2008. Pediatrics 127:1034–1042
7. Bright Futures Steering Committee Medical Home Initiatives for Children With Special Needs Project Advisory (2006) Identifying infants and young children with developmental disorders in the medical home: an algorithm for developmental surveillance and screening. Pediatrics 118:405–420
8. Chauhan A, Sahu JK, Jaiswal N et al (2019a) Prevalence of autism spectrum disorder in Indian children: a systematic review and meta-analysis. Neurol India 67:100–104
9. Chauhan A, Singh M, Jaiswal N et al (2019b) Prevalence of cerebral palsy in Indian children: a systematic review and meta-analysis. Indian J Pediatr 86:1124–1130
10. Chiarotti F, Venerosi A (2020) Epidemiology of autism spectrum disorders: a review of worldwide prevalence estimates since 2014. Brain Sci 10:1–21
11. Child Development Centre, Kerala, India. https://www.cdckerala.org/. Accessed 23 July 2020
12. Child developmental screening checklist. Centre for Disease Control and Prevention (CDC). https://www.cdc.gov/ncbddd/actearly/pdf/checklists/Checklists-with-Tips_Reader_508.pdf. Accessed 23 July 2020
13. Das SK, Biswas A (2006) A random sample survey for prevalence of major neurological disorders in Kolkata. Indian J Med Res 124:163–172

14. Durkin M (2002) The epidemiology of developmental disabilities in low income countries. Ment Retard Dev Disabil Res Rev 8:206–211
15. Durkin MS (2019) Increasing prevalence of developmental disabilities among children in the US: a sign of progress? Pediatrics 144:e20192005
16. Early child development measurement inventory [Internet]. World Bank 2017. World Bank Open Knowledge Repository. https://openknowledge.worldbank.org/handle/10986/29000?loc aleattribute=en&locale-attribute=es. Accessed 23 July 2020
17. Elsabbagh M, Divan G, Koh Y et al (2012) Global prevalence of autism and other pervasive developmental disorders. Autism Res 5:160–179
18. Faraone SV, Asherson P, Banaschewski T et al (2015) Attention-deficit/hyperactivity disorder. Nat Rev Dis Primers 1:1–23
19. Fernald LCH, Prado E, Kariger P, Raikes A (2017) A toolkit for measuring early childhood development in low- and middle-income countries. International Bank for Reconstruction and Development/The World Bank, Washington
20. Fiest KM, Sauro KM, Wiebe S et al (2017) Prevalence and incidence of epilepsy: a systematic review and meta-analysis of international studies. Neurology 88:296–303
21. Fischer VJ, Morris J, Martines J (2014) Developmental screening tools: feasibility of use at primary healthcare level in low- and middle-income settings. J Health Popul Nutr 32:314–326
22. Forsgren L, Beghi E, Oun A, Sillanpaa M (2005) The epidemiology of epilepsy in Europe—a systematic review. Eur J Neurol 12:245–253
23. Gladstone M (2010) A review of the incidence and prevalence, types and aetiology of childhood cerebral palsy in resource-poor settings. Ann Trop Pediatr 30:181–196
24. Global Research on Developmental Disabilities Collaborators (2018) Developmental disabilities among children younger than 5 years in 195 countries and territories, 1990–2016: a systematic analysis for the Global Burden of Disease Study. Lancet Glob Health 6:e1100–e1121
25. Gogate P, Gilbert C, Zin A (2011) Severe visual impairment and blindness in infants: causes and opportunities for control. Middle East Afr J Ophthalmol 18:109–115
26. Goldsmith S, Mcintyre S, Smithers-Sheedy H et al (2016) An international survey of cerebral palsy registers and surveillance systems. Dev Med Child Neurol 58:11–17
27. Gourie-Devi M, Gururaj G, Satishchandra P, Subbakrishna DK (2004) Prevalence of neurological disorders in Bangalore, India: a community-based study with a comparison between urban and rural areas. Neuroepidemiology 23:261–268
28. Government of India, Ministry of Health and Family Welfare (2019) Answers data of Rajya Sabha questions for session 240/year wise physical status Rashtriya Bal Swasthya Karyakram (RBSK) during 2014–15 and 2015–16. https://data.gov.in/node/3978901/download. Accessed 2 Mar 2018
29. Gulati S, Aneja S, Juneja M et al (2014) INCLEN diagnostic tool for neuromotor impairments (INDT-NMI) for primary care physician: development and validation. Indian Pediatr 51:613–619
30. Himpens E, Van den Broeck C, Oostra A et al (2008) Review: prevalence, type, distribution, and severity of cerebral palsy in relation to gestational age: a meta-analytic review. Dev Med Child Neurol 50:334–340
31. Hirtz D, Thurman DJ, Gwinn-Hardy K et al (2007) How common are the "common" neurologic disorders? Neurology 68:326–337
32. Hossain MD, Ahmed HU, Uddin MMJ et al (2017) Autism spectrum disorders (ASD) in South Asia: a systematic review. BMC Psychiatry 17:1–7
33. Huang J, Zhu T, Qu Y, Mu D (2016) Prenatal, perinatal and neonatal risk factors for intellectual disability: a systemic review and meta-analysis. PLoS ONE 11:1–12
34. Juneja M, Mishra D, Russell P et al (2014) INCLEN diagnostic tool for autism spectrum disorder (INDT-ASD): development and validation. Indian Pediatr 51:359–365
35. Kliegman ST, Blum G et al (2016) Nelson textbook of pediatrics, 20th edn. Elsevier Inc
36. Kong L, Fry M, Al-Samarraie M et al (2012) An update on progress and the changing epidemiology of causes of childhood blindness worldwide. J AAPOS 16:501–507

37. Korver AMH, Smith RJH, Van Camp G et al (2018) Congenital hearing loss. Nat Rev Dis Primers 3:1–37
38. Korzeniewski SJ, Slaughter J, Lenski M et al (2018) The complex aetiology of cerebral palsy. Nat Rev Neurol 14:528–543
39. Lakhan R, Ekúndayò OT, Shahbazi M (2015) An estimation of the prevalence of intellectual disabilities and its association with age in rural and urban populations in India. J Neurosci Rural Pract 6:523–528
40. Lord C, Brugha T, Charman T et al (2020) Autism spectrum disorder. Nat Rev Dis Primers 6:1–23
41. Maenner M, Shaw K, Bajo J (2020) Prevalence of autism spectrum disorder among children aged 8 years—autism and developmental disabilities monitoring network, 11 sites, United States, 2016. MMWR Surveill Summ
42. Mani KS, Rangan G, Srinivas HV et al (1998) The Yelandur study: a community-based approach to epilepsy in rural South India—epidemiological aspects. Seizure 7:281–288
43. Marlow M, Servili C, Tomlinson M (2019) A review of screening tools for the identification of autism spectrum disorders and developmental delay in infants and young children: recommendations for use in low- and middle-income countries. Autism Res 12:176–199
44. Maulik PK, Darmstadt GL (2007) Childhood disability in low- and middle-income countries: overview of screening, prevention, services, legislation, and epidemiology. Pediatrics 120:1–55
45. Maulik PK, Mascarenhas MN, Mathers CD et al (2011) Prevalence of intellectual disability: a meta-analysis of population-based studies. Res Dev Disabil 32:419–436
46. Mckenzie K, Milton M, Smith G, Ouellette-Kuntz H (2016) Systematic review of the prevalence and incidence of intellectual disabilities: current trends and issues. Curr Dev Disord Rep 3:104–115
47. Mukherjee S, Aneja S, Krishnamurthy V, Srinivasan R (2014a) Incorporating developmental screening and surveillance of young children in office practice. Indian Pediatr 51:621–635
48. Mukherjee S, Aneja S, Russell P et al (2014b) INCLEN diagnostic tool for attention deficit hyperactivity disorder (INDT-ADHD). development and validation. Indian Pediatr 51:457–462
49. Nair M, Nair H, Mini A et al (2013a) Development and validation of Language Evaluation Scale Trivandrum for children aged 0–3 years—LEST (0–3). Indian Pediatr 50:463–467
50. Nair MKC, Nair GSH, George B et al (2013b) Development and validation of Trivandrum development screening chart for children aged 0–6 years [TDSC (0–6)]. Indian J Pediatr 80:248–255
51. Nair MKC, Princly P, Leena ML et al (2014) CDC Kerala 17: early detection of developmental delay/disability among children below 3 y in Kerala—a cross sectional survey. Indian J Pediatr 81:S156–S160
52. Nair MKC, Leela LM, George B et al (2016) CDC Kerala—the untold story. Indian J Pediatr 83:426–433
53. National Sample Survey Organization. Ministry of statistics and programme implementation. Government of India. https://mospi.nic.in/nsso. Accessed 23 July 2020
54. Office of the Registrar General & Census Commissioner, India. Ministry of Home Affairs, Government of India. https://censusindia.gov.in. Accessed 23 July 2020
55. Oskoui M, Coutinho F, Dykeman J et al (2013) An update on the prevalence of cerebral palsy: a systematic review and meta-analysis. Dev Med Child Neurol 55:509–519
56. Poornima M, Polinedi A, Praveen Kumar PTV et al (2013) Prenatal, perinatal and neonatal risk factors of autism spectrum disorder: a comprehensive epidemiological assessment from India. Res Dev Disabil 34:3004–3013
57. Radhakrishnan K, Pandian JD, Santhoshkumar T et al (2000) Prevalence, knowledge, attitude, and practice of epilepsy in Kerala, South India. Epilepsia 41:1027–1035
58. Ramesh K, Mishra D, Gulati S et al (2014) INCLEN diagnostic tool for epilepsy (INDT-EPI) for primary care physicians: development and validation. Indian Pediatr 51:539–543
59. Scheffer IE, Berkovic S, Capovilla G et al (2017) ILAE classification of the epilepsies: position paper of the ILAE Commission for Classification and Terminology. Epilepsia 58:512–521

60. Setting up district early intervention centres. Operational guidelines (2014) Rashtriya Bal Swasthya Karyakram, Ministry of Heath & Family Welfare. Government of India. https://nhm.gov.in/images/pdf/programmes/RBSK/Operational_Guidelines/Operational%20Guidelines_RBSK.pdf. Accessed 23 July 2020

61. Shetty PH, Naik RK, Saroja AO, Punith K (2011) Quality of life in patients with epilepsy in India. J Neurosci Rural Pract 2:33–38

62. Sirven JI (2015) Epilepsy: a spectrum disorder. Cold Spring Harb Perspect Med 5:1–16

63. Skounti M, Philalithis A, Galanakis E (2007) Variations in prevalence of attention deficit hyperactivity disorder worldwide. Eur J Pediatr 166:117–123

64. Sridharan R, Murthy BN (1999) Prevalence and pattern of epilepsy in India. Egypt J Med Hum Genet 40:631–636

65. Thomas R, Sanders S, Doust J et al (2015) Prevalence of attention-deficit/hyperactivity disorder: a systematic review and meta-analysis. Pediatrics 135:e994–e1001

66. Tsai LY (2014) Impact of DSM-5 on epidemiology of autism spectrum disorder. Res Autism Spectr Disord 8:1454–1470

67. Wijnhoven TMA, De Onis M, Onyango AW et al (2004) Assessment of gross motor development in the WHO multicentre growth reference study. Food Nutr Bull 25:S37–S45

68. Willcutt EG (2012) The prevalence of DSM-IV attention-deficit/hyperactivity disorder: a meta-analytic review. Neurotherapeutics 490–499

69. Williams J, Higgins J, Brayne C (2006) Systematic review of prevalence studies of autism spectrum disorders. Arch Dis Child 91:8–15

70. Winter S, Autry A, Boyle C, Yeargin-Allsopp M (2002) Trends in the prevalence of cerebral palsy in a population-based study. Pediatrics 110:1119–1125

71. Zablotsky B, Black LI, Maenner MJ et al (2020) Prevalence and trends of developmental disabilities among children in the US: 2009–2017. Pediatrics 144:1–21

Chapter 8
Magnitude and Characteristics of Children with Congenital Disabilities in India

Amruta Chutke

Abstract Non-fatal birth defects and developmental disorders cause congenital disabilities, that is disabilities that are present since birth. There is limited data on the number of survivors with congenital disabilities in India. This article describes the magnitude and characteristics of children with congenital disabilities in the country. It uses data from the National Sample Survey (NSS) 2002. The survey collected data on the time of onset of disability, which provided an opportunity to categorize disability as congenital, that is disability since birth, or acquired, that is disability that occurred later in life. The data indicated that 58% of disabilities among children below 18 years of age were of congenital origin. Among the different types of disabilities, 88% of speech, 85% of multiple, 78% of cognitive, and 63% of visual impairments were of congenital origin. Congenital disability prevalence was four-fold higher at birth as compared to acquired disability and achieved its highest prevalence in the age group of 15–19 years. The impact of congenital disability was considerable, as the person-years lived with disability since birth was nearly double than that due to acquired causes. Severity of disability was more among children born with disabilities. These data suggest the need for further research, and the need to link disability services to maternal and child health services in order to address the needs of children born with disabilities.

Keywords Children with disability · Congenital · Acquired · Birth defects · Vision disorders · Hearing disorders · Speech disorders

This article describes the magnitude and the socio-demographic profile of children with congenital disabilities in India. Childhood disability may be acquired, caused by injuries, chronic conditions like cancer, infections such as poliomyelitis and leprosy [1], vitamin deficiencies causing rickets or blindness, poor nutrition, poverty, and lack of access to needed medical care and rehabilitation [2, 3]. Congenital disability, that is disability since birth, results from a diverse group of congenital disorders, collectively termed birth defects and developmental disabilities. Examples include common

A. Chutke (✉)
Department of Community Medicine, Bharati Vidyapeeth DTU Medical College, Pune, India

© Springer Nature Singapore Pte Ltd. 2021
A. Kar (ed.), *Birth Defects in India*,
https://doi.org/10.1007/978-981-16-1554-2_8

childhood conditions like cerebral palsy, cleft palate, congenital deafness, congenital cataract, intellectual disabilities, spina bifida, and congenital talipes equinovarus.

Congenital disabilities are highly incapacitating conditions, causing cognitive, speech, hearing, vision, and locomotor impairments [4]. They have serious public health implications as children have special medical and rehabilitation needs, majority of which are lifelong in nature [5]. Children require special education and skills to ensure participation in society [6]. Congenital disorders contribute to considerable number of Years Lived with Disability (YLD) [7–10]. Families confront substantial economic burden [11]. Thus, living with disabilities since birth has several health, social and economic consequences for individuals and families, especially in resource-constrained settings where organized disability services are not in place.

Disability data are primarily available from high-income countries [12–15]. The overall magnitude of children with disabilities (CWD) is not well characterized in low and middle-income countries (LMICs). Available data suggest that disability disproportionately affects children in these countries. The World Health Organization enumerated that nearly 80% of the 100–200 million children living with disability worldwide were from LMICs [16]. The report estimated 93 million (5.1%) children below14 years of age with moderate disability and 13 million (0.7%) children with severe disability in 2004. These estimates indicated that there may be 1.86 billion disabled children below 15 years of age in 2010. The World Bank review of 13 LMICs identified a wide range of disability prevalence, ranging from 0.49 to 3.2% among school-going children [17]. A cross-sectional survey of 900,000 children aged 0–17 years from 30 countries participating in the Plan International Sponsorship Programme in 2012 estimated the prevalence of disability in the range of 0.4–3% [18].

There is even less data on children born with congenital disabilities (CWCD). Among available studies, a study from rural Pakistan reported disability prevalence in children less than two years of age was 5.5 out of 1000, while the prevalence was 5.4 out of 1000 for children between two to five years of age. Almost 56% of disability was present since birth and cerebral palsy was the most common disability among children under five years of age [19]. Another study conducted in rural Cambodia reported 40% of caregivers recalled that disability was present since birth [20]. A study done in Rwanda identified that more than one-third of musculoskeletal impairments were due to congenital deformity and neurological causes [21]. A study in Ethiopia identified a smaller proportion (15%) of individuals reporting disability during infancy, with 5.7% describing the cause of disability to be congenital [22]. An Indian study from the state of Gujarat showed that 60% of individuals had disability during infancy [23]. A 30-country study reported that about 80–90% of speech, multiple and cognitive disabilities was present since birth [18]. Another study from Uttar Pradesh, India reported that speech disability was primarily of congenital origin [24].

Data on birth defects and developmental disability survivors is an extremely important child health indicator, as these children require appropriate services. Global data shows that epidemiological transition has resulted in a shift in the major causes of childhood morbidity and mortality and emergence of chronic and disabling conditions of childhood [25–28]. As health service activities improve survival of premature and low birth weight infants, the likelihood of survival with developmental disabilities increase. With decrease in other causes of neonatal mortality (such as prematurity, intrapartum complications, and neonatal sepsis), the visibility of birth defects as contributors to child mortality will increase. Modell and colleagues offer an indicator, the ratio of increase in disability to proportion of congenital disorder deaths [29]. This is an important indicator for service planning, as disability survivors have to be provided appropriate services. It is also an indicator that identifies the need for interventions to prevent these conditions.

In India, disability statistics are available from the decadal Census [30] and from the National Sample Surveys (NSS) [31, 32]. The decadal Census enumerates the total Indian population and collects disability data through a single question to measure the magnitude of disability in the country. Disability is categorized as impairment in vision, hearing, speech, and movement, and mental retardation, mental illness, and multiple disabilities. Additionally, the National Sample Survey Office (NSSO) has conducted two disability surveys in the last two decades. The NSS 58th round (2002) [31] and the 76th round (2018) [32] measured household, socio-demographic, and disability characteristics from a nationally representative stratified sample of individuals across the country.

The NSS is the only source from which data on congenital disability may be obtained, as it includes a question on the time of onset of disability. The data is respondent reported, and not further validated with other records. The time of onset of disability provides the opportunity to categorize children into two groups, those born with disabilities (congenital disability) and those who acquired the disability later in life (acquired disability). The NSS collected data on the severity of disability, self-care, treatment received, and education, providing an insight into the lives of affected children. This article describes the epidemiology of children with disabilities (CWD) and children with congenital disabilities (CWCD) under 18 years of age in India.

A note of caution on the absolute numbers in this article is warranted. The data are from the period 2002 as disaggregated data from the 2018 survey was not available at the time of writing. Furthermore, despite the similarity in categorization of disability types between the Census and the NSS data sets, there are substantial differences in prevalence estimates, mainly due to differences in the definitions used in these two surveys [33]. Nevertheless, the NSS data provide an insight into the characteristics of CWCD in India.

Magnitude of CWD

The magnitude of disability across all age groups in India is reflected in the sheer number of disabled persons in the country. The NSS in 2002 estimated 18.49 million (1.80%) disabled individuals. Disabled children made up 1.14% that is 5.27 million

Table 8.1 Disability prevalence by residence and gender

Residence	Numbers in 00,000 (%)		
	Male	Female	Total
Rural	25.36	16.15	41.51 (79%)
Urban	6.56	4.66	11.22 (21%)
Total	31.92 (61%)	20.81 (39%)	52.73

children with disability under 18 years of age in the country. Nearly 0.5% (that is 0.54 million) children under five years of age were estimated to be disabled. Disability was a larger problem in rural areas, with 75% children with disabilities (over 4 million) being resident in rural areas (Table 8.1). Disability was higher in boys (about 3.2 million, 61%) than girls (2 million, 39%) (Table 8.1).

Prevalence of Childhood Disability by Type of Disability

The disability rate for children below 18 years of age was 114 per 10,000 children. The rates were highest for locomotor disability (66 per 10,000), followed by multiple disability (16 per 10,000), cognitive disability (12 per 10,000), speech (10 per 10,000), and hearing and visual impairments (5 each per 10,000) (Table 8.2).

Table 8.2 Age-specific prevalence of childhood disabled by type of disability

Type of disability	Childhood disability prevalence (per 10,000) and absolute numbers (in 0000)	Childhood congenital disability prevalence (per 10,000) and absolute numbers (in 0000)	Childhood acquired disability prevalence (per 10,000) and absolute numbers (in 0000)	Rate ratio (95% CI) (congenital/acquired disability rate)
Cognitive	11 (45.89)	8 (35.88)	2 (10.01)	3.58 (3.56–3.61)*
Visual	5 (2.18)	3 (1.37)	2 (0.81)	1.7 (1.68–1.71)*
Hearing	5 (2.09)	2 (0.94)	3 (1.15)	0.82 (0.81–0.82)*
Speech	10 (4.32)	9 (3.79)	1 (0.52)	7.26 (7.20–7.33)*
Locomotor	63 (26.52)	28 (11.71)	35 (14.81)	0.79 (0.79–0.79)*
Multiple	16 (6.64)	13 (5.61)	2 (1.03)	5.47 (5.43–5.50)*
Total	110 (46.33)	64 (27.01)	46 (19.32)	1.4 (1.40–1.40)*

*p value <0.05

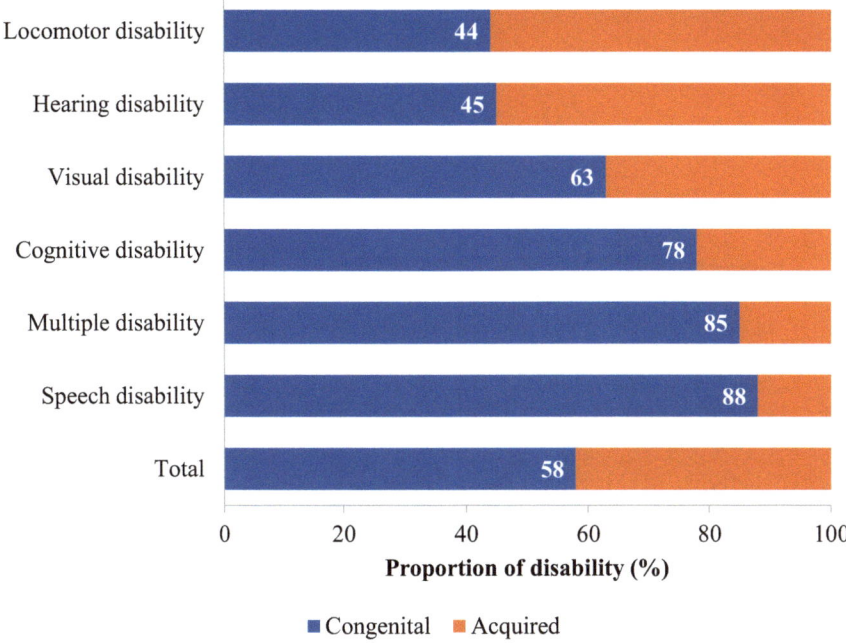

Fig. 8.1 Proportion of disability since birth

Magnitude of CWCD

The data indicated that 58% of children were born with disability (2.70 million children, prevalence 64 per 10,000), while the remaining 42%, (1.93 million children, prevalence 49 per 10,000) had acquired disability (Table 8.2). The data indicated that 88% of speech disability, 85% of multiple disability, 78% of cognitive disability, 63% of visual disability, 45% of hearing, and 44% of locomotor disability were reported to have been present since birth (Fig. 8.1).

The rate ratio of congenital to acquired disability was significantly higher for speech, multiple, cognitive, and visual disability (RR > 1, $p < 0.05$) indicating these conditions to be more likely among children due to congenital disorders. Hearing and locomotor disability were more likely to be acquired (RR < 1, $p < 0.05$) (Table 8.2).

Age-Specific Prevalence Per 10,000

Figure 8.2 shows the age-specific prevalence of childhood congenital and acquired disability. As compared to acquired disability, congenital disability was four-fold higher (39 vs. 9 per 10,000) among children under five years of age, and achieved highest prevalence in the age group of 15–19 years (77 vs. 66 per 10,000). After the

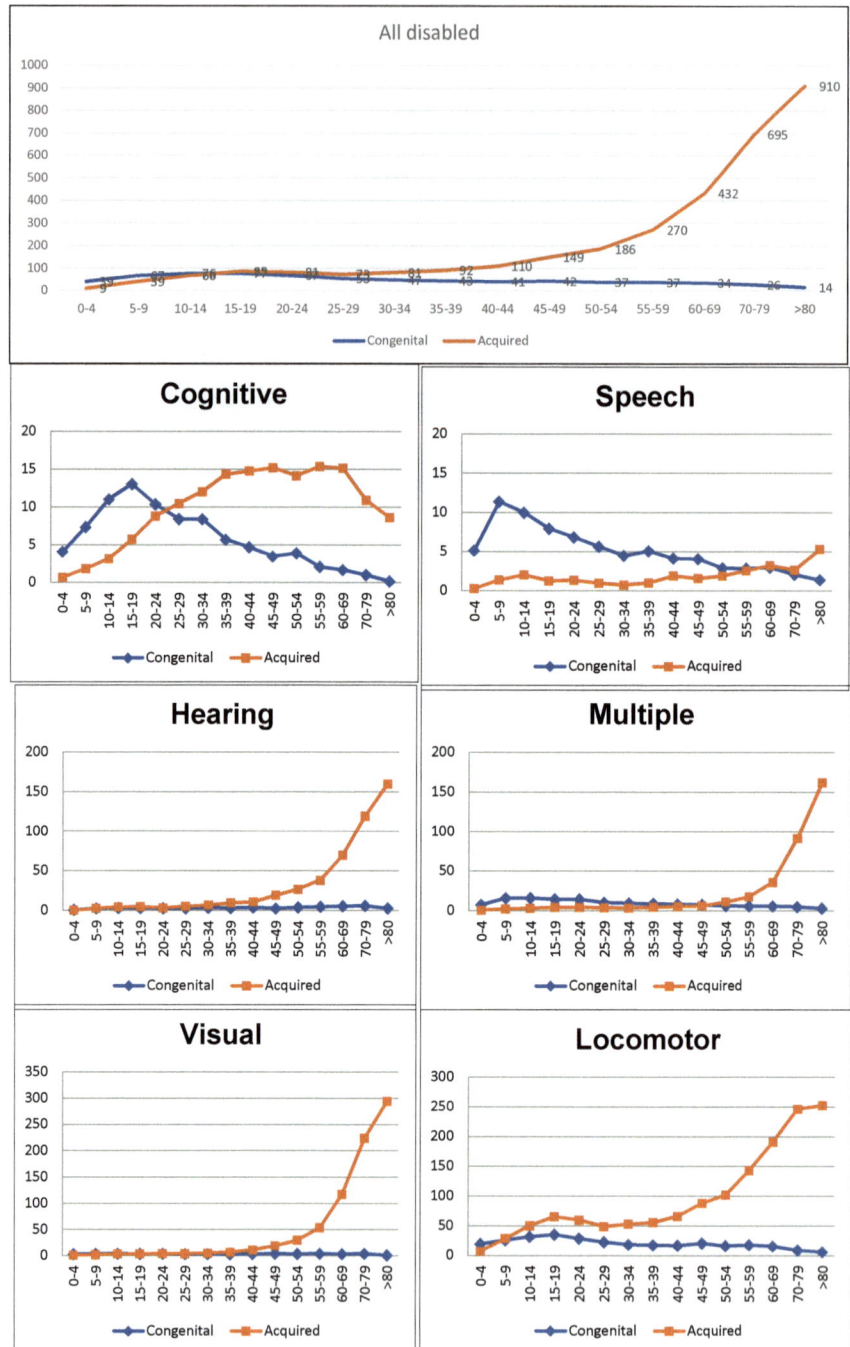

Fig. 8.2 Age-specific prevalence by disability type. X-axis indicates age groups and Y-axis indicates rate per 10,000 population.

age of 19 years, congenital disability rates declined (67 vs. 81 per 10,000), reflecting reduced survival of CWCD.

Congenital locomotor disability prevalence was highest in the age group of 15–19 years (35 per 10,000), and remained as high as 16 per 10,000 till the age group of 60–69 years (Fig.8.2). The next highest prevalence was for multiple disability, which had the highest prevalence in the age group of 5–9 years (16 per 10,000). The prevalence of congenital speech disability and cognitive impairment were similar. Congenital speech impairment was highest in the age group of 5–9 years (11 per 10,000), while cognitive impairment was the highest in the age group of 15–19 years (13 per 10,000). Congenital visual impairment rate was highest in the age group of 5–9 years (4 per 10,000) but the age-specific hearing impairment prevalence remained more or less constant (around 3 per 10,000) over age groups (Fig. 8.2).

Duration of Disability

The mean age years lived with congenital disability was significantly higher (10.05 years, 95% CI 10.04–10.06) than that of children with acquired disability (7.66 years, 95% CI 7.66–7.67) ($p < 0.001$) (Table 8.3). The person-years lived with congenital disability was 27.15 million which was nearly double than that of childhood disability due to acquired causes (14.80 million). The person-years lived with congenital disability was highest for locomotor disability (11.62 million), followed by multiple disability (5.52 million), cognitive impairment (3.90 million), speech disability (3.74 million), visual impairment (1.33 million), and hearing impairment

Table 8.3 Years lived with disability

Type of disability	Acquired disability		Congenital disability		P value
	Mean number of years lived with disability (95% CI)	Person-years lived	Mean number of years lived with disability (95% CI)	Person-years lived	
Cognitive	7.14 (7.10–7.17)	0.71	10.87 (10.85–10.89)	3.90	*0.0001
Visual	6.02 (6.00–6.05)	0.49	9.69 (9.66–9.73)	1.33	*0.0001
Hearing	6.11 (6.08–6.14)	0.70	11.08 (11.03–11.12)	1.04	*0.0001
Speech	6.87 (6.83–6.92)	3.74	9.87 (9.85–9.89)	3.74	*0.0001
Locomotor	7.96 (7.96–7.97)	11.79	9.92 (9.91–9.93)	11.62	*0.0001
Multiple	7.25 (7.22–7.27)	0.74	9.84 (9.83–9.86)	5.52	*0.0001
Total	7.66 (7.66–7.67)	14.80	10.05 (10.04–10.06)	27.15	*0.0001

*$p < 0.001$

(1.04 million). The person-years lived with speech and locomotor disability due to congenital or acquired cause was almost similar (Table 8.3).

Characteristics of CWCD

Table 8.4 shows that CWCD were more likely to report higher proportion of parental consanguinity (17% and 12% among CWCD and acquired disability respectively), belong to families with smaller household size, and appear to have more than one child with a congenital disability. CWCD were more likely to be male, uneducated, belong to socially deprived groups (scheduled castes, other backward castes), and among families with lower average monthly expenditure.

Severity of Disability

Disability appeared more severe in CWCD. Table 8.5 shows that the risk of no perception to light was nearly double in CWCD as compared to those with acquired disability (61% congenital to 48% acquired). Children who were unable to speak were fifteen times more likely to have been born with this impairment (31% vs. 14%). The risk of being able to communicate in single words was nine times higher among children with congenital disabilities. Children born with communication impairments had a seven times higher risk of speaking unintelligibly, four times higher risk of stammering, had nearly six times increased risk of speaking with abnormal voice as compared to children reporting acquiring this disability. The risk of profound hearing impairment was two and a half times elevated for congenital disability (30% vs. 10%). Locomotor disability due to acquired causes was higher, but the proportion of limb deformity was higher among CWCD (64% vs. 58%).

Self-care and Utilization of Services

Another reflection of the severity of congenital disability was the dependence on others for self-care. The proportion was significantly higher (17.5%) among CWCD as compared to children with acquired disability (9.37%) (Table 8.6). A higher proportion of CWCD (25.8%) was without treatment as compared to children with acquired disabilities (9.76%). In the sample, 11.78% CWCD had not tried an aid/appliance or the aid/appliance was not available as compared to 8.9% of children with acquired disabilities. However, less than 1% of CWCD and even lower proportions of children with acquired disabilities were enrolled in special schools.

Table 8.4 Characteristics of CWCD

	Characteristics	Congenital n in 00,000 (%)	Acquired n in 00,000 (%)	OR (95% CI)
i	*Sex*			
	Male	15.83 (58.61)	11.82 (61.22)	*1.12 (1.11–1.12)
	Female	11.18 (41.39)	7.49 (38.78)	Ref
ii	*Place of residence*			
	Rural	21.28 (78.78)	15.33 (79.36)	*1.04 (1.03–1.04)
	Urban	5.73 (21.22)	3.99 (20.64)	Ref
iii	*Social group*			
	Schedule tribe	2.34 (8.67)	1.38 (7.17)	*0.95 (0.94–0.95)
	Schedule caste	6.09 (22.55)	4.71 (24.37)	*1.24 (1.23–1.24)
	Other backward castes	11.08 (41.01)	8.53 (44.18)	*1.23 (1.23–1.24)
	Others	7.50 (27.77)	4.69 (24.28)	Ref
iv	*Consanguinity*			
	Yes	4.38 (17.01)	2.32 (12.43)	*0.69 (0.69–0.70)
	No	21.34 (82.99)	16.38 (87.57)	Ref
v	*Household size*			
	1–4 members	6.93 (25.65)	3.51 (18.16)	Ref
	5–9 members	17.35 (64.21)	13.49 (69.83)	*1.54 (1.53–1.54)
	Members 10+	2.74 (10.13)	2.32 (12.01)	*1.67 (1.66–1.69)
vi	*Education of the individual (6–17 years)*			
	No education	34.14 (61.14)	63.61 (53.79)	*0.60 (0.59–0.60)
	Primary	17.94 (32.12)	42.88 (36.26)	*0.77 (0.76–0.77)
	Secondary and higher	3.76 (6.74)	11.76 (9.95)	Ref
vii	*Average monthly expenditure*			
	Low (<4000)	23.38 (86.54)	16.68 (86.38)	*1.43 (1.40–1.45)
	Middle (4000–8000)	3.22 (11.92)	2.42 (12.54)	*1.50 (1.48–1.53)
	High (>8000)	0.41 (1.54)	0.21 (1.08)	Ref
viii	*No. of disabled children in the household*			
	1	23.04 (85.27)	17.10 (88.53)	Ref
	2	3.30 (12.22)	2.02 (10.45)	*2.54 (2.50–2.58)
	≥3	0.68 (2.51)	0.20 (1.03)	*2.09 (2.06–2.13)

*$p < 0.001$

Table 8.5 Severity of disability

Severity of physical disability (n in 0000)	Disabled children		
	Prevalence for congenital disability per 100,000 (n in 0000)	Prevalence for acquired disability per 100,000 (n in 0000)	Rate ratio (congenital to acquired)
Visual disability n = 21.66			
No light perception	1.96 (8.29)	0.90 (3.80)	*2.18 (2.15–2.21)
Light perception, cannot count fingers upto 1 m, normally uses spectacles	0.17 (0.71)	0.17 (0.70)	1.01 (0.98–1.05)
Cannot count fingers upto 1 m, normally does not use spectacles	0.53 (2.26)	0.34 (1.42)	*1.59 (1.56–1.62)
Cannot count 1–3 m, use spectacles	0.16 (0.67)	0.17 (0.73)	*0.92 (0.89–0.95)
Cannot count 1–3 m, do not use spectacles	0.42 (1.77)	0.31 (1.30)	*1.36 (1.33–1.39)
Speech disability n = 43.08			
Cannot speak	2.79 (11.79)	0.17 (0.74)	*15.98 (15.61–16.37)
Speak single words	1.81 (7.63)	0.20 (0.84)	*9.12 (8.92–9.33)
Speak unintelligibly	1.61 (6.81)	0.23 (0.95)	*7.14 (6.99–7.30)
Stammers	1.77 (7.49)	0.44 (1.85)	*4.04 (3.97–4.10)
Speaks with abnormal voice	0.62 (2.61)	0.11 (0.45)	*5.77 (5.59–5.95)
Others	0.36 (1.54)	0.09 (0.37)	*4.13 (3.99–4.29)

(continued)

Conclusions

In conclusion, the NSS survey provides an opportunity to understand the magnitude and quality of survival of children born with disabilities in India. The data suggests that 58% of disability among children below the age of 18 years was caused by congenital causes. The numbers of CWCD increased due to accrual of birth cohorts till the end of the first decade of life, or during the first half of the second decade of life. Subsequently, the prevalence declined, implying higher mortality at these ages. As reported in other studies, CWCD were likely not to have received treatment, nor were children enrolled in schools [34, 35]. One of the major impacts of disability was that the person-years lived with disabilities were nearly two-fold higher in children with congenital disabilities, not necessarily because the children lived longer, but because the onset of disability was since birth.

Table 8.5 (continued)

Severity of physical disability (n in 0000)	Disabled children		
	Prevalence for congenital disability per 100,000 (n in 0000)	Prevalence for acquired disability per 100,000 (n in 0000)	Rate ratio (congenital to acquired)
Hearing disability n = 20.89			
Profound	0.66 (2.78)	0.27 (1.12)	*2.48 (2.42–2.53)
Severe	0.89 (3.76)	1.26 (5.33)	*0.71 (0.70–0.72)
Moderate	0.68 (2.87)	1.19 (5.03)	*0.57 (0.56–0.58)
Locomotor disability n = 202.36			
Paralysis	2.34 (9.91)	4.44 (18.75)	*0.53 (0.52–0.53)
Deformity of limb	17.76 (75.10)	20.16 (85.25)	*0.88 (0.88–0.88)
Loss of limb	1.00 (4.22)	1.84 (7.80)	*0.54 (0.53–0.55)
Dysfunction of limb joints	4.02 (17.02)	5.33 (22.55)	*0.75 (0.75–0.76)
Others	2.55 (10.78)	3.15 (13.31)	*0.81 (0.80–0.82)

n—number of disabled children
*p value <0.05

Table 8.6 Self-care and service utilization

Characteristics n (in 00,000)	Congenital n (in 00,000) (%)	Acquired n (in 00,000) (%)	OR
Extent of disability n = 42.30			
Cannot take self-care even with aid/appliance	4.26 (17.5)	1.73 (9.37)	*0.46 (0.46–0.46)
Can take self-care with only aid/appliance	2.43 (10.17)	2.28 (12.37)	*1.06 (1.05–1.06)
Aid/appliance not tried/not available	2.81 (11.78)	1.64 (8.90)	*0.66 (0.65–0.66)
Can take self-care without aid/appliance	14.38 (60.20)	12.77 (69.36)	Ref
Services for disabled children n = 46.33			
Attending special school	0.15 (0.57)	0.02 (0.10)	*0.14 (0.13–0.14)
No treatment	6.97 (25.80)	1.89 (9.76)	*0.30 (0.30–0.30)
Yes: undergoing treatment: consulting doctor	1.85 (6.85)	0.11 (5.81)	*0.67 (0.66–0.67)
Yes: taken: otherwise	1.01 (3.73)	0.84 (4.33)	*0.91 (0.91–0.92)
Yes: taken: consulting doctor	17.03 (63.05)	15.45 (80.01)	Ref

*p < 0.001

The data has some limitations, including definitions, categorization, and validation of respondent reported information. The survey data cannot be compared with the Census disability data. Issues related to stigma, lack of diagnosis, poor awareness, and perception of disability may contribute to non-reporting of disability and underestimation [33]. Despite these limitations, the data remains an invaluable source of information on the lives of children born with disabilities. The use of standard definitions, tools such as the UN Washington Group on Disability Statistics Short Set questions [34], which can yield globally comparable data would be effective in further enhancing the quality, applicability, and utility of the data.

References

1. Narang T, Kumar B (2019) Leprosy in children. Indian J Paediatr Dermatol 20(1):12–24
2. Groce NE, Kerac M, Farkas A, Schultink W, Bieler RB (2013) Inclusive nutrition for children and adults with disabilities. Lancet Glob Health 1(4):e180–e181
3. DeCesaro A, Hemmeter J (2009) Unmet health care needs and medical out-of-pocket expenses of SSI children. J Vocat Rehabil 30(3):177–199
4. Perrin EC, Cole CH, Frank DA, Glicken SR, Guerina N, Petit K et al (2003) Criteria for determining disability in infants and children: failure to thrive. Evid Rep Technol Assess (Summ) 72:1
5. Boulet SL, Boyle CA, Schieve LA (2009) Health care use and health and functional impact of developmental disabilities among US children, 1997–2005. Arch Pediatr Adolesc Med 163(1):19–26
6. Imms C, Reilly S, Carlin J, Dodd K (2008) Diversity of participation in children with cerebral palsy. Dev Med Child Neurol 50(5):363–369
7. Vos T, Allen C, Arora M, Barber RM, Bhutta ZA, Brown A et al (2016) Global, regional, and national incidence, prevalence, and years lived with disability for 310 diseases and injuries, 1990–2015: a systematic analysis for the Global Burden of Disease Study 2015. Lancet 388(10053):1545–1602
8. Higashi H, Barendregt JJ, Kassebaum NJ, Weiser TG, Bickler SW, Vos T (2015) The burden of selected congenital anomalies amenable to surgery in low and middle-income regions: cleft lip and palate, congenital heart anomalies and neural tube defects. Arch Dis Child 100(3):233–238
9. McKenna MT, Michaud CM, Murray CJ, Marks JS (2005) Assessing the burden of disease in the United States using disability-adjusted life years. Am J Prev Med 28(5):415–423
10. Groce NE (2018) Global disability: an emerging issue. Lancet Glob Health 6(7):e724–e725
11. Gupta VB (2007) Comparison of parenting stress in different developmental disabilities. J Dev Phys Disabil 19(4):417–425
12. McDermott V. Disability in the United Kingdom 2016 facts and figures. Papworth Trust. Available at https://www.papworthtrust.org.uk/about-us/publications/papworth-trust-disability-facts-and-figures-2018.pdf. Accessed 13 Aug 2020
13. Madden R, Collins P (2004) Children with disabilities in Australia. Australian Institute of Health and Welfare, Canberra. Available at https://www.aihw.gov.au/getmedia/a792fd34-61db-4f62-91f8-9aa8e41f8207/cda.pdf.aspx?inline=true. Accessed 13 Aug 2020
14. Blackburn CM, Spencer NJ, Read JM (2010) Prevalence of childhood disability and the characteristics and circumstances of disabled children in the UK: secondary analysis of the Family Resources Survey. BMC Pediatr 10(1):21
15. Banks J, Maître B, McCoy S (2015) Insights into the lives of children with disabilities: findings from the 2006 National Disability Survey. Economic and Social Research Institute (ESRI). Available at https://www.esri.ie/system/files?file=media/file-uploads/2015-08/BKMNEXT274.pdf. Accessed 13 Aug 2020

16. World Health Organization (2011) World report on disability 2011. World Health Organization. Available at https://www.who.int/disabilities/world_report/2011/en/index.html. Accessed 25 Mar 2018
17. Filmer D (2008) Disability, poverty, and schooling in developing countries: results from 14 household surveys. World Bank Econ Rev 22(1):141–163
18. Kuper H, Monteath-van Dok A, Wing K, Danquah L, Evans J, Zuurmond M et al (2014) The impact of disability on the lives of children; cross-sectional data including 8,900 children with disabilities and 898,834 children without disabilities across 30 countries. PLoS ONE 9(9):e107300
19. Ibrahim SH, Bhutta ZA (2013) Prevalence of early childhood disability in a rural district of Sind, Pakistan. Dev Med Child Neurol 55(4):357–363
20. VanLeit B, Channa S, Rithy P (2007) Children with disabilities in rural Cambodia: an examination of functional status and implications for service delivery. Asia Pac Disabil Rehabil J 18(2):33–48
21. Rischewski D, Kuper H, Atijosan O, Simms V, Jofret-Bonet M, Foster A et al (2008) Poverty and musculoskeletal impairment in Rwanda. Trans R Soc Trop Med Hyg 102(6):608–617
22. Fitaw Y, Boersma JM (2006) Prevalence and impact of disability in north-western Ethiopia. Disabil Rehabil 28(15):949–953
23. Morris A, Sharma G, Sonpal D (2005) Working towards inclusion: experiences with disability and PRA. General section, 5. Available at https://pubs.iied.org/pdfs/G02138.pdf. Accessed 13 Aug 2020
24. Srivastava DK, Khan JA, Pandey S, Pillai DS, Bhavsar AB (2014) Awareness and utilization of rehabilitation services among physically disabled people of rural population of a district of Uttar Pradesh, India. Int J Med Sci Public Health 3:1157–1160
25. Liu L, Oza S, Hogan D, Perin J, Rudan I, Lawn JE et al (2015) Global, regional, and national causes of child mortality in 2000–13, with projections to inform post-2015 priorities: an updated systematic analysis. Lancet 385(9966):430–440
26. Were WM, Daelmans B, Bhutta Z, Duke T, Bahl R, Boschi-Pinto C et al (2015) Children's health priorities and interventions. BMJ 351:h4300
27. Santosa A, Wall S, Fottrell E, Högberg U, Byass P (2014) The development and experience of epidemiological transition theory over four decades: a systematic review. Glob Health Action 7(1):23574
28. Perrin JM, Anderson LE, Van Cleave J (2014) The rise in chronic conditions among infants, children, and youth can be met with continued health system innovations. Health Aff 33(12):2099–2105
29. Modell B, Darlison MW, Malherbe H, Moorthie S, Blencowe H, Mahaini R et al (2018) Congenital disorders: epidemiological methods for answering calls for action. J Community Genet 9:335–340
30. Office of the Registrar General & Census Commissioner, India. Ministry of Home Affairs, Government of India. Census 2011. Available at https://censusindia.gov.in. Accessed 25 Mar 2018
31. National Sample Survey Organization, Ministry of Statistics and Programme Implementation, Government of India (2002) Disabled persons in India NSS 58th round (July–December 2002). Available at https://mospi.nic.in/nsso. Accessed 25 Mar 2018
32. Ministry of Statistics and Programme Implementation (2018) Persons with disabilities in India NSS 76th round, July to August 2018. Available at https://www.mospi.gov.in/sites/default/files/publication_reports/Report_583_Final_0.pdf. Accessed 26 May 2020
33. Mitra S, Sambamoorthi U (2006) Measurement of disability: disability estimates in India—what the census and NSS tell us. Econ Polit Wkly 41(38):23–29
34. Madans JH, Loeb ME, Altman BM (2011) Measuring disability and monitoring the UN Convention on the Rights of Persons with Disabilities: the work of the Washington Group on Disability Statistics. BMC Public Health 11(Suppl 4):S4
35. Human Resources and Skill Development, Canada (HRSDC) (2011) Disability in Canada: a 2006 profile. Available at https://www.hrsdc.gc.ca/eng/disability_issues/reports/disability_profile/2011/index.html. Accessed 15 May 2017

Part III
Prevention

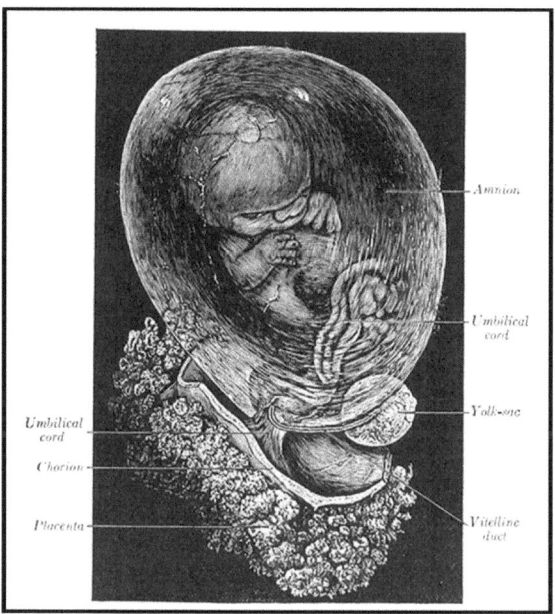

Foetus of about eight weeks, enclosed in the amnion. Magnified a little over two diameters. (Drawn from stereoscopic photographs lent by Prof. A. Thomson, Oxford.) Henry Gray (1918) Anatomy of the Human Body. Revised by Warren H. Lewis Anatomy of the Human Body Publisher Leah and Febiger Philadephia and New York 1918. Plate number 30 Available from https://commons.wikimedia.org/wiki/File:Gray30.png.

Prevention of birth defects is challenged by limited knowledge of the aetiology of most of these conditions. Majority of birth defects are however caused by gene-environmental interactions. The environmental component provides the opportunity for introducing public health interventions for prevention. The first article in this part (9. "Preventing Congenital Anomalies Through Existing Maternal Health Services in India") describes several modifiable factors that can be targeted through primary prevention activities of the existing maternal and child health programme in India. The second article in this part (10. "Neural Tube Defects and Folate Status in India") reviews the prevalence of folate insufficiency and the prevalence of neural tube defects in the country. The last article in this section (11. "Haemoglobinopathies: Genetic Services in India") discusses the national guidelines for prevention and control of haemoglobinopathies in India.

Chapter 9
Preventing Congenital Anomalies Through Existing Maternal and Child Health Services in India

Anita Kar, Prajkta Bhide, and Pooja Gund

Abstract Congenital anomalies (birth defects) affect 2–3% of pregnancies. In India, birth defects are responsible for over 50,000 neonatal deaths annually. Significant numbers of children survive with disabilities. Prevention has a major role in alleviating suffering and in lessening health and welfare service costs. Although congenital anomalies have a genetic component, several are caused by gene-environmental and environmental factors. These include maternal nutritional, health and lifestyle factors, and teratogenic factors such as infectious, chemical and physical exposures during pregnancy. Primary prevention activities target these environmental risks and exposures. The timing of health promotion messages is important, as risk factors need to be addressed prior to conception. Awareness on birth defect prevention and specific protection measures such as folate supplementation and rubella immunization need to target women in the preconception and inter-conception periods. Congenital anomalies are mostly prevented through secondary prevention activities. These involve detection of affected pregnancies, followed by elective termination of the pregnancy. Many of the risk factors for congenital anomalies are non-specific and shared with other adverse pregnancy outcomes like stillbirths, miscarriage, prematurity and low birth weight. A package of interventions targeting these shared risk factors can mainstream primary prevention of birth defects into general maternal health services. Studies to test the impact of primary and secondary prevention for congenital anomalies are scarce. The available data suggest that nearly 60% of birth defects can be prevented. The RMNCHA+ (Reproductive, Maternal, Neonatal, Child and Adolescent) programme in India has several opportunities for implementing services for prevention of birth defects. The existing programmatic deficiency is that maternal health services initiate in the antenatal period, so that interventions during the preconception period remain unaddressed. Currently, the RMNCHA+ lacks services to inform women about birth defects and their prevention.

A. Kar (✉) · P. Bhide
Birth Defects and Childhood Disability Research Centre, Pune 411020, India

P. Gund
Indian Institute of Public Health, Hyderabad 500033, India

© Springer Nature Singapore Pte Ltd. 2021
A. Kar (ed.), *Birth Defects in India*,
https://doi.org/10.1007/978-981-16-1554-2_9

Keywords Congenital malformations · Congenital anomalies · Prevention ·
Preconception · RMNCH + A · India

Background

Congenital anomalies are structural birth defects, where deviations in the process of
organogenesis cause defective development of anatomical structures of the foetus.
They include common malformations like congenital heart defects, orofacial clefts,
spina bifida, anorectal anomalies and a host of conditions, some of which can be
detected during the antenatal period. Most are diagnosed in the neonatal period, or in
early childhood [1, 2]. Several malformations are not compatible with life and result
in spontaneous abortions, stillbirths or early neonatal deaths. Majority of non-fatal
anomalies require immediate surgical interventions, or surgery during childhood
[3]. Survivors have disabilities, and several have disabling medical complications.
Medical care, physical and social rehabilitation services need to be instituted for care
of children and support of caregivers.

Congenital anomalies affect 2–3% of pregnancies and are the fourth largest cause
of neonatal deaths globally [4]. Congenital anomaly birth prevalence in India ranges
between 18.44 and 23.05 per 1000 births, implying that the country is likely to have
over 500,000 congenital anomaly affected births each year. Estimates of congenital anomaly neonatal mortality range from 54,000 to 99,000 (Chap. 6). Mortality,
morbidity, disability and the associated health care and welfare costs make prevention
of congenital anomalies a major public health responsibility.

Aetiology of Congenital Anomalies

Prevention activities target modifiable risk factors of diseases and disorders. The aetiology of 66–80% of congenital anomalies are unknown [5] (Table 9.1). However,
like all non-communicable conditions, several congenital anomalies are caused
by *gene-environmental* factors. Others are caused by *environmental* factors, and
wholly *genetic* factors. The term 'environment' refers to all physical, chemical and
biological factors external to the human host, including behaviours that are potentially modifiable [5, 6]. These environmental risk factors and exposures form the
target of public health activities [7]. It is noteworthy that measurement of environmental risk exposures are from industrialized settings, and not from resource-
constrained settings, where poor nutrition, environmental exposures and infections
are more prevalent.

There are differences in estimates of the genetic and environmental contribution
in the aetiology of congenital anomalies (Table 9.1). Stevenson and Hall [1] reported
that the aetiology of 40–60% of congenital anomalies is unknown. Genetic causes
(single gene mutations and chromosomal abnormalities) were associated with the

causation of 15–25% of congenital anomalies. Multifactorial (i.e. genetic and environmental) causation could be ascribed to 20–25% of congenital anomalies. Environmental factors were associated with 8–12% of congenital malformation-affected births. A study investigating the aetiology of 5504 birth defects occurring among 270,878 births to women registered with the Utah Birth Defects Network reported that the aetiology could not be identified for 80% of congenital anomalies [8]. Among the remaining 20% of cases for which an aetiology could be identified, majority were of genetic or chromosomal origin. About 5% of birth defects were of environmental origin, primarily poorly controlled, pre-existing diabetes.

The Modell Global Database of Congenital Disorders (MGDb) reported reference birth defect rates for congenital anomalies (Table 9.1). Estimates do not include conditions that are exclusively of environmental origin (e.g. anomalies associated with congenital infections) [9]. The MGDb estimates that anomalies associated with chromosomal and single-gene disorders account for 0.4% of anomalies, while environmental causes were responsible for 0.23% of anomalies. Data from the British Columbia Health Surveillance Registry estimated that the genetic aetiology of congenital anomalies was 0.79% [10] (Table 9.1). Despite these differences in estimates, all data indicate that malformations are mostly caused by gene-environmental interactions.

Teratogens

Teratogens are common environmental exposures that are known causes of congenital anomalies. A teratogen refers to an agent that alters the growth or structure of the developing embryo or foetus, thereby causing a malformation [11–13]. Rubella infection during pregnancy was the first teratogen to be identified. Congenital rubella syndrome presents with a triad of defects (cataracts, heart anomalies and deafness). The next teratogen to be identified was thalidomide. Its association with severe birth defects was recognized in 1961 and provided the first evidence that drugs could cross the placental barrier and affect foetal development (Chap. 3).

Teratogens are not causative, rather, teratogenic agents are associated with the risk of development of a congenital anomaly. The exposure dose and the duration of exposure, as well as the developmental stage at the time of exposure are important considerations (Box 9.1). There are critical periods of sensitivity to agents. Developing organ systems during this window of teratogenic exposure are usually affected. The embryonic period between implantation to around 60 days post-conception is the most vulnerable period as this is when organogenesis occurs. Exposures at this stage increases the risk of anomalies. The risk decreases significantly in the subsequent trimesters. Teratogenic effects are dependent on the chemical property of the teratogen, the route of exposure, maternal/foetal bioactivation, placental transport and other factors. Several of these characteristics are dependent on the maternal and foetal genotype. For example, embryonic exposure to cigarette smoke is a risk factor for orofacial clefts. The risk is enhanced in presence of the transforming growth factor

Table 9.1 Aetiology of congenital anomalies: genetic and environmental factors

Stevenson and Hall [1]	Causes of anomalies in liveborn infants (%)
Unknown	40–60
Genetic	15–25
Chromosome	10–15
Single gene	2–10
Multifactorial	20–25
Environmental	8–12
Maternal diseases	6–8
Uterine/placental	2–3
Drugs/chemicals	0.5–1
Twinning	0.5–1
Feldkamp et al. [8]	5504 birth defects among 270,878 (2%) births to women resident in Utah
Unknown	79.8% (4390)
Definite cause	20.2% (1114 out of 5504)
Genetic causes Chromosomal or genetic cause	94.4% (1052 out of 1114)
Family history (similarly affected first degree relative)	4.8% (266)
Non-genetic causes Teratogens (mostly poorly controlled pregestational diabetes)	4.1% (46)
Twinning (conjoined or acardiac twins)	1.4% (16)
Moorthie et al. [9]	Global reference estimates from the Modell Database of Congenital Disorders (MGDb)
Congenital anomalies	27.0/1000
Genetic (Anomalies associated with chromosomal and single gene disorders such as Down syndrome, microdeletion and genetic syndromes)	4.0/1000
Non-genetic	23.0/1000
Environmental (anomalies primarily attributed to environmental factors like congenital rubella syndrome and foetal alcohol syndrome)	0.32/1000
Miscellaneous: association with rare syndromes or other conditions e.g. craniosynostosis, twins	1.5/1000
Non-syndromic, i.e. anomalies with multi-factorial or unknown primary cause	21.2/1000
Isolated anomalies	19.8/1000

(continued)

Table 9.1 (continued)

Stevenson and Hall [1]	Causes of anomalies in liveborn infants (%)
Multiple anomalies	1.4/1000
Baird et al. [10]	British Columbia Health Surveillance Registry, population based registry with multiple sources of ascertainment, with the aim of estimating the population load from genetic disease; population born in British Columbia between 1952 and 1983
Single-gene disorders	3.6/1000
Autosomal dominant	1.4/1000
Autosomal recessive	1.7/1000
X-linked recessive disorders	0.5/1000
Chromosomal anomalies	1.8/1000
Multifactorial disorders	46.4/1000
Precise mechanism unknown	1.2/1000
Congenital anomalies with genetic aetiology	79/1000

α (*TGFA*) gene variant, or a nitric oxide synthase (*NOS3*) gene variant. The effect of the teratogen may range from no observable effect to total lethality. Awareness about teratogens and avoidance of teratogenic exposure form an important component of birth defects prevention communication.

Box 9.1 Wilson's six principles of teratology [12]

1. Teratogens affect normal development, and cause mortality, medical complications and malformations.
2. Teratogens affect the regulation of genetic pathways determining developmental events.
3. Teratogens exert their effects in a dose-dependent manner. The effect ranges from no effect to lethality.
4. The chemical properties of the teratogen and its route of exposure determine the teratogenic effect.
5. The maternal and foetal genotype and the species characteristics determines the sensitivity to the teratogen.
6. The developmental stage at the time of exposure determines the organ systems that are affected.

Environmental Risk Factors for Congenital Anomalies

Other environmental risk factors include teratogenic medications during pregnancy, behavioural factors such as smoking or alcohol intake during pregnancy, nutritional factors, maternal infections and chronic health conditions, physical agents including ionizing radiation (such as X-rays), environmental chemical exposures, maternal reproductive history, and socio-demographic factors [5, 14]. It is important to note that establishing associations between exposures/risk factors and congenital anomalies is challenged by the rarity of the conditions, and by issues relating to categorization of the anomaly and measurement of exposure. Methods of case ascertainment and categorization, challenges in measuring dose and duration of exposures, and confounding factors such as diet, lifestyle and socio-economic factors influence data on the strength of association [15]. Qualitative and quantitative synthesis of data using systematic reviews and meta-analyses are alternative methodological approaches in understanding the strengths and uncertainties in associations between environmental factors and congenital anomalies.

1.   Teratogenic medication

Several types of medications are teratogenic [16]. These include antiepileptic drugs, folate antimetabolites, antiblastic agents, anticoagulants, retinoic acid derivatives, ACE inhibitors and others (Table 9.2).

Antiepileptic drugs are known to be strongly teratogenic. A systematic review and meta-analysis of 96 studies that included 58,461 patients identified a higher risk of congenital anomalies as compared to controls. The odds of congenital malformation were 3.04 for ethosuximide, 2.93 for valproate, 1.9 for topiramate, 1.83 for phenobarbital, 1.67 for phenytoin, 1.37 for carbamazepine, but appeared relatively safer for newer generation anti-epileptic drugs like lamotrigine (OR 0.96) and levetiracetam (OR 0.72) [17].

Antifolate metabolites are also associated with increased risk of congenital anomalies. Among 527 women exposed to at least one folic acid antagonist in the first trimester of pregnancy (349 exposed to dihydrofolate reductase inhibitors, 346 to trimethoprim/sulfamethoxazole, two to methotrexate and one to sulfasalazine) and 179 to one or more 'other' folic acid antagonists (112 to carbamazepine, 35 to valproic acid, 21 to phenobarbital, 14 to phenytoin, eight to lamotrigine and one to primidone and cholestyramine), the odds of congenital anomalies increased by 2.43. The odds for neural tube defects was 6.5, and that for cardiovascular defects was increased by 1.76 [18].

Thalidomide, a strongly teratogenic drug, is associated with severe disabilities. Thalidomide embryopathy affects several organ systems, but its most disabling presentation is phocomelia, a striking limb deformity [19]. The drug was banned in 1961, but re-licenced for management of leprosy reactions and multiple myeloma. Reports of thalidomide embryopathy from Brazil were caused by unregulated use of thalidomide for management of type II leprosy reactions [20]. Prenatal misoprostol exposure is associated with an increased risk of congenital anomalies namely

Table 9.2 Some common teratogenic drugs

Antihypertensive	ACE inhibitors	Renal dysplasia
Anti-inflammatory	Chloroquine (higher doses 250–500 mg/day)	Chorioretinitis, deafness
Hormonal	Diethylstilbestrol	Uterine anomalies, vaginal adenocarcinoma
Anti-acne	Retinoids	Ear and eye defects, hydrocephalus
Antibiotics	Streptomycin	Deafness
	Tetracycline	Dental enamel hypoplasia
Antinauseant	Thalidomide	Phocomelia, cardiac and ear abnormalities
Antiepileptic	Lithium	Elevated risk of major congenital anomalies
	Phenytoin phenobarbital	Impaired growth, motor development, mortality; foetal phenytoin syndrome
	Valproic acid	Neural tube defects, cardiac defects, foetal valproate syndrome with characteristic facies, limb abnormalities, lip/cleft palate, urinary tract defects
Anticoagulants	Warfarin	Nasal hypoplasia, stippled calcification of epiphyses, risk of foetal Warfarin syndrome high between 6 and 9 weeks of gestation
Antineoplastic		Increased risk of major congenital anomalies

Möbius sequence (OR 25.31, 95% CI 11.11–57.66) and terminal transverse limb defects (OR 11.86, 95% CI 4.86–28.90) [21]. There is uncertainty in the data on opioid use and congenital anomalies due to variabilities of study design, and study quality and measurement issues [22]. Studies have documented associations with oral clefts, ventricular septal defects, atrial septal defects and clubfoot.

Associations between antibiotic use and congenital anomalies were established in a large population-wide cohort study in Denmark. The study did not find any increased risk of congenital anomalies among women with history of first-trimester use of ten commonly prescribed antibiotics (doxycyclin, amoxicillin, pivmecillinam, dicloxacillin, sulfamethizole, erythromycin, roxithromycin, azithromycin, ciprofloxacin, nitrofurantoin) [23]. Exposure to macrolide antibiotics during the first trimester of pregnancy was associated with an increased risk of major anomalies, including cardiovascular anomalies. Exposure in any trimester of pregnancy was associated with an increased risk of genital anomalies [24].

Maternal hyperthyroidism is treated with antithyroid drugs which are thionamides that block the synthesis of thyroid hormones. The available drugs are methimazole, carbimazole and propylthiouracil. From the early 1970s, use of methimazole in early pregnancy has been associated with birth defects like aplasia cutis, choanal atresia, oesophageal atresia, omphalocele and omphalomesenteric duct anomalies [25]. The teratogenic effects of antithyroid medication and birth defects have been corroborated in a large Danish study that identified that maternal thyroid dysfunction in early pregnancy per se was not responsible for birth defects. As compared to the risk of birth defects in the non-exposed group (6.7%), the risk of anomalies was 9.6% in the methimazole/carbimazole group, and 8.3% in women managed with propylthiouracil [25, 26].

The use of asthma and allergy medication during pregnancy or asthma during pregnancy is not associated with an increased risk of congenital malformation [27]. Further information on teratogenicity of drugs is available from the EUROMEDICAT (Euromedicat: Safety of medication use in pregnancy in relation to risk of congenital anomalies) [28].

2. Behavioural factors

Active smoking and alcohol consumption during pregnancy are risk factors for several types of common congenital anomalies. A systematic review of 172 studies that included 173,687 cases and 11,674,332 controls identified that smoking was associated with an increased risk for anomalies of the cardiovascular, musculoskeletal and gastrointestinal systems, orofacial clefts and cryptorchidism. There was a reduced risk for hypospadias and skin defects. There was no association with genitourinary, respiratory or central nervous system anomalies [29]. Foetal alcohol syndrome, a birth defect where children are born with microcephaly, facial dysmorphism and developmental delay, is strongly associated with alcohol intake during pregnancy. Data indicate a fourfold increase in birth defects associated with alcohol exposure during the first trimester [30].

3. Nutritional factors

Maternal folate deficiency is one of the major risk factors for neural tube defects. There is a clear relationship between maternal early pregnancy folate levels and risk of neural tube defects, with studies indicating that inadequate intake of folic acid increases the risk of neural tube defects in the offspring by two- to eightfold [31–35]. Folate-rich diet and periconceptional folate supplementation has been associated with reduction of neural tube defects and remains a very important component of a birth defects prevention programme.

Maternal obesity is associated with various adverse maternal and foetal outcomes. Compared to women with healthy weights, overweight and obese women are at a higher risk of developing pregnancy complications. Maternal overweight and obesity increase the risk of several types of anomalies, like CHDs, neural tube defects and orofacial clefts [36].

4. Maternal infections

Congenital TORCH infections which include toxoplasmosis, rubella, cytomegalovirus, herpes simplex, and other infections such as syphilis, varicellazoster, measles virus, human immunodeficiency virus, and recently Zika virus are associated with severe congenital anomalies, especially if the infection occurs prior to the 9th week of pregnancy. Cytomegalovirus (CMV) infection is a major cause of congenital hearing loss. Congenital rubella syndrome (CRS) causes intrauterine growth restriction, intracranial calcifications, microcephaly, cataracts, cardiac anomalies and neurologic disease. Congenital varicella syndrome (CVS) results in spontaneous abortion, chorioretinitis, cataract, limb atrophy, cerebral cortical atrophy and/or neurological disability. Zika virus infection in early pregnancy is associated with congenital Zika syndrome (CZS), characterized by microcephaly and neurological damage.

5. **Maternal chronic conditions**

Poorly controlled pregestational diabetes is associated with a two- to ninefold increase in the risk of cardiac anomalies, neural tube defects and other anomalies. There is a proportional relationship with the degree of maternal hyperglycaemia and congenital anomaly risk [37]. Maternal hypertension, irrespective of treatment status, increases the risk of cardiac anomalies by almost two times (RR 1.8; 95% CI 1.5–2.2). This association was observed for both treated (RR 2.0, 95% CI 1.5–2.7) and untreated hypertension (RR 1.4, 95% CI 1.2–1.7) [38]. Maternal asthma is associated with an increased risk of congenital anomalies (RR 1.11, 95% CI 1.02–1.21) and specifically cleft lip with or without cleft palate (RR 1.30, 95% CI 1.01–1.68) [39]. Maternal phenylketonuria also increases the risk of congenital anomalies like cardiac anomalies and microcephaly, with the risk increasing with higher phenylalanine levels. Strict dietary compliance to a phenylalanine restricted diet initiated before conception can prevent adverse outcomes [40].

6. **Physical exposures**

Fever and hyperthermia in the first trimester have been associated with an increased risk of cardiac anomalies, neural tube defects, orofacial clefts, and renal anomalies. However, a large Danish cohort study did not detect any association between fever in the first trimester and risk of congenital anomalies among live births. Stillbirths and pregnancy terminations were not included and this might have biased the results of the study [41]. Ionizing radiation such as X-rays has teratogenic and mutagenic effects. Effects of in utero exposure depend on the amount of exposure and stage of foetal development.

7. **Environmental chemical exposures**

Various studies have noted a positive association between pesticide exposure and congenital anomalies [42]. An extensive review of epidemiological studies that have examined the association between various environmental chemical pollutants such as drinking water contaminants (heavy metals and nitrates, chlorinated and aromatic solvents, and chlorination by-products), residence near waste disposal sites and contaminated land, pesticide exposure in agricultural areas, air pollution and industrial pollution sources, food contamination, and chemical disasters however was unable to establish strong associations [43]. Maternal exposure to air pollution during pregnancy has been associated with increase in the risk of specific cardiac anomaly subtypes. Increasing nitrogen dioxide (NO_2) and sulphur dioxide (SO_2) concentrations was associated with an increased risk of coarctation of the aorta, and Tetralogy of Fallot, and increased risk of atrial septal defects for PM_{10} [44]. Chlorination disinfection by-products were reported to be associated with an increased risk of some types of congenital anomalies [45, 46]. Some teratogenic exposures have very obvious manifestations. For example, foetal methylmercury syndrome (Minamata disease) was recognized in Japan in the 1950s as a cluster of cases of cerebral palsy and microcephaly. It is associated with maternal exposure to methylmercury. In most cases, epidemiological studies to determine associations between exposures and congenital anomalies remain challenging. The need for careful study designs and meta-analysis of published studies become important in deciding how the findings can be translated into public health measures [47].

8.   Maternal reproductive history

Maternal gravidity is a significant risk factor for some congenital anomalies. Data from the Polish Registry of Congenital Anomalies identified a relationship between gravidity and higher odds of congenital heart defects (OR = 1.22, 95% CI 1.09, 1.36), cleft lip with or without cleft palate (OR = 1.21, 95% CI 1.09, 1.36) [48]. History of pregnancy loss has been associated with an increased risk of congenital anomalies. In a population-based case–control study, women with previous pregnancy loss were at higher odds of having a congenital anomaly affected child, with the risk of serious birth defects increasing from 2.5% for women with no prior pregnancy loss to 4.2% for women with three or more pregnancy losses [49]. This increased risk of congenital anomalies among women experiencing pregnancy loss was not observed in the Polish study [48].

Birth defects are associated with a high recurrence risk. In a population-based study in Norway, mothers of infants with birth defects were two and a half times more likely to have a second infant with any defect. The increased risk of the same defect occurring in the second child (RR 7.6, 95% CI 6.5–8.8) was higher than that of a different defect occurring (RR 1.5, 95% CI 1.3–1.7) in the child [50]. Multiple births is another risk factor, with the risk of congenital anomalies increasing by as much as 50% among multiple births [51].

9.   Socio-demographic factors

Maternal age is a risk factor for birth defects. Down syndrome (a chromosomal anomaly) birth prevalence, for example, shows a J-shaped curve when plotted by maternal age. The risk increases from 1/1400 births at age 20–24 to 1/25 births in women above 45 years of age. The rates of all cytogenetic abnormalities rise from 2 per 1000 at younger maternal ages, to 2.6 per 1000 at age 30, 5.6 per 1000 at age 35 to 15.8 per 1000 at age 40 and 53.7 per 1000 at age 45 [52]. Structural anomalies like neural tube defects, (OR 1.81, 95% CI 1.30–2.52), orofacial clefts (OR 1.88, 95% CI 1.30–2.73) and female genital defects (OR 1.57, 95% CI 1.12–2.19) were more common among teenage mothers (14–19 years), while older mothers (35–40 years) were at an increased risk of having an offspring with cardiac anomalies (OR 1.12,

95% CI 1.03–1.22) [53]. The evidence for paternal age as a risk factor for congenital anomalies is inconclusive.

Certain maternal and paternal occupations are associated with an increased risk of congenital anomalies in the offspring. Maternal occupations such as janitors/cleaners, scientists and electronic equipment operators were significantly associated with one or more birth defects in the National Birth Defects Prevention Study [54]. Even in industrialized countries, socioeconomic disadvantage increases the risk of congenital anomalies [55].

Consanguineous unions increase the risk of congenital heart defects, with the risk being highest for first cousin unions. Consanguinity also increases the risk of recurrence of anomalies. The risk of recurrence is higher for siblings of children born to first cousin parents compared to those of non-consanguineous unions. Among first cousin unions, the risk of a sibling of an affected child was almost double (recurrence risk 6.8%) compared to infants of consanguineous unions, where the older sibling did not have a birth defect. This can be attributed to the increased homozygosity among offspring of consanguineous unions [56, 57].

Shared Risk Factors with Other Adverse Pregnancy Outcomes

Several of the risk factors for congenital anomalies are shared with other adverse foetal pregnancy outcomes (APOs) like low birth weight (LBW), preterm births, stillbirths and miscarriage (Fig. 9.1). Shared risk factors provide the opportunity for using a package of interventions to target all these conditions. It is important

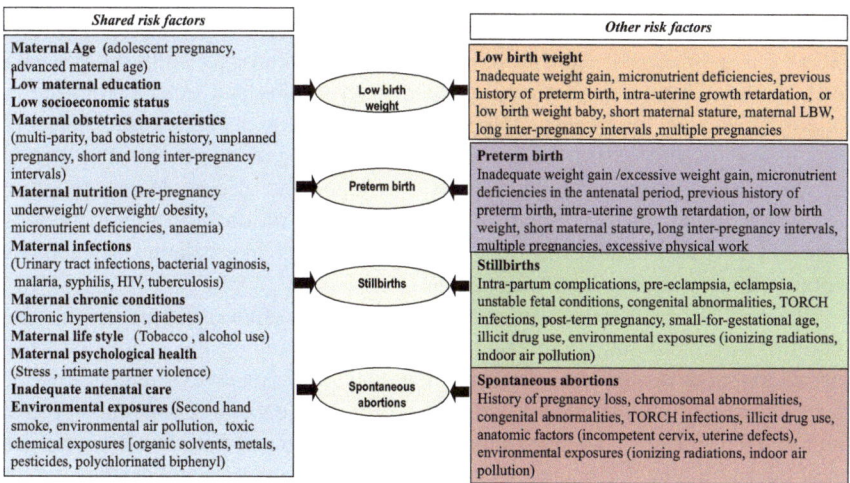

Fig. 9.1 Shared risk factors for APOs

to note that several of these interventions have to be implemented in the preconception period, that is before conception [58]. In India and low- and middle-income countries (LMICs) in general, interventions usually initiate once the woman presents for antenatal care, so that the benefits of preconception interventions do not reach beneficiaries.

Universally known *preconception* nutritional risk factors associated with APOs are anaemia and low birth weight, prepregnancy underweight and preterm birth, prepregnancy overweight and pre-eclampsia, and preconception folic acid use and reduced risk of neural tube defects [59]. Maternal nutrition in the antenatal period is also an important determinant of APOs [60]. Overweight and obesity are associated with several complications, including congenital anomalies. Correcting iron-deficiency anaemia, and folate deficiency reduces the risk of several adverse pregnancy outcomes, including the risk of neural tube defects [61, 62].

Prevention and management of maternal infections and morbidities before conception is critically important for a healthy pregnancy outcome [63]. Maternal infections (including TORCH infections) are known risk factor for poor pregnancy outcomes. In addition, sexually transmitted diseases (syphilis, bacterial vaginosis) are associated with pregnancy loss and foetal death. Uncontrolled and untreated chronic illnesses (pre-gestational diabetes, chronic hypertension, asthma) during pregnancy are known risk factors for poor maternal and child health outcomes [64].

Smoking [65] and alcohol use [66] in the preconception and antenatal period are known determinants of APOs. Tobacco use, in the form of smoking during and into pregnancy is highly prevalent among women of reproductive age in industrialized countries, while smokeless tobacco use is prevalent among women of reproductive age group in LMICs [67]. Smoking during pregnancy increases the risk of stillbirths, while periconception alcohol consumption was associated with an increased risk of small for gestational age babies and preterm births [67].

Preconception interventions are only possible if pregnancies are planned. Ensuring access to family planning services is important, as advanced maternal age (age >35 yrs) is a strong determinant of APO such as miscarriage, stillbirth, perinatal death, preterm birth, low birth weight, Down syndrome and other chromosomal abnormalities, and pregnancy complications including pre-eclampsia, placental abruption and foetal growth restriction [68]. Providing access to family planning services not only prevent pregnancies at older ages, but also adolescent pregnancies, which are known to significantly increase the risk of low birth weight, preterm birth and neonatal death, and maternal complications like eclampsia, puerperal endometritis, and systemic infections [69].

Socioeconomic inequalities interact with other risk factors and aggravate the likelihood of adverse maternal and foetal outcomes, continuing the vicious circle of poverty and poor health outcomes. Low maternal education level is associated with increased risk of stillbirths and perinatal mortality. Low socioeconomic status is a major determinant of adverse outcomes and is associated with low birth weight, small for gestational age babies, spontaneous abortion and neonatal deaths [70].

Prevalence of Risk Factors for Congenital Anomalies and APOs in Indian Women

Data on the prevalence of risk factors for congenital anomalies are available from different sources. In 2015–16, the National Family Health Survey 4 (NFHS-4) reported data from 572,000 households from 640 districts across India [71]. The survey data was collected through personal interviews of 699,686 women between the ages of 15 and 49 years, and 122,051 men aged 15–54 years [72]. The survey identified that 21% of women were overweight or obese (Body Mass Index (BMI) $\geq$ 25.0 kg/m^2), while 23% women had BMI below normal (<18.5 kg/m^2). High blood sugar level (>140 mg/dl) was present in 5.8%, and 2.8% women had very high (>160 mg/dl) blood sugar levels. Nearly 7% women had slightly above normal blood pressure (systolic 140–159 mm of Hg and/or diastolic 90–99 mm of Hg), 1.4% had moderately high (systolic 160–179 mm of Hg and/or diastolic 100–109 mm of Hg), and 0.7% women had very high (systolic $\geq$ 180 mm of Hg and/or diastolic $\geq$ 110 mm of Hg) blood pressure. The prevalence of anaemia was 53% among non-pregnant women (15–49 years) and 50% among pregnant women in 2015–2016. About 12% women were married to a blood relative with 9% being first cousin marriages [72].

Studies have reported the prevalence of specific risk factors for congenital anomalies. A study reported 3.8% prevalence of pregestational diabetes, with the main risk factors being age more than 25 years, increased Body Mass Index, family history of diabetes, history of stillbirth, gestational diabetes in the previous pregnancy and thyroid disease [73]. A cross-sectional study identified 13% prevalence of hypothyroidism among 2599 pregnant women in the first trimester of pregnancy [74]. The prevalence of TORCH infection was high in women with bad obstetric history. In an analysis of 380 serum samples of pregnant women with bad obstetric history, there was 27% rubella, 11% Toxoplasma, 8% CMV and 4% HSV II IgM positivity [71]. Data from five sentinel sites located in urban areas, but catering to district population identified rubella as a persistent problem in India. During 2016–18, the sentinel sites enrolled 645 patients with suspected congenital rubella syndrome, of which 21% were laboratory confirmed cases. Congenital heart defects were identified in 79% patients, eye signs were present in 60% and hearing impairment among 39% patients [75].

The practice of self-prescription and over the counter (OTC) drug use is also common in India. Among 483 women enrolled in a study, 13% of women reported OTC drug use. Highest use was in the first (39%) and second (44%) trimesters, and among urban women from higher economic strata. The OTC drug was acquired mostly by describing the symptoms to the pharmacist. Most women were unaware of the teratogenic effects of drugs [76]. Another study examined the prescriptions received by 1163 women attending an antenatal clinic and reported that Category A drugs were most frequently prescribed (76%), followed by Category B (17.5%), C (5%), D (2%) and Category X (0.17%). The latter drugs were more commonly prescribed in in-patient settings [77].

A study that specifically examined periconceptional risk factors for birth defects and other adverse foetal outcomes among 2107 women in early pregnancy identified that one-fifth of women (18%) were below 19 years of age, while only 13 women (0.6%) were above the age of 35 years [78]. Twenty percent women reported consanguineous marriage. Smokeless tobacco (SLT) use was 7% in the preconception period, of which 4% continued SLT use into pregnancy. In terms of nutritional risk factors, 27% were underweight, while 14% were overweight or obese. Only seven (0.3%) women reported using preconception folic acid. These were prescribed by doctors when the women had presented for some other complaint [78].

Prevalence of any maternal acute illness during the periconception period was 8%. Only 2% reported chronic conditions, of which thyroid disorders were the highest, affecting 26 women. Other chronic conditions included five women with clinically diagnosed epilepsy and two women with psychiatric disorders. Three women reported a congenital condition (two women with congenital heart disease, while one woman had a diagnosis of spina bifida). In addition to this, three women were beta thalassemia carriers. Of 28 women who reported using non-allopathic medicines, 26 reported using ayurvedic 'tonics'. Two percent of the women reported using OTC analgesics [78].

Exposure to occupational and household chemical exposures was high, as 17% women reported participating in agricultural work prior to marriage, involving direct contact with pesticides and fertilizers. Seventy-six percent reported exposure to household cleaners and 64% reported exposure to insect repellents. Women from families below the poverty line (OR 1.3, 95% CI 1.0–1.6) and with low education levels (OR 1.4, 95% CI 1.1–1.6) were more likely to report five or more risk factors [78].

Prevention

Disease prevention activities are implemented at different phases of the natural history of a disease or disorder (Fig. 9.2). Primary prevention is implemented prior to disease onset, with the objective of removing/managing the risk factors of disease. Primary prevention measures for birth defects have to be implemented *before pregnancy*, in the preconception period. These activities involve addressing the modifiable risk factors for congenital disorders, using methods of health promotion (providing women knowledge on prevention of birth defects), or specific protection such as folic acid supplementation and rubella immunization.

For implementing primary prevention services for birth defects, it is imperative to enhance knowledge about pregnancy planning, so that there is time to correct/manage preconception risk factors (such as folate deficiency). Required health service activities include strengthening of family planning services and improving awareness and access to contraceptives, so that pregnancy planning is possible. Education on

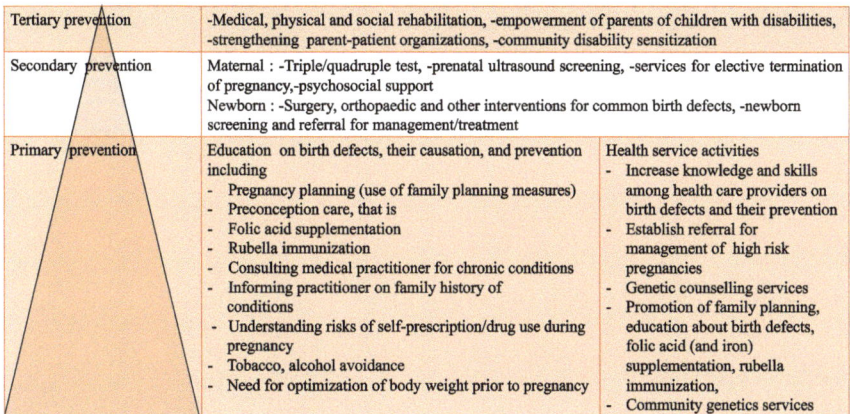

Tertiary prevention	-Medical, physical and social rehabilitation, -empowerment of parents of children with disabilities, -strengthening parent-patient organizations, -community disability sensitization	
Secondary prevention	Maternal : -Triple/quadruple test, -prenatal ultrasound screening, -services for elective termination of pregnancy,-psychosocial support Newborn : -Surgery, orthopaedic and other interventions for common birth defects, -newborn screening and referral for management/treatment	
Primary prevention	Education on birth defects, their causation, and prevention including - Pregnancy planning (use of family planning measures) - Preconception care, that is - Folic acid supplementation - Rubella immunization - Consulting medical practitioner for chronic conditions - Informing practitioner on family history of conditions - Understanding risks of self-prescription/drug use during pregnancy - Tobacco, alcohol avoidance - Need for optimization of body weight prior to pregnancy	Health service activities - Increase knowledge and skills among health care providers on birth defects and their prevention - Establish referral for management of high risk pregnancies - Genetic counselling services - Promotion of family planning, education about birth defects, folic acid (and iron) supplementation, rubella immunization, - Community genetics services

Fig. 9.2 Primary secondary and tertiary prevention of congenital anomalies and their complications

common birth defects, their causes, risk factors and exposures is necessary. Important birth defects prevention messages would include information on rubella vaccination, folic acid supplementation, tobacco and alcohol cessation prior to conception, avoidance of occupational and household teratogenic exposures, compulsory preconception check-up especially for women with diagnosed medical conditions and on prescription medications, women above 35 years of age, women with history of pregnancy loss, or a history of a birth defect-affected pregnancy or a family history of a disorder (Fig. 9.2).

Birth defect prevention messages can be incorporated into a larger package of messages aimed at prevention of several other APOs. These messages would include awareness about early pregnancy registration and full antenatal care, increasing knowledge about diet during pregnancy, iron and folic acid supplementation [79], healthy lifestyle and personal hygiene, and the benefits of achieving optimal weight prior to conception. For underweight women, enrolling in supplemental feeding programmes would address the issue of underweight.

Secondary prevention for congenital anomalies is the most effective method of prevention, but fraught with ethical dilemmas. Public health activities focus on screening and early detection of high-risk pregnancies (Fig. 9.2). Screening for birth defects includes serum screening and ultrasonographic examination during pregnancy. Table 9.3 lists the serum screening tests typically used for screening for Down syndrome [80]. First trimester testing is offered between 10 and 13 weeks 6 days, and combines serum screening for two markers (free beta–human chorionic gonadotropin (hCG) and pregnancy-associated plasma protein A) in combination with ultrasonographic examination for nuchal translucency (ultrasonographic detection of collection of fluid under the skin behind the foetal neck). Risk is estimated using a combination of measures using maternal age, past pregnancy history, number of foetuses in the current gestation, weight, race, serum markers and nuchal

Table 9.3 Screening for aneuploidy [80]

Screening test	Gestational age at screening (in weeks)	Detection rate for trisomy 21 (%)	Analytes and/or measurements
First-trimester screen	10–13	82–87	Nuchal translucency Papp-A hCG
Triple screen	15–22	69	5hCG AFP uE3
Quad screen	15–22	81	hCG AFP uE3 DIA
Cell-free DNA	Any age after 9–10 weeks	99	Molecular evaluation of cell-free foetal DNA within maternal serum

AFP alpha-fetoprotein; *DIA* dimeric inhibin A; *hCG* human chorionic gonadotropin; *Papp-A* pregnancy-associated plasma protein A; *uE3* unconjugated estriol

translucency measurement. The Quad screen is offered between 15 and 22 weeks of gestation. It measures four markers (hCG, alpha-fetoprotein (AFP), inhibin A, and unconjugated estriol). Risk is estimated by combining these measurements with patients' age, race, weight, number of foetuses in the current gestation, diabetes status, and gestational age. Cell-free DNA, referred to as non-invasive prenatal test (NIPT) screens for risk using foetal DNA in maternal serum [81]. Ultrasonographic examination remains the most widely used test. Screening for congenital anomalies typically involves trans-abdominal ultrasonography between 18 and 23 weeks of pregnancy.

Diagnostic testing is done to confirm a specific disorder. It is routinely offered to women above 35 years of age, those with a history of a genetic disorder or a congenital anomaly-affected pregnancy. Chorionic villus sampling is conducted after 9 weeks of gestation. Amniocentesis is done after 15 weeks of gestation. The retrieved samples are tested using a range of cytogenetic and molecular tests [82]. Non-directive counselling is required for assisting parents in making the difficult decision of carrying the pregnancy to term, or for termination of the pregnancy.

Tertiary prevention measures are directed at newborns with the goal of reducing disabilities or complications of disease. Surgical services for newborns or infants can correct/address several congenital malformations and reduce mortality. Services for early detection and referral for surgical care for congenital heart defects can reduce morbidity and mortality. Orthopaedic interventions can correct conditions like clubfoot that can cause movement difficulties if untreated. Medical services and rehabilitation have been discussed elsewhere in the book (Chap. 12). Newborn screening (NBS) aims at presymptomatic detection of newborns in order to treat and reduce the complications of congenital conditions. Universal newborn hearing screening is an important strategy for early identification of children with hearing impairment. Early intervention has shown to improve language outcomes and school performance [83]. NBS screening panels are available for common conditions like phenylketonuria, congenital hypothyroidism, congenital adrenal hypoplasia, cystic fibrosis and others [84].

Effectiveness of Preventive Measures

Although primary prevention for congenital anomalies has been advocated [85, 86], the data on the effectiveness of strategies to prevent birth defects is scarce. Lassi et al. estimated a 30% risk reduction for congenital malformations (RR 0.30 95% CI, 0.22–0.41) and perinatal mortality (RR 0.3 95% CI, 0.19–0.53) in women who received preconception care (preconception counselling, preconception glycaemic control) [87].

A single study attempted to measure the proportion of birth defects that can be prevented using primary and secondary interventions [88, 89]. Using data from the Hungarian registry, the authors reported that primary prevention (referral of families with known single gene disorders for genetic counselling, care of high risk pregnancies, rubella immunization and alcohol avoidance) could reduce around 41% of congenital anomalies. Secondary prevention was more effective in either prevention or early correction of anomalies. Secondary prevention involved prenatal ultrasonographic examination and pregnancy termination, neonatal screening and specific surgical treatments. The authors concluded that nearly 60% of congenital anomalies can be prevented. In their analysis, secondary prevention strategies were more realistic than primary prevention.

Opportunities for Prevention in Public Health Programmes: RMNCHA+ Programme in India

Maternal health services in India are offered through the RMNCH + A (Reproductive, Maternal, Neonatal, Child and Adolescent Health) programme [90]. The RMNCH + A strategy proposes a continuum of care approach (designated by the 'Plus') that links maternal and child health services, to adolescent services, family planning and nutrition-based interventions. Services are either home, community or facility based. The components of the RMNCH + A strategy are listed in Table 9.4. The focus on a continuum of care services, provides the opportunity to introduce birth defects prevention services as a package of preconception and antenatal services. However, preconception services are yet to be initiated in India. There are at present no messages to alert women about congenital anomalies and methods of prevention.

Figure 9.3 shows the measures that need to be incorporated into the RMNCH +

Table 9.4 Key components of the RMNCH + A strategy

Reproductive	1. Focus on birth spacing	2. Ensuring access to pregnancy test kits	3. Strengthening abortion care services	4. Maintaining quality of female (tubal ligation) and male sterilization (vasectomy) services	5. Home delivery of contraceptives
Maternal health	1. Early registration, ensuring all ANC visits	2. Detection of high-risk pregnancies, including severely anaemic women, appropriate management	3. Well-trained staff at delivery points	4. Maternal, infant and child death review for corrective measures	5. Identification of villages with high home births
Newborn health	1. Early initiation of exclusive breastfeeding	2. Home-based newborn care	3. Essential newborn care and resuscitation service at all delivery points	4. Special newborn care units	5. Community-level dispensation of gentamycin for women nearing labour
Child health	1. Complementary feeding, IFA supplementation	2. Diarrhoea management with ORS and zinc nutrition	3. Management of pneumonia	4. Full immunization coverage	5. RBSK, community-based screening and referral for management of children with diseases, nutritional deficiencies, birth defects and developmental delays and disabilities
Adolescent health	1. Address teenage pregnancy, increase contraceptive use	2. Community-based adolescent health services through peer educators	3. Strengthening of adolescent reproductive and sexual health (ARSH) clinics	4. National Iron plus programme, implemented as WIFS (weekly iron and folic acid supplementation)	5. Promotion of menstrual hygiene

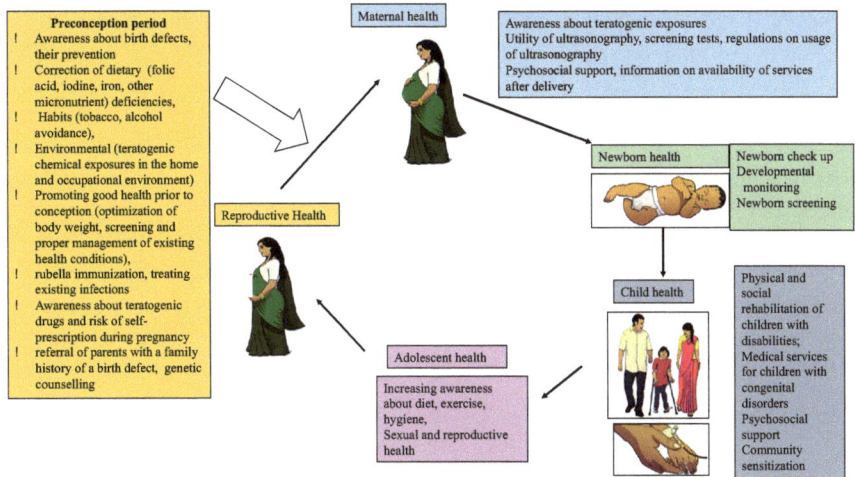

Fig. 9.3 **Birth defect prevention activities in maternal and child health services**

A service for prevention of birth defects. Firstly, there is need to initiate messaging about preparing for pregnancy, by targeting women in the preconception and inter-conception period. Messages should increase awareness about common birth defects, and their prevention, including correction of dietary (folic acid, iodine, iron, other micronutrient) deficiencies, messaging on habits (tobacco, alcohol avoidance, dangers of self-medication during pregnancy), environmental risk factors (terato-genic chemical exposures in the home and occupational environment), promoting good health prior to conception (optimization of body weight, consulting a medical practitioner while planning a pregnancy, proper management of existing health condi-tions and rubella immunization). Sensitive messaging, psychosocial support in case of an adverse ultrasonography report and empowerment of parents with information on availability of services after delivery are important. It is imperative to remember that counselling should be non-directive, so that parents have an opportunity to independently decide on whether to continue or terminate the pregnancy.

It is equally important to introduce the practice of thorough examination of the newborn as proposed in the Rashtriya Bal Swasthya Karyakram (Chap. 12). Newborn screening, for example screening for hearing impairment and developmental moni-toring, can lead to early identification of newborns with birth defects and devel-opmental disabilities, providing the opportunity for early intervention. Low-cost newborn screening services can be implemented, as has been done in the states of Goa, Kerala and Chandigarh where the infant mortality rate has reduced to less than 10. The major impediment to the NBS programme is the lack of specialists and infrastructure for treatment/management of patients [91]. The expanded RMNCH + A strategy has to be supported with enhancing the competency of healthcare providers, and devel-oping requisite skills to manage birth defects. Birth defect prevention services in India also have to be reinforced with appropriate policies and regulations to control exposure to environmental pollution, and ensuring occupational health regulations to prevent workplace exposures. The pharmacovigilance service [92] needs to be strengthened, with stronger post-marketing surveillance.

In conclusion, birth defect prevention activities are challenged by the fact that the aetiology of most of these conditions is unknown. However, several conditions have modifiable/actionable risk factors that are shared with other adverse birth outcomes like miscarriage, stillbirth, preterm birth and infants born with low birth weight. Birth defect prevention strategies can therefore be integrated into a package of interventions that have the potential of addressing all these adverse events. Interventions need to initiate before conception, so that services have to   expand to the preconception period, and target women intending a pregnancy. In India and majority of LMICs, maternal health programmes like the RMNCHA+ focus on antenatal care. They lack messages alerting women on birth defects and their prevention. As this discussion shows, health promotion messages informing women on how they can prevent birth defects are not resource intensive, and can easily be implemented as a part of ongoing maternal health services. Birth defect prevention messages have the potential for addressing several adverse pregnancy outcomes, thereby furthering activities intended towards the achievement of the Sustainable Development Goal, maternal and child health targets.

References

1. Stevenson RE, Hall JG (2015) Introduction. In: Stevenson RE, Hall JG, Everman DB, Solomon BD (eds) Human anomalies and related anomalies, No. 66. Oxford University Press
2. WHO/CDC/ICBDSR (2014) Birth defects surveillance: a manual for programme managers. World Health Organization, Geneva
3. Sitkin NA, Ozgediz D, Donkor P et al (2015) Congenital anomalies in low- and middle-income countries: the unborn child of global surgery. World J Surg 39(1):36–40
4. United Nations Inter-agency Group for Child Mortality Estimation (UN IGME) (2017) Levels & trends in child mortality: report 2017. Estimates developed by the UN Inter-Agency Group for Child Mortality Estimation, United Nations Children's Fund, New York. https://www.uni cef.org/publications/files/Child_Mortality_Report_2017.pdf. Accessed 6 May 2020
5. Taruscio D, Arriola L, Baldi F et al (2014) European recommendations for primary prevention of congenital anomalies: a joined effort of EUROCAT and EUROPLAN projects to facilitate inclusion of this topic in the National Rare Disease Plans. Public Health Genomics 17(2):115–123
6. Brent RL (2004) Environmental causes of human congenital anomalies: the pediatrician's role in dealing with these complex clinical problems caused by a multiplicity of environmental and genetic factors. Pediatrics 113(Supplement 3):957–968
7. Whitworth M, Dowswell T (2009) Routine pre-pregnancy health promotion for improving pregnancy outcomes. Cochrane Database Syst Rev 4:CD007536
8. Feldkamp ML, Carey JC, Byrne JL et al (2017) Etiology and clinical presentation of birth defects: population based study. BMJ 357:j2249
9. Moorthie S, Blencowe H, Darlison MW et al (2018) Estimating the birth prevalence and pregnancy outcomes of congenital anomalies worldwide. J Community Genet 9(4):387–396
10. Baird PA, Terence AW, Newcombe HB et al (1988) Genetic disorders in children and young adults: a population study. Am J Hum Genet 42(5):677–693

11. Alwan S, Chambers CD (2015) Identifying human teratogens: an update. J Pediatr Genet 4(2):39–41
12. Tantibanchachai C (2014) Teratogens. Embryo Project Encyclopedia. https://embryo.asu.edu/pages/teratogens. Accessed 20 May 2020
13. Holmes LB (2011) Human teratogens: update 2010. Birth Defects Res A 91(1):1–7
14. Taruscio D, Baldi F, Carbone P (2017) Primary prevention of congenital anomalies: special focus on environmental chemicals and other toxicants, maternal health and health services and infectious diseases. In: Posada de la Paz M, Taruscio D, Groft S (eds) Rare diseases epidemiology: update and overview. Advances in experimental medicine and biology, vol 1031. Springer, Cham
15. Tinker SC, Gilboa S, Reefhuis J et al (2015) Challenges in studying modifiable risk factors for birth defects. Curr Epidemiol Rep 2(1):23–30
16. Andersen JT, Futtrup TB (2020) Drugs in pregnancy. Adv Drug React Bull 321(1):1243–1246
17. Veroniki AA, Cogo E, Rios P et al (2017) Comparative safety of anti-epileptic drugs during pregnancy: a systematic review and network meta-analysis of congenital anomalies and prenatal outcomes. BMC Med 15(1):1–20
18. Matok I, Gorodischer R, Koren G et al (2009) Exposure to folic acid antagonists during the first trimester of pregnancy and the risk of major anomalies. Br J Clin Pharmacol 68(6):956–962
19. Vargesson N (2015) Thalidomide-induced teratogenesis: history and mechanisms. Birth Defects Res C Embryo Today 105:140–156
20. Castilla EE, Ashton-Prolla P, Barreda-Mejia E et al (1996) Thalidomide, a current teratogen in South America. Teratology 54(6):273–277
21. Dal Pizzol TDS, Knop FP, Mengue SS (2006) Prenatal exposure to misoprostol and congenital anomalies: systematic review and meta-analysis. Reprod Toxicol 22(4):666–671
22. Lind JN, Interrante JD, Ailes EC et al (2017) Maternal use of opioids during pregnancy and congenital anomalies: a systematic review. Pediatrics 139(6):e20164131
23. Damkier P, Brønniche LM, Korch-Frandsen JF et al (2019) In utero exposure to antibiotics and risk of congenital anomalies: a population-based study. Am J Obstet Gynecol 221(6).648.e1-648.e15
24. Fan H, Gilbert R, O'Callaghan F et al (2020) Associations between macrolide antibiotics prescribing during pregnancy and adverse child outcomes in the UK: population based cohort study. BMJ 368:m331
25. Andersen SL, Olsen J, Wu CS (2013) Birth defects after early pregnancy use of antithyroid drugs: a Danish nationwide study. J Clin Endocrinol Metab 98(11):4373–4381
26. Li X, Gui-Yang L, Jian-Li M et al (2015) Risk of congenital anomalies associated with antithyroid treatment during pregnancy: a meta-analysis. Clinics (Sao Paulo) 70(6):453–459
27. Schatz M, Zeiger RS, Harden K et al (1997) The safety of asthma and allergy medications during pregnancy. J Allergy Clin Immunol 100(3):301–306
28. Final report summary—EUROMEDICAT (EUROmediCAT: safety of medication use in pregnancy in relation to risk of congenital anomalies). https://cordis.europa.eu/project/id/260598/reporting. Accessed 20 May 2020
29. Hackshaw A, Rodeck C, Boniface S (2011) Maternal smoking in pregnancy and birth defects: a systematic review based on 173 687 malformed cases and 11.7 million controls. Hum Reprod Update 17(5):589–604
30. O'Leary CM, Nassar N, Kurinczuk JJ et al (2010) Prenatal alcohol exposure and risk of birth defects. Pediatrics 126(4):e843–e850
31. Rasmussen SA, Erickson JD, Reef SE et al (2009) Teratology: from science to birth defects prevention. Birth Defects Res A Clin Mol Teratol 85(1):82–92
32. Wald N, Sneddon J, Densem J et al (1991) Prevention of neural tube defects: results of the Medical Research Council Vitamin Study. Lancet 338(8760):131–137
33. Czeizel AE, Dudás I (1992) Prevention of the first occurrence of neural-tube defects by periconceptional vitamin supplementation. N Engl J Med 327(26):1832–1835
34. Berry RJ, Li Z, Erickson JD et al (1999) Prevention of neural-tube defects with folic acid in China. N Engl J Med 341(20):1485–1490

35. De-Regil LM, Peña-Rosas JP, Fernández-Gaxiola AC (2015) Effects and safety of periconceptional oral folate supplementation for preventing birth defects. Cochrane Database Syst Rev (12):CD007950
36. Persson M, Cnattingius S, Villamor E et al (2017) Risk of major congenital anomalies in relation to maternal overweight and obesity severity: cohort study of 1.2 million singletons. BMJ 357:j2563
37. Gabbay-Benziv R, Reece EA, Wang F (2015) Birth defects in pregestational diabetes: defect range, glycemic threshold and pathogenesis. World J Diabetes 6(3):481–488
38. Ramakrishnan A, Lee LJ, Mitchell LE et al (2015) Maternal hypertension during pregnancy and the risk of congenital heart defects in offspring: a systematic review and meta-analysis. Pediatr Cardiol 36(7):1442–1451
39. Murphy VE, Wang G, Namazy JA et al (2013) The risk of congenital anomalies, perinatal mortality and neonatal hospitalisation among pregnant women with asthma: a systematic review and meta-analysis. BJOG 120(7):812–822
40. Levy HL, Guldberg P, Güttler F et al (2001) Congenital heart disease in maternal phenylketonuria: report from the Maternal PKU Collaborative Study. Pediatr Res 49(5):636–642
41. Sass L, Urhoj SK, Kjærgaard J et al (2017) Fever in pregnancy and the risk of congenital anomalies: a cohort study. BMC Pregnancy Childbirth 17(1):413
42. Kalliora C, Mamoulakis C, Vasilopoulos E (2018) Association of pesticide exposure with human congenital abnormalities. Toxicol Appl Pharmacol 346:58–75
43. Dolk H, Vrijheid M (2003) The impact of environmental pollution on congenital anomalies. Br Med Bull 68(1):25–45
44. Vrijheid M, Martinez D, Manzanares S et al (2011) Ambient air pollution and risk of congenital anomalies: a systematic review. Environ Health Perspect 119:598–606
45. Nieuwenhuijsen MJ, Martinez D, Grellier J et al (2009) Chlorination disinfection byproducts in drinking water and congenital anomalies; review and meta-analyses. Environ Health Perspect 117:1486–1493
46. Hwang BF, Jaakkola JJK, Guo HR (2008) Water disinfection by-products and the risk of specific birth defects: a population-based cross-sectional study in Taiwan. Environ Health 7:1–11
47. Nieuwenhuijsen MJ, Dadvand P, Grellier J et al (2013) Environmental risk factors of pregnancy outcomes: a summary of recent meta-analyses of epidemiological studies. Environ Health 12:6
48. Materna-Kiryluk A, Więckowska B, Wiśniewska K et al (2011) Maternal reproductive history and the risk of isolated congenital anomalies. Paediatr Perinat Epidemiol 25(2):135–143
49. Khoury MJ, Erickson JD (1993) Recurrent pregnancy loss as an indicator for increased risk of birth defects: a population-based case-control study. Paediatr Perinat Epidemiol 7(4):404–416
50. Lie RT, Wilcox AJ, Skjaerven R (1994) A population-based study of the risk of recurrence of birth defects. N Engl J Med 331(1):1–4
51. Mastroiacovo P, Castilla EE, Arpino C et al (1999) Congenital anomalies in twins: an international study. Am J Med Genet 83:117–124
52. Hook EB (1981) Rates of chromosome abnormalities at different maternal ages. Obstet Gynecol 58(3):282–285
53. Reefhuis J, Honein MA (2004) Maternal age and non-chromosomal birth defects, Atlanta—1968–2000: teenager or thirty-something, who is at risk? Birth Defects Res A Clin Mol Teratol 70(9):572–579
54. Lin S, Herdt-Losavio ML, Chapman BR (2013) Maternal occupation and the risk of major birth defects: a follow-up analysis from the National Birth Defects Prevention Study. Int J Hyg Environ Health 216(3):317–323
55. Spencer NJ, Blackburn CM, Read JM (2015) Disabling chronic conditions in childhood and socioeconomic disadvantage: a systematic review and meta-analyses of observational studies. BMJ Open 5(9):e007062
56. Shieh JT, Bittles AH, Hudgins L (2012) Consanguinity and the risk of congenital heart disease. Am J Med Genet A 158(5):1236–1241
57. Stoltenberg C, Magnus P, Skrondal A et al (1999) Consanguinity and recurrence risk of birth defects: a population-based study. Am J Med Genet 82(5):423–428

58. Dean SV, Lassi ZS, Imam AM et al (2014) Preconception care: closing the gap in the continuum of care to accelerate improvements in maternal, newborn and child health. Reprod Health 11(3):1–8
59. Dean S, Lassi Z, Imam A et al (2014) Preconception care: nutritional risks and interventions. Reprod Health 11(Suppl 3):S3
60. Goldstein R, Abell S, Ranasinha S et al (2017) Association of gestational weight gain with maternal and infant outcomes. JAMA 317(21):2207–2225
61. Haider B, Olofin I, Wang M et al (2013) Anaemia, prenatal iron use, and risk of adverse pregnancy outcomes: systematic review and meta-analysis. BMJ 346:f3443
62. Ahankari A, Leonardi-Bee J (2015) Maternal hemoglobin and birth weight: systematic review and meta-analysis. Int J Med Sci Public Health 4:435–445
63. Lassi Z, Imam A, Dean S et al (2014) Preconception care: preventing and treating infections. Reprod Health 11(Suppl 3):S4
64. Lassi Z, Imam A, Dean S et al (2014) Preconception care: screening and management of chronic disease and promoting psychological health. Reprod Health 11(Suppl 3):S5
65. Lange S, Probst C, Rehm J et al (2018) National, regional, and global prevalence of smoking during pregnancy in the general population: a systematic review and metaanalysis. Lancet Glob Health 6:e76-76
66. Nykjaer C, Alwan N, Greenwood D et al (2014) Maternal alcohol intake prior to and during pregnancy and risk of adverse birth outcomes: evidence from a British cohort. J Epidemiol Community Health 68:542–549
67. England L, Kim S, Tomar S et al (2010) Non-cigarette tobacco use among women and adverse pregnancy outcomes. Acta Obstet Gynecol Scand 89:454–464
68. Kenny L, Lavender T, Mc Namee R et al (2013) Advanced maternal age and adverse pregnancy outcome: evidence from a large contemporary cohort. PLoS ONE 8(2):e56583
69. Althabe F, Moore J, Gibbons L et al (2015) Adverse maternal and perinatal outcomes in adolescent pregnancies: The Global Network's Maternal Newborn Health Registry study. Reprod Health 12(Suppl 2):S8
70. Kramer M, Séguin L, Lydon J et al (2000) Socio-economic disparities in pregnancy outcome: why do the poor fare so poorly? Paediatr Perinat Epidemiol 14:194–210
71. Turbadkar D, Mathur M, Rele M (2003) Seroprevalence of TORCH infection in bad obstetric history. Indian J Med Microbiol 21(2):108–110
72. International Institute for Population Sciences. Macro international national family health survey (NFHS-4) 2015–16. https://www.rchiips.org/nfhs/nfhs4.shtml. Accessed 22 Mar 2020
73. Renji SR, Lekshmi ST, Chellamma N (2017) Prevalence of pre-gestational diabetes among the antenatal women attending a tertiary care center. Int J Reprod Contracept Obstet Gynecol 6(3):797–801
74. Dhanwal DK, Bajaj S, Rajput R et al (2016) Prevalence of hypothyroidism in pregnancy: an epidemiological study from 11 cities in 9 states of India. Indian J Endocrinol Metab 20(3):387–390
75. Murhekar M, Verma S, Singh K et al (2020) Epidemiology of Congenital Rubella Syndrome (CRS) in India, 2016–18, based on data from sentinel surveillance. PLoS Negl Trop Dis 14(2):e0007982
76. Shruti G, Sree PK, Rao YV (2015) Drug use pattern of over-the-counter and alternative medications in pregnancy: a cross sectional descriptive study. Natl J Physiol Pharm Pharmacol 5(3):195–199
77. Dhar M, Komaram RB (2017) Assessment of drug utilization pattern and teratogenicity risk among pregnant women attending a tertiary care hospital, Andhra Pradesh. Int J Pharm Sci Res 8(12):5291–5297
78. Gund P, Bhide P, Kar A (2016) Prevalence of periconception risk factors for adverse pregnancy outcomes in a cohort of urban Indian women: implications for preconception health education. J Women's Health Care 5(296). ISSN 2167-0420
79. Peña-Rosas JP, Viteri FE (2009) Effects and safety of preventive oral iron or iron + folic acid supplementation for women during pregnancy. Cochrane Database Syst Rev (4):CD004736

80. Carlson LM, Vora NL (2017) Prenatal diagnosis: screening and diagnostic tools. Obstet Gynecol Clin North Am 44(2):245–256
81. Grace MR, Hardisty E, Dotters-Katz SK et al (2016) Cell-free DNA screening: complexities and challenges of clinical implementation. Obstet Gynecol Surv 71(8):477–487
82. Berisha SZ, Shetty S, Prior TW et al (2020) Cytogenetic and molecular diagnostic testing associated with prenatal and postnatal birth defects. Birth Defects Res 112(4):293–306
83. Thompson DC, McPhillips H, Davis RL et al (2001) Universal newborn hearing screening: summary of evidence. JAMA 286(16):2000–2010
84. Pitt JJ (2010) Newborn screening. Clin Biochem Rev 31(2):57–68
85. Taruscio D, Mantovani A, Carbone P et al (2015) Primary prevention of congenital anomalies: recommendable, feasible and achievable. Public Health Genomics 18(3):184–191
86. Kerber KJ, de Graft-Johnson JE, Bhutta ZA et al (2007) Continuum of care for maternal, newborn, and child health: from slogan to service delivery. Lancet 370(9595):1358–1369
87. Lassi Z, Salam R, Haider B et al (2013) Folic acid supplementation during pregnancy for maternal health and pregnancy outcomes. Cochrane Database Syst Rev 3(3):CD006896
88. Czeizel AE (2005) Birth defects are preventable. Int J Med Sci 2(3):91–92
89. Czeizel AE, Intödy Z, Modell B (1993) What proportion of congenital abnormalities can be prevented? Br Med J 306:499–503
90. Taneja G, Sridhar VSR, Mohanty JS (2019) India's RMNCH + A strategy: approach, learnings and limitations. BMJ Glob Health 4(3):e001162
91. Mookken T (2020) Universal implementation of newborn screening in India. Int J Neonatal Screen 6(2):24
92. Kalaiselvan V, Srivastava S, Singh A, Gupta SK (2019) Pharmacovigilance in India: present scenario and future challenges. Drug Saf 42(3):339–346

Chapter 10
Neural Tube Defects and Folate Status in India

Prajkta Bhide

Abstract Prevention of neural tube defects (NTDs) by folic acid supplementation in the preconception period is a well-established strategy for prevention of these common and debilitating birth defects. Results from two systematic reviews and meta-analyses estimated the pooled birth prevalence of NTDs in India to be in the range of 4.1–4.5 per 1000 births. A cohort study that included data on pregnancy terminations (that was missing in the earlier meta-analyses) reported a lower prevalence of 27.44 per 10,000 births. Evidence indicates that the risk of NTDs increases in the presence of folate deficiency. The Comprehensive National Nutrition Survey identified a high population prevalence of folate deficiency in India, being 23%, 28% and 37% among pre-school children, school-age children and adolescents, respectively. However, data on the proportion of women with adequate periconception folate stores is relatively scarce. In a study on women in the periconception period, folate deficiency was identified in 24.3% and possible deficiency in 20.7% women. Overall, Indian studies to measure NTD prevalence and folate deficiency suffer from various methodological issues. Difficulty in assessing individual study quality, variation in the method of biochemical assessment, and lack of use of standard threshold values for defining deficiency make comparisons across studies difficult. Existing government folic acid supplementation services target adolescent girls and pregnant women but do not include preconception folic acid supplementation. The reported high numbers of NTDs and high prevalence of folate deficiency argue for including preconception folic acid supplementation as a component of maternal health services in India.

Keywords Neural tube defects · India · Folate deficiency · Preconception

P. Bhide (✉)
Birth Defects and Childhood Disability Research Centre, Pune 411020, India

© Springer Nature Singapore Pte Ltd. 2021   235
A. Kar (ed.), *Birth Defects in India*,
https://doi.org/10.1007/978-981-16-1554-2_10

Neural Tube Defects

Neural tube defects are malformations of the central nervous system that occur when the normal process of neural tube closure fails. The normal closure of the neural tube occurs soon after conception and is completed by 28 days post conception. Anencephaly and spina bifida are the dominant NTDs, while craniorachischisis, encephalocele and inencephaly are less prevalent [1]. Neural tube defects have a high burden in low- and middle-income countries (LMICs), largely owing to the high number of births [2]. However, the lack of quality data in resource-poor settings, in part due to absence of robust surveillance systems, has contributed to the invisibility of the problem at the national level in most LMICs. Estimates suggest that annually more than 300,000 NTD affected births occur globally [3]. Lo et al. analysed NTD data from 18 countries in six World Health Organization (WHO) regions and estimated the total median live birth NTD burden to be 1.67/1000 (inter-quartile range (IQR) 0.98–3.49) live births. This estimate increased to 2.55/1000 (IQR 1.56–3.91) when stillbirths and terminations of pregnancy were added to the analyses. The authors predicted that about 190,000 neonates were born each year with NTDs in LMICs alone [4]. A more extensive review and meta-analysis of the global NTD data by Zaganjor et al. using data from 75 countries showed higher rates of NTDs in the South-East Asian region (15.8 per 10,000 births, IQR 1.9–66.2) compared to other world regions [5].

Magnitude of NTDs in India

In India, primary sources of prevalence statistics for congenital anomalies like NTDs are ad hoc studies. Neural tube defects are externally visible anomalies that are relatively easy to diagnose on physical examination and are commonly reported in prevalence studies. However, a large variation in the reported birth prevalence rates has been observed, ranging from 0.7 [6] to 18.2 [7] per 1000 births. The variation in reported prevalence rates can be attributed to various methodological reasons and lack of uniformity among the studies including type of surveillance, case definition and method of case ascertainment, inclusion of pregnancy outcomes and age of diagnosis, among others. For example, the inclusion of data on pregnancy terminations increased the reporting of NTD cases and certain syndromes by 5% or more [8]. A study in Lucknow showed significantly higher prevalence of NTDs in a government hospital associated with a medical college compared to another government hospital that was not associated with any teaching institute [9]. Other studies have reported higher rates among stillbirths and preterm births [10].

Two systematic reviews and meta-analyses estimated the prevalence of NTDs to be in the range of 4.1–4.5 per 1000 births [11, 12]. While the meta-analyses may be considered as the best evidence, they are restricted by the quality of available studies that are included. In both cases, majority of the studies were hospital-based and as such would not represent the true population prevalence. In India, hospitals typically have undeterminable catchment areas due to high patient mobility and out-of-pocket expenditure. High level of heterogeneity was also observed among the studies, highlighting the absence of uniformity in the adopted methodologies.

In contrast, a cohort study (Pune Urban Birth Outcomes study, PUBOS) that measured the birth prevalence of congenital anomalies among 2107 women recruited in early pregnancy and followed till pregnancy outcomes, reported NTD birth prevalence of 27.44 per 10,000 births [13]. This study adopted internationally recognized standard methodology including classification criteria [14] and active case ascertainment. Another important consideration was the inclusion of terminations of pregnancies for foetal anomalies, which was missing in the previous meta-analyses. Despite this additional inclusion data, the NTD prevalence in the cohort study was lower than that estimated from the meta-analyses. In the study, all NTDs were identified in the prenatal period, and the pregnancies were terminated. Although the study identified a lower prevalence than the meta-analyses, it does not undermine the magnitude of these conditions in India. The study estimated that in India, NTDs would affect around 70,233 births each year [13].

Risk Factors

Neural tube defects are complex traits having multifactorial etiologies. However, less than half of observed NTDs can be attributed to known causes [15]. Known risk factors for NTDs include low maternal folate levels [16, 17], maternal obesity and high pre-pregnancy weight [18], maternal illnesses such as diabetes [17], hyperthermia [19] and epilepsy [17], and the use of anti-folate drugs like valproic acid and carbamazepine during pregnancy [20]. Isolated NTDs have a recurrence risk of 3% when one prior pregnancy is affected [21]. Epidemiological studies have reported a higher risk for NTD affected children among households with low socio-economic status [22] and low maternal education [23]. Environmental exposure to pesticides and occupational exposures to organic solvents and glycol ether have also been implicated in the occurrence of NTDs [17, 24]. The maternal methylenetertahydrofolate reductase (*MTHFR*) gene C677T polymorphism increases the possibility of NTD in offspring through its influence on homocysteine concentrations [25].

Despite their complex aetiology, NTDs are preventable if known risk factors are addressed on a timely basis. Primary prevention with maternal periconception folic acid supplementation either alone, or as a part of multivitamins containing folic acid has been demonstrated to lower the incidence of NTDs by 41–62% [16, 26]. Women in the periconception period, either planning to or capable of getting pregnant are recommended daily folic acid supplements as dietary intake is typically inadequate. However, there is an acute lack of data on the prevalence of known risk factors including folate deficiency among pregnant women. Evidence regarding the prevalence of folate deficiency in India is inconclusive; nonetheless, folate intake

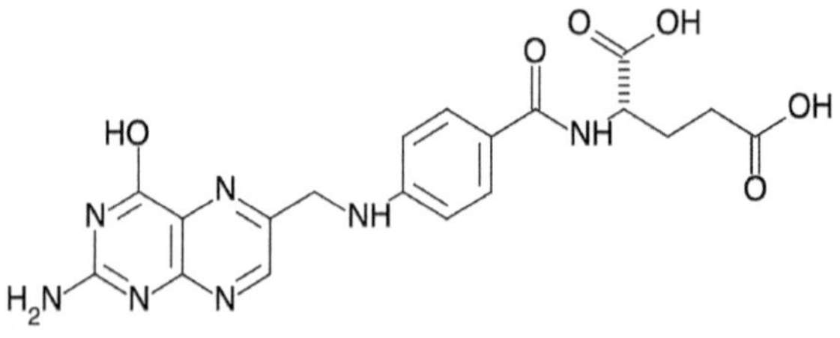

Folic acid

Fig. 10.1 Structure of folic acid *Source*: https://commons.wikimedia.org/wiki/File:Folic_acid_s tructure.svg

below the estimated average requirement (EAR) has been identified often among women in resource-poor settings [27].

Folate

Folate is the collective name for a group of water soluble B vitamin forms that are crucial for growth and maintenance of health. Folate plays a critical role in DNA synthesis and cell proliferation through its involvement in the de novo synthesis of purine and thymidine nucleotides. It is also necessary for the remethylation of homocysteine to methionine. In addition, folate can also affect gene expression as it is required for various cellular methylation processes [28]. The synthetic form of folate is folic acid which has greater bioavailability (~85%) compared to naturally occurring folates (~50%) (Fig. 10.1) [29]. Prolonged folate deficiency is character-ized by megaloblastic anaemia (where erythrocytes are lower in number but larger (macrocytic)), and increased plasma homocysteine levels which is a known risk factor for cardiovascular diseases.

Assessment of Folate Status

Folate status in the population can be assessed in various ways: through dietary assess-ment, assessment of supplement usage and biomarker assessment such as serum and red blood cell (RBC) folate concentrations. The traditional microbiological assay is still considered the "gold standard" for the quantification of serum and RBC folate and the basis for cut-offs to determine folate status [30]. *Lactobacillus rhamnosus* (previously *L. casei*) is used for the measurement of folate in the microbiological assay as the bacteria can grow using many different forms of folate. The growth of the organism is proportional to the amount of folate present in the serum and the turbidity of the inoculated medium on measurement yields the folate concentra-tion. The microbiological assay has certain clear advantages over the other methods of assessment (namely protein binding assay and chromatography-based MS/MS assay). It has the sensitivity for measurement of folate in small sample volumes. The method is relatively inexpensive and so it is suitable for low-resource settings with in-house control of performance. It has scalability with 96-well plate technology and its ability to measure all biologically active forms of folate is also an advantage [29].

Measurement of serum folate is comparatively easier than RBC folate as the long chain polyglutamate forms present in RBCs need to be converted to monoglutame forms (only forms in plasma) prior to analysis of RBC folate [29]. However, serum folate concentrations are typically lower than RBC concentrations. While RBC folate levels serve to indicate long-term folate status, serum folate concentrations vary with recent folate intake.

Cut-Offs for Determining Folate Status

In 1968, the World Health Organization Scientific Group Meeting on Nutritional Anaemias for the first time proposed serum folate concentration <3 ng/ml (<6.8 nmol/L) as cut-off for clinical deficiency, that is, the appearance of macrocytic anaemia based on earlier studies [31]. Serum folate concentrations between 3 and 6 ng/ml (6.8–13.4 nmol/L) were indicative of possible deficiency and 6–20 ng/ml (13.5–45.3 nmol/L) represented normal range. For indicating RBC folate deficiency, a single value of <100 ng/ml (<226.5 nmol/L) was proposed as cut-off. These thresholds used macrocytic anaemia as a haematological indicator of deficiency and were subsequently affirmed in later WHO consultations in 1972 [32] and 1975 [33]. Although megaloblastic anaemia is a characteristic sign of folate deficiency, it was also noted that severe vitamin B-12 deficiency can lead to the same.

Serum and RBC folate cut-off values were revised in 2005 based on metabolic indicators to identify folate deficiency [34]. While serum and RBC folate are direct measures of folate status, another way to determine status is by measuring surrogate markers (metabolites) that may increase as a result of deficiency. One such metabolite is total homocysteine (tHcy) that can serve as a functional indicator of folate status [35, 36]. New cut-off values of 4 ng/ml (10 nmol/L) for serum folate and <151 ng/ml (<340 nmol/L) for RBC folate based on the metabolic indication of increased plasma tHcy were proposed [37] and recommended by the WHO in 2005 [34]. However, tHcy is non-specific for folate as it also depends on other B vitamin concentrations and is elevated in cases of vitamin B-12 inadequacy as well as vitamin B-6 and riboflavin deficiency. It was also recognized that these thresholds are neither appropriate nor adequate for assessing folate status in pregnancy and for the prevention of NTDs [34].

The "optimal" population threshold of RBC folate concentrations was estimated by Crider et al. to be approximately 1000 nmol/L [38]. These results concurred with a single prospective study in the Irish population that indicated the lowest risk of NTDs to be at RBC folate concentrations of 906 nmol/L (400 ng/ml) [39]. The WHO has estimated that for the prevention of NTDs, RBC folate concentrations >400 ng/ml

(906 nmol/L) are considered necessary at the population level [40]. The various cut-offs for folate measurement are summarized in Table 10.1.

Prevalence of Folate Deficiency in India

Together with iron, folate deficiency is considered among the most prevalent micronutrient deficiencies worldwide. However, the true magnitude of the deficiency is not known due to lack of representative data and inconsistent use of measurement cut-offs [41]. Great variation in the prevalence of folate deficiency has been reported worldwide: Switzerland, 2–5% [42]; Belgium, 39% [43]; Japan, 0.5% [44]; Venezuela, 36.3%; Canada, 2.4%; Nepal, 12%; Sweden, 43.1% [41]. Available data also indicates that folate intake was below the EAR in all studies of pregnant women in resource-poor settings [27].

In India, dietary assessment studies have consistently reported a gross deficiency in nutritional folate intake among pregnant women in rural areas [45–47]. In contrast, measurement of serum and RBC folate concentrations has revealed varying results. A community-based study from rural India reported no RBC folate deficiency (<283 nmol/L) among 80 pregnant women at 28 weeks gestation, but all these women had consumed iron and folic acid supplements [48]. Two hospital-based studies reported RBC folate deficiencies (<283 nmol/L) of 14% ($n = 316$) [49] and 22.2% ($n = 351$) [50] among urban pregnant women in the first trimester consuming varying levels of folic acid supplements. A single study measuring serum folate levels among 163 pregnant women in the second trimester reported a deficiency of 1.2% [51]. However, the women in the study had received variable doses of folic acid supplements in early pregnancy. Studies in late pregnancy (>28 weeks) reported wide variation in the prevalence of serum folate deficiency, ranging from 4.4% ($n = 774$) [52] to 26.3% ($n = 266$) [45]. All these reported values are indicative of severe folate deficiency (<3 ng/ml) and none of the three abovementioned studies reported the proportion of women with possible folate deficiency. A cross-sectional study in a rural area in North India reported 3.2% prevalence of severe folate deficiency and 26.2% prevalence of possible deficiency among 95 pregnant women spanning 8–38 weeks of gestation [53].

Table 10.1 Cut-offs for measurement of folate deficiency

Clinical indication of deficiency - appearance of macrocytic anaemia		
Serum folate [31]	Deficiency	<3 ng/ml (<6.8 nmol/L)
	Possible deficiency	3–6 ng/ml (6.8–13.4 nmol/L)
	Normal	6–20 ng/ml (13.5–45.3 nmol/L)
Metabolic indication of deficiency - elevated tHcy concentrations		
Serum folate [34]		<10 nmol/L (4 ng/mL)
Red blood cell folate [34]		<340 nmol/L (151 ng/mL)

Table 10.2 Studies measuring folate levels in India

Study	Type of study and location	Population	Number of women	Per cent deficient	Cut-off	Assessment
Dietary intake						
Pathak et al. [45]	Community-based villages near Delhi	Pregnant women (>28 weeks)	283	99	400 µg/day	24-h recall method and food frequency questionnaire
Gautam et al. [46]	Community-based rural Rajasthan	Pregnant women (12–20 weeks)	114	99.1	400 µg/day	24-h recall method and food frequency questionnaire
Singh et al. [47]	Hospital-based Bangalore city	Pregnant women (<13 weeks)	1626	30	500 µg/day	Food frequency questionnaire
Dwarkanath et al. [49]	Hospital-based Tamil Nadu	Pregnant women (>24 weeks)	1000	73	6 µg/L	Microbiological assay
Serum/plasma folate measurement						
Yusufji et al. [46]	Hospital-based Tamil Nadu	Pregnant women (>24 weeks)	1000	73	6 µg/L	Microbiological assay
Pathak et al. [45]	Community-based rural Haryana	Pregnant women (>28 weeks)	266	26.3	3 ng/mL	Radioimmunoassay
Khanduri et al. [57]	Hospital-based New Delhi	Non-pregnant women	50	12	3 ng/mL	Competitive enzyme immunoassay
Yajnik et al. [58]	Hospital-based Mysore city	Non-pregnant women	40	25	3 ng/mL	Radioimmunoassay
Krishnaveni et al. [52]	Hospital-based Mysore city	Pregnant women (>24 weeks)	774	4.4	7 nmol/L	Microbiological assay
Katre et al. [51]	Urban and rural Maharashtra	Pregnant women (>24 weeks)	536	4.1	7 nmol/L	Microbiological assay

(continued)

Table 10.2 (continued)

Study	Type of study and location	Population	Number of women	Per cent deficient	Cut-off	Assessment
Veena et al. [59]	Hospital-based Mysore city	Non-pregnant women	109	2	6.8 nmol/L	Radioimmunoassay
Menon et al. [60]	Community-based rural and tribal Maharashtra	Pregnant women (<16 weeks)	584	24.3	3 ng/mL	Microbiological assay
Bhide and Kar [55]	Hospital-based Pune city	Pregnant women (<16 weeks)	584	24.3	3 ng/mL	Microbiological assay
Saxena et al. [53]	Community-based rural Uttarakhand	Pregnant women (8–38 weeks)	95	29.4	5.9 ng/mL	Competitive immunoassay
RBC folate measurement						
Yajnik et al. [48]	Community-based rural Maharashtra	Pregnant women (>28 weeks)	80	0	283 nmol/L	Radioimmunoassay
Samuel et al. [50]	Hospital-based Bangalore city	Pregnant women (<14 weeks)	351	22.2	283 nmol/L	Chemiluminescence immunoassay
Dwarkanath et al. [49]	Hospital-based Bangalore city	Pregnant women (<13 weeks)	316	14	283 nmol/L	Chemiluminescence immunoassay
Lukose et al. [61]	Hospital-based Bangalore city	Pregnant women (<13 weeks)	358	26	283 nmol/L	Chemiluminescence immunoassay

The single instance of nationally representative data in the country, the Comprehensive National Nutrition Survey aimed to assess micronutrient deficiencies among pre-schoolers (0–4 years), school-age children (5–9 years) and adolescents (10–19 years) [54]. A high prevalence of folate deficiency was detected: 23% among pre-school children, 28% among school-age children and 37% among adolescents [54].

In the PUBOS cohort, serum folate concentrations were analysed in 584 women using the microbiological assay [55]. The recruited sample had a mean gestational age of 11 ± 3 weeks at the time of sample collection. The mean age of the women who constituted the sample was 22 ± 3 years. More than half (62%) of the women reported ≤ 10 years of completed education. Severe folate deficiency (serum folate concentration <3 ng/ml) was detected in 142 women (24.3% (95% CI 21%–27.9%)) and possible deficiency (3–5.9 ng/ml) in 121 women (20.7% (95% CI 17.6%–24.2%)). Fifty-five per cent (321) women had normal folate concentrations (≥ 6 ng/ml). As serum folate concentrations are an indicator of recent folate uptake, the study highlighted the dietary deficiency of this key nutrient among women in the periconception period [55]. Table 10.2 summarizes the data on folate deficiency among women in India.

Determinants of Folate Deficiency

Inadequate intake is the main cause of folate deficiency. Overall, fewer than 50% women in most countries take periconceptional folic acid supplements [62]. Studies have shown many lifestyle as well as demographic factors to be associated with low folate concentrations. These include factors like alcoholism, smoking and use of anti-folate drugs in addition to factors like women's age, body mass index (BMI), educational level and socio-economic status [29, 63, 64]. The methylenetetrahydrofolate reductase (*MTHFR*) 677C > T polymorphism has also been associated with blood folate concentrations, showing lower concentrations across the genotypes CC > CT > TT [65]. In a large-scale cohort study of Japanese women, younger maternal age, lower educational level, lower annual income and smoking increased the risk of low folate levels up to two times [44]. A national survey from Belgium reported increased folate concentrations among women who did not smoke, had high education levels, and had planned their pregnancy [43]. Data from the United States National Health and Nutrition Examination Survey attests to the inverse association of BMI with serum folate concentrations [66].

There are limited studies evaluating the determinants of folate deficiency in India. In rural North India among the 114 pregnant women who were studied, dietary folate intake was significantly lower among women from low-income families and those who were younger, poorly educated and primiparous [46]. Similarly, in another study, among 283 rural North Indian pregnant women, an increased risk of serum folate deficiency (by 2–3 times) was associated with lower socio-economic status, poor education, higher parity and inter-pregnancy interval of less than 30 months [45].

Data from the PUBOS cohort was analysed to evaluate the association between potential determinants and folate concentrations [55]. Women with lower educational levels ($\leq$10 years of completed education) emerged as having a higher likelihood of both severe folate deficiency (OR 1.74, 95% CI 1.14–2.64) as well as possible folate deficiency (OR 1.59, 95% CI 1.02–2.46) compared to women with >10 years of completed education. Folate deficiency was also more likely to occur among multigravid women than primigravid women (OR 1.73, 95% CI 1.23–2.44). Women having moderate or severe anaemia were more likely to have folate deficiency (OR 1.91, 95% CI 1.07–3.43). In contrast, prenatal folic acid supplement use lowered the odds of folate deficiency (OR 0.17, 95% CI 0.07–0.39) and possible deficiency (OR 0.34, 95% CI 0.17–0.69) in the women. Women consuming prenatal folic acid supplements were less likely to have any folate deficiency (OR 0.24, 95% CI 0.14–0.43).

Maternal Methylenetetrahydrofolate Reductase (*MTHFR*) 677C > T Polymorphism

The enzyme methylenetetrahydrofolate reductase (MTHFR) plays a vital role in one-carbon metabolism, in turn affecting cell division and other metabolic reactions. MTHFR brings about the reduction of 5,10-methylenetetrahydrofolate to 5-methyltetrahydrofolate (5-Methyl-THF). This irreversible reaction resulting in the formation of 5-Methyl-THF is critical for the further processes of remethylation of homocysteine to methionine. The MTHFR enzyme is encoded by the *MTHFR* gene located on the short arm of chromosome 1 (1p36.22). A common polymorphism in the *MTHFR* gene (677C > T, rs1801133) can affect enzymatic activity. A change at nucleotide 677 (C > T) in exon 4 of the *MTHFR* gene leads to the substitution of alanine with valine in codon 222, resulting in a thermolabile form of the enzyme. In individuals homozygous (TT) for the minor allele, the enzyme activity is reduced by 30%, and by 10% in heterozygous (CT) individuals [67]. Whenever MTHFR is inhibited, this results in a lack of available methyl groups due to a reduced synthesis of 5-Methyl-THF, which in turn reduces the remethylation of homocysteine and leads to increasing homocysteine concentrations in blood. This resultant hyperhomocysteinemia increases predisposition to several conditions, including cardiovascular diseases [68, 69] and cancers [70] among other conditions.

The MTHFR 677C > T polymorphism has also been associated with maternal and pregnancy-related conditions, including preeclampsia [71], hypertension in pregnancy [72] and recurrent pregnancy loss [73]. Among the best studied associations of the *MTHFR* 677C > T polymorphism are its associations with NTDs. Meta-analyses have provided evidence for the association of the maternal TT genotype with an increased risk of NTDs in the offspring [25, 74, 75]. The *MTHFR* 677C > T polymorphism has also been associated with lower blood folate concentrations.

Decreasing folate concentrations are associated with the genotypes CC > CT > TT [65].

The prevalence of *MTHFR* TT genotype varies from <1% among Africans to more than 20% among certain populations like Italians and Hispanics [76]. There is a wide variation in the T allele frequency among world populations: 24–40% among Europeans [77], 26–37% in Japanese [78] and around 11% among African Americans [79]. Studies have reported a lower prevalence of the risk allele in Indian population in comparison to other populations (minor allele prevalence ranges from 10 to 13%). The Indian Genome Variation initiative study reported a T allele frequency of 13% among 1871 individuals belonging to 32 large and 23 isolated endogamous groups of India [80]. Naushad et al. reported the T allele frequency to be 10% among 1818 individuals from South India [81]. A third study also reported the T allele frequency of 10% for 23 different endogamous populations of India comprising of various castes, tribes and religious groups [82]. In the PUBOS cohort, overall genotypic frequencies of the *MTHFR* 677C > T polymorphism were 76% for the CC genotype, 23% for the CT genotype and 1% for the TT genotype and the C and T allele frequencies were 87% and 13%, respectively [55].

Summary and Conclusion

There are few studies assessing folate deficiency in early pregnancy in India and there is a clear absence of nationally representative data on folate status among women in the reproductive ages. Results from the cohort study (PUBOS) identified that among an urban population of women using public ante-natal care services, nearly one-fourth (24%) of the women had severe folate deficiency. In addition, one-fifth (21%) of the women had possible folate deficiency and were consequently more likely to become folate deficient in the near future. The study also identified that folate deficient women were in addition likely to be iron deficient. Women with poor education and multigravid women were most likely to present with folate deficiency. These findings coupled with the significant prevalence of NTDs in the country highlight the need for preconception folic acid supplementation. Preconception folic acid supplementation is a cost-effective strategy to prevent the occurrence of NTDs [83]. Additional benefits of folic acid supplementation extend to the reduction of congenital heart defects [84] and possibly preterm births [85]. Existing government folic acid supplementation services target adolescent girls and pregnant women. The latter group are provided folic acid supplements at registration for ante-natal care, and one hundred days of iron and folic acid during the second trimester of pregnancy. However, compliance is an issue and the services do not include preconception folic acid supplementation. The reported high numbers of NTDs and prevalence of folate deficiency argue for exploring feasible health service approaches for including preconception folic acid supplementation as a component of maternal health services in India.

References

1. Moore C (2006) Classification of neural tube defects. In: Wyszynski D (ed) Neural tube defects from origin to treatment. Oxford University Press, New York
2. World Health Organization (2016) Congenital anomalies. https://www.who.int/news-room/fact-sheets/detail/congenital-anomalies. Accessed 3 July 2020
3. Christianson A, Howson C, Modell B (2006) Global report on birth defects: the hidden toll of dying and disabled children. New York
4. Lo A, Polšek D, Sidhu S (2014) Estimating the burden of neural tube defects in low- and middle-income countries. J Glob Health 4:010402
5. Zaganjor I, Sekkarie A, Tsang B et al (2016) Describing the prevalence of neural tube defects worldwide: a systematic literature review. PLoS ONE 11:e0151586
6. Taksande A, Vilhekar K, Chaturvedi P, Jain M (2010) Congenital malformations at birth in Central India: a rural medical college hospital based data. Indian J Hum Genet 16:159–163
7. Sharma J, Gulati N (1992) Potential relationship between dengue fever and neural tube defects in a northern district of India. Int J Gynaecol Obstet 39:291–295
8. Ethen M, Canfield M (2002) Impact of including elective pregnancy terminations before 20 weeks gestation on birth defect rates. Teratology 66:S32–S35
9. Sharma A, Upreti M, Kamboj M et al (1994) Incidence of neural tube defects at Lucknow over a 10 year period from 1982–1991. Indian J Med Res 99:223–226
10. Verma M, Chhatwal J, Singh D (1991) Congenital malformations—a retrospective study of 10,000 cases. Indian J Pediatr 58:245–252
11. Bhide P, Sagoo GS, Moorthie S et al (2013) Systematic review of birth prevalence of neural tube defects in India. Birth Defects Res Part A Clin Mol Teratol 97:437–443
12. Allagh KP, Shamanna BR, Murthy GVS et al (2015) Birth prevalence of neural tube defects and orofacial clefts in India: a systematic review and meta-analysis. PLoS ONE 10:e0118961–e0118961. https://doi.org/10.1371/journal.pone.0118961
13. Bhide P, Gund P, Kar A (2016) Prevalence of congenital anomalies in an Indian maternal cohort: healthcare, prevention, and surveillance implications. PLoS ONE 11:e0166408. https://doi.org/10.1371/journal.pone.0166408
14. EUROCAT (2005) EUROCAT guide 1.3: instruction for the registration of congenital anomalies
15. Agopian AJ, Tinker SC, Lupo PJ et al (2013) Proportion of neural tube defects attributable to known risk factors. Birth Defects Res A Clin Mol Teratol 97:42–46. https://doi.org/10.1002/bdra.23138
16. Berry RJ, Li Z, Erickson JD et al (1999) Prevention of neural-tube defects with folic acid in China. China-U.S. collaborative project for neural tube defect prevention. N Engl J Med 341:1485–1490. https://doi.org/10.1056/NEJM199911113412001
17. Wyszynski D (2006) Maternal exposure to selected environmental factors and risk of neural tube defects in the offspring. In: Wyszynski D (ed) Neural tube defects from origin to treatment. Oxford University Press, New York
18. Stothard K, Tennant P, Bell R, Rankin J (2009) Maternal overweight and obesity and the risk of congenital anomalies: a systematic review and meta-analysis. JAMA 301:636–650
19. Moretti M, Bar-Oz B, Fried S, Koren G (2005) Maternal hyperthermia and the risk for neural tube defects in offspring: systematic review and meta-analysis. Epidemiology 16:216–219. https://doi.org/10.1097/01.ede.0000152903.55579.15
20. Alsdorf R, Wyszynski D (2005) Teratogenicity of sodium valproate. Expert Opin Drug Saf 4:345–353. https://doi.org/10.1517/14740338.4.2.345
21. Cowchock S, Ainbender E, Prescott G et al (1980) The recurrence risk for neural tube defects in the United States: a collaborative study. Am J Med Genet 5:309–314
22. Wasserman C, Shaw G, Selvin S et al (1998) Socioeconomic status, neighborhood social conditions, and neural tube defects. Am J Public Health 88:1674–1680
23. Farley T, Hambidge S, Daley M (2002) Association of low maternal education with neural tube defects in Colorado, 1989–1998. Public Health 116:89–94

24. Ren A, Qiu X, Jin L et al (2011) Association of selected persistent organic pollutants in the placenta with the risk of neural tube defects. Proc Natl Acad Sci USA 108:12770–12775
25. Yadav U, Kumar P, Yadav S et al (2015) Polymorphisms in folate metabolism genes as maternal risk factor for neural tube defects: an updated meta-analysis. Metab Brain Dis 30:7–24
26. Blencowe H, Cousens S, Modell B, Lawn J (2010) Folic acid to reduce neonatal mortality from neural tube disorders. Int J Epidemiol 39. https://doi.org/10.1093/ije/dyq028
27. Torheim L, Ferguson E, Penrose K, Arimond M (2010) Women in resource-poor settings are at risk of inadequate intakes of multiple micronutrients. J Nutr 140:2051S–2058S. https://doi.org/10.3945/jn.110.123463
28. Beaudin A, Abarinov E, Malysheva O et al (2012) Dietary folate, but not choline, modifies neural tube defect risk in Shmt1 knockout mice. Am J Clin Nutr 95:109–114
29. Bailey L, Stover P, McNulty H et al (2015) Biomarkers of nutrition for development—folate review. J Nutr 145:1636S–1680S. https://doi.org/10.3945/jn.114.206599
30. Yetley E, Coates P, Johnson C (2011) Overview of a roundtable on NHANES monitoring of biomarkers of folate and vitamin B-12 status: measurement procedure issues. Am J Clin Nutr 94:297S–302S
31. World Health Organization (1968) Nutritional anaemias. Report of a WHO scientific group. Geneva
32. World Health Organization (1972) Nutritional anemias. Report of a WHO group of experts. Geneva
33. World Health Organization (1975) Control of nutritional anaemia with special reference to iron deficiency. Report of an IAEA/USAID/WHO joint meeting. Geneva
34. de Benoist B (2008) Conclusions of a WHO technical consultation on folate and vitamin B12 deficiencies. Food Nutr Bull
35. Green R (2008) Indicators for assessing folate and vitamin B-12 status and for monitoring the efficacy of intervention strategies. Am J Clin Nutr 94:666S–672S. https://doi.org/10.3945/ajcn.110.009613
36. Selhub J, Jacques PF, Wilson PW, Rush DRI (1993) Vitamin status and intake as primary determinants of homocysteinemia in an elderly population. JAMA 270:2693–2698
37. Selhub J, Jacques P, Dallal G et al (2008) The use of blood concentrations of vitamins and their respective functional indicators to define folate and vitamin B12 status. Food Nutr Bull 29:S67–S73
38. Crider KS, Devine O, Hao L et al (2014) Population red blood cell folate concentrations for prevention of neural tube defects: Bayesian model. BMJ 349:g4554. https://doi.org/10.1136/bmj.g4554
39. Daly L, Kirke P, Molloy A et al (1995) Folate levels and neural tube defects. Implications for prevention. JAMA 274:1698–1702
40. World Health Organization (2015) Guidelines for optimal serum and red blood cell folate concentrations in women of reproductive age for prevention of neural tube defects. Geneva
41. McLean E, de Benoist B, Allen L (2008) Review of the magnitude of folate and vitamin B12 deficiencies worldwide. Food Nutr Bull 29:S38–S51
42. Hess S, Zimmermann M, Brogli S, Hurrell R (2001) A national survey of iron and folate status in pregnant women in Switzerland. Int J Vitam Nutr Res 71:268–273. https://doi.org/10.1024/0300-9831.71.5.268
43. Vandevijvere S, Amsalkhir S, Van OH, Moreno-Reyes R (2012) Determinants of folate status in pregnant women: results from a national cross-sectional survey in Belgium. Eur J Clin Nutr 66111:1172–1177. https://doi.org/10.1038/ejcn.2012.111
44. Yila TA, Araki A, Sasaki S et al (2016) Predictors of folate status among pregnant Japanese women: the Hokkaido Study on Environment and Children's Health, 2002–2012. Br J Nutr 115:2227–2235. https://doi.org/10.1017/S0007114516001628
45. Pathak P, Kapil U, Kapoor S et al (2004) Prevalence of multiple micronutrient deficiencies amongst pregnant women in a rural area of Haryana. Indian J Pediatr 71:1007–1014
46. Gautam V, Taneja D, Sharma N et al (2008) Dietary aspects of pregnant women in rural areas of Northern India. Matern Child Nutr 4:86–94. https://doi.org/10.1111/j.1740-8709.2007.00131.x

47. Singh MB, Fotedar R, Lakshminarayana J (2009) Micronutrient deficiency status among women of desert areas of western Rajasthan, India. Public Health Nutr 12:624–629. https://doi.org/S1368980008002395 (pii)\r10.1017/S1368980008002395 (doi)
48. Yajnik C, Deshpande S, Panchanadikar A et al (2005) Maternal total homocysteine concentration and neonatal size in India. Asia Pac J Clin Nutr 14:179–181
49. Dwarkanath P, Barzilay J, Thomas T et al (2013) High folate and low vitamin B-12 intakes during pregnancy are associated with small-for-gestational age infants in South Indian women: a prospective observational cohort study. Am J Clin Nutr 98:1450–1458. https://doi.org/10.3945/ajcn.112.056382
50. Samuel T, Duggan C, Thomas T et al (2013) Vitamin B(12) intake and status in early pregnancy among urban South Indian women. Ann Nutr Metab 62:113–122. https://doi.org/10.1159/000345589
51. Katre P, Bhat D, Lubree H et al (2010) Vitamin B12 and folic acid supplementation and plasma total homocysteine concentrations in pregnant Indian women with low B12 and high folate status. Asia Pac J Clin Nutr 19:335–343
52. Krishnaveni G, Hill J, Veena S et al (2009) Low plasma vitamin B12 in pregnancy is associated with gestational "diabesity" and later diabetes. Diabetologia 52:2350–2358. https://doi.org/10.1007/s00125-009-1499-0
53. Saxena V, Naithani M, Singh R (2017) Epidemiological determinants of Folate deficiency among pregnant women of district Dehradun. Clin Epidemiol Glob Health 5:21–27
54. Ministry of Health and Family Welfare, Government of India, UNICEF, Population Council (2019) Comprehensive national nutrition survey national report 2016–2018. New Delhi
55. Bhide P, Kar A (2018) Prevalence and determinants of folate deficiency among urban Indian women in the periconception period. Eur J Clin Nutr. https://doi.org/10.1038/s41430-018-0255-2
56. Yusufji MV, Baker S (1973) Iron, folate, and vitamin B12 nutrition in pregnancy: a study of 1000 women from southern India. Bull World Health Organ 48:15–22
57. Khanduri U, Sharma A, Joshi A (2005) Occult cobalamin and folate deficiency in Indians. Natl Med J India 18:182–183
58. Yajnik C, Lubree H, Thuse N et al (2007) Oral vitamin B12 supplementation reduces plasma total homocysteine concentration in women in India. Asia Pac J Clin Nutr 16:103–109
59. Veena S, Krishnaveni G, Srinivasan K et al (2010) Higher maternal plasma folate but not vitamin B-12 concentrations during pregnancy are associated with better cognitive function scores in 9- to 10-year-old children in South India. J Nutr 140:1014–1022
60. Menon K, Skeaff S, Thomson C et al (2011) Concurrent micronutrient deficiencies are prevalent in nonpregnant rural and tribal women from central India. Nutrition 27:496–502
61. Lukose A, Ramthal A, Thomas T, Bosch R, Kurpad AV, Duggan CSK (2014) Nutritional factors associated with antenatal depressive symptoms in the early stage of pregnancy among urban South Indian women. Matern Child Health J 18:161–170
62. Ray JG, Singh G, Burrows RF (2004) Evidence for suboptimal use of periconceptional folic acid supplements globally. BJOG 111:399–408. https://doi.org/10.1111/j.1471-0528.2004.00115.x
63. Pfeiffer C, Sternberg M, Schleicher R, Rybak M (2013) Dietary supplement use and smoking are important correlates of biomarkers of water-soluble vitamin status after adjusting for sociodemographic and lifestyle variables in a representative sample of U.S. adults. J Nutr 143:957S–965S. https://doi.org/10.3945/jn.112.173021
64. Pfeiffer C, Sternberg M, Fazili Z et al (2015) Folate status and concentrations of serum folate forms in the US population: national health and nutrition examination survey 2011–2. Br J Nutr 1–13. https://doi.org/10.1017/S0007114515001142
65. Tsang B, Devine O, Cordero A et al (2015) Assessing the association between the methylenetetrahydrofolate reductase (MTHFR) 677C>T polymorphism and blood folate concentrations: a systematic review and meta-analysis of trials and observational studies. Am J Clin Nutr 101:1286–1294

66. Tinker S, Hamner H, Berry R et al (2012) Does obesity modify the association of supplemental folic acid with folate status among nonpregnant women of childbearing age in the United States? Birth Defects Res Part A Clin Mol Teratol 94:749–755. https://doi.org/10.1002/bdra. 23024

67. Frosst P, Blom HJ, Milos R et al (1995) A candidate genetic risk factor for vascular disease: a common mutation in methylenetetrahydrofolate reductase. Nat Genet 10:111–113. https://doi.org/10.1038/ng0595-111

68. Husemoen L, Skaaby T, Jørgensen T et al (2014) MTHFR C677T genotype and cardiovascular risk in a general population without mandatory folic acid fortification. Eur J Nutr 53:1549–1559

69. Li P, Qin C (2014) Methylenetetrahydrofolate reductase (MTHFR) gene polymorphisms and susceptibility to ischemic stroke: a meta-analysis. Gene 535:359–364. https://doi.org/10.1016/j.gene.2013.09.066

70. Tang M, Wang S, Liu B et al (2014) The methylenetetrahydrofolate reductase (MTHFR) C677T polymorphism and tumor risk: evidence from 134 case-control studies. Mol Biol Rep 41:4659–4673

71. Wang X, Wu H, Qiu X (2013) Methylenetetrahydrofolate reductase (MTHFR) gene C677T polymorphism and risk of preeclampsia: an updated meta-analysis based on 51 studies. Arch Med Res 44:159–168

72. Yang B, Fan S, Zhi X et al (2014) Associations of MTHFR gene polymorphisms with hypertension and hypertension in pregnancy: a meta-analysis from 114 studies with 15411 cases and 21970 controls. PLoS One 9. https://doi.org/10.1371/journal.pone.0087497

73. Chen H, Yang X, Lu M (2015) Methylenetetrahydrofolate reductase gene polymorphisms and recurrent pregnancy loss in China: a systematic review and meta-analysis. Arch Gynecol Obstet 293:283–290. https://doi.org/10.1007/s00404-015-3894-8

74. Yan L, Zhao L, Long Y et al (2012) Association of the maternal MTHFR C677T polymorphism with susceptibility to neural tube defects in offsprings: evidence from 25 case-control studies. PLoS ONE 7:e41689

75. Zhang T, Lou J, Zhong R et al (2013) Genetic variants in the folate pathway and the risk of neural tube defects: a meta-analysis of the published literature. PLoS ONE 8:e59570

76. Botto LD, Yang Q (2000) 5,10-methylenetetrahydrofolate reductase gene variants and congenital anomalies: a HuGE review. Am J Epidemiol 151:862–877. https://doi.org/10.1093/oxfordjournals.aje.a010290

77. van der Put N, Eskes T, Blom H (1997) Is the common 677CrT mutation in the methylenetetrahydrofolate reductase gene a risk factor for neural tube defects? A meta-analysis. QJM 90:111–115

78. Papapetrou C, Lynch S, Burn J, Edwards Y (1996) Methylenetetrahydrofolate reductase and neural tube defects. Lancet 348:58

79. Stevenson R, Schwartz C, Du Y-Z, Adams MJ (1997) Differences in methylenetetrahydrofolate reductase genotype frequencies between whites and blacks. Am J Hum Genet 60:229–230

80. Indian Genome Variation Consortium (2008) Genetic landscape of the people of India: a canvas for disease gene exploration. J Genet 87:3–20. https://doi.org/10.1007/s12041-008-0002-x

81. Naushad S, Krishnaprasad C, Devi A (2014) Adaptive developmental plasticity in methylene tetrahydrofolate reductase (MTHFR) C677T polymorphism limits its frequency in South Indians. Mol Biol Rep 41:3045–3050

82. Saraswathy K, Asghar M, Samtani R et al (2012) Spectrum of MTHFR gene SNPs C677T and A1298C: a study among 23 population groups of India. Mol Biol Rep 39:5025–5031. https://doi.org/10.1007/s11033-011-1299-8

83. MRC Vitamin Study Research Group (1991) Prevention of neural tube defects: results of the Medical Research Council Vitamin Study. Lancet 338:131–137

84. Czeizel A, Dudás I, Vereczkey A, Bánhidy F (2013) Folate deficiency and folic acid supplementation: the prevention of neural-tube defects and congenital heart defects. Nutrients 5:4760–4775

85. Greenberg J, Bell S, Guan Y, Yu Y (2011) Folic acid supplementation and pregnancy: more than just neural tube defect prevention. Rev Obstet Gynecol 4:52–59

Chapter 11
Haemoglobinopathies: Genetic Services in India

Sumedha Dharmarajan

Abstract Genetic testing is an integral component of a birth defects service. Using the example of haemoglobinopathies, the most common single-gene disorder, this chapter presents the magnitude of these disorders and describes their consequences on patients and their families. Haemoglobinopathies impact the public health system, because of the chronicity and expensive treatment modalities. Genetic testing and screening are key prevention tools. This chapter discusses the need for development of genetic services in India by describing the magnitude of these conditions and tracks the history of development of services for haemoglobinopathies till the launch of the national guidelines on prevention and control of haemoglobinopathies in India. The relevance of these guidelines and their similarity to global guidelines have been discussed. The need for the programme to be ethical and culturally sensitive in order for it to be successful has also been deliberated. Monitoring of such a programme is important, and the available indicators for monitoring the programme have been examined.

Keywords Haemoglobinopathies · Beta thalassemia · Sickle cell anaemia · Public health · India

Thalassemia, sickle cell anaemia, muscular dystrophy and haemophilia are examples of common genetic (monogenic or single gene) disorders. Single-gene disorders are caused by mutations in a single gene whose function is of critical importance in one or more metabolic pathway(s) of the body. Genetic disorders can be classified based on their pattern of inheritance that is whether they are sex-linked or autosomal, recessive or dominant. Clinical manifestation of the disorder depends on the number of copies of the mutation (recessive or dominant), location (X-linked, Y-linked or autosomal), penetrance, epistatic interactions, presence of secondary modifiers and others. Though individually monogenic disorders are less prevalent than other common diseases and disorders, collectively they affect a significant proportion of the global population. The chronic, often life-threatening nature of the conditions adversely impacts the quality of life of patients and their families. In

S. Dharmarajan (✉)
Birth Defects and Childhood Disability Research Centre, Pune, India

© Springer Nature Singapore Pte Ltd. 2021
A. Kar (ed.), *Birth Defects in India*,
https://doi.org/10.1007/978-981-16-1554-2_11

251

low-income countries, subsidized care is limited or may not be available for all types of genetic disorders. This limited or absence of services causes physical, economic and emotional distress to patients and caregivers. Where available, these services may primarily be available through private medical services which impose considerable financial burden on families.

From a public health perspective, the expenditure needed for patient management, the different clinical types of genetic disorders, their varied diagnoses and treatment modalities make it difficult to view genetic services through the prism of public health services, especially in resource limited settings. Genetic disorders are accorded a low priority in these regions primarily because their management is long term and expensive, and there are more urgent public health challenges in the form of infectious diseases, malnutrition and perinatal complications that are major contributors to child mortality [1]. Public health activities to address prevalent genetic disorders have however started emerging in low- and low-middle-income countries (LMICs). This article focuses on the guidelines for a national prevention and control programme for haemoglobinopathies in India. These guidelines form a first step towards development of a public health genetic service in India.

Haemoglobinopathies

Thalassemias are inherited haemoglobin disorders which together with sickle cell anaemias and other haemoglobin disorders are referred to as haemoglobinopathies. Thalassemias are caused by mutations in the alpha or beta globin gene, resulting in a quantitative deficiency in the synthesis of alpha or beta globin, causing reduced haemoglobin in red blood cells, decreased red blood cell production and anaemia [2]. Sickle cell anaemia is caused by a structural alteration in the beta globin subunit of haemoglobin. Haemoglobin polymerization leading to red cell damage causes vaso-occlusion, leading to infarction, anaemia, inflammation, hypercoagulability, oxidative stress and vascular endothelial dysfunction [3]. Mutations causing severe alpha thalassemia are less common in India, but the beta thalassemias pose a considerable public health challenge in the country [4].

Beta thalassemia and sickle cell anaemia are autosomal recessive disorders, with asymptomatic carriers (heterozygotes) (referred to as beta thalassemia trait and sickle cell trait, respectively) transmitting the mutation to the next generation. Individuals homozygous for beta thalassemia (beta thalassemia major/intermedia) are present with chronic, haemolytic anaemia. Heterozygote carriers are mostly asymptomatic, presenting with mild anaemia. The clinical manifestations of thalassemia and their management are summarized in Box 11.1 [5–12]. Sickle cell disease is marked by periodic episodes of pain (crises), organ ischemia and severe systemic complications. Organ failure is the primary cause of death [13]. Pain, infections and anemia result in frequent hospitalization [14]. Stroke is common in children with sickle cell disease, but there is limited data from India [13].

Box 11.1 Clinical manifestations and management of beta thalassemias

- **Genotype**
  Thalassemia major/intermedia: β°/β°, β°/β^+, β^+/β^+
  Thalassemia minor: β°/β^N, β^+/β^N
  Prevalent mutations in India
  IVS 1–5 (G → C), 619 bp del, IVS 1–1 (G → T), Cd 41/42 (−TCTT), Cd 8/9 (+G), Cd 15 (G → A), Cd 30 (G → C), Cap site +1 (A → C), Cd 5 (−CT), Cd 16 (−C) [5].
- **Pattern of inheritance**: Autosomal recessive.
- **Phenotype**
  Beta thalassemia major, beta thalassemia intermedia, beta thalassemia minor/trait/carrier.
- **Clinical manifestations in under-transfused patients with beta thalassemia major**
  Chronic anaemia, pallor, growth retardation, poor musculature, craniofacial abnormalities often referred to as thalassemic facies (frontal bossing, prominent malar eminence, depression of the bridge of the nose, mongoloid slant of the eye, hypertrophy of the maxillae), genu valgum and hepatosplenomegaly [6].
- **Beta thalassemia intermedia**
  Craniofacial deformities, gallstones, leg ulcers, thrombosis leading to pulmonary embolism.
- **Beta thalassemia minor**
  Clinically asymptomatic, may have mild anaemia.
- **Management**

  - Blood transfusions
    Frequency increases with age and may range from 3 to 20 transfusions in a year (unpublished data).
  - Iron chelation therapy
    Desferrioxamine, deferiprone and desirox to be prescribed at the discretion of the clinician [7].

- **Quality of life**
  Eighty-six percent of parents reported emotional stress, anxiety and depression [8].
  Physical, psychological, social, school functioning and environmental domains were affected, and discontinuation of schooling was reported [9–12].

Patients with beta thalassemia major are treated with repeated blood transfusions to correct anaemia and with iron chelation that removes the excess iron that accumulates in the body due to repeated transfusions. The importance of iron chelation therapy is evidenced in the survival rates of patients who are chelated and those who are not [2]. Survival of patients without iron chelation is 35 years [15], whereas well-transfused and well-chelated patients survive up to 45 years [16]. The Standardized Mortality Ratio (SMR) was estimated to reduce from 28.9 to 13.5 after introduction of iron chelation [17]. The current iron chelation therapies recommend initiation of therapy at the age of 2–3 years. Hydroxyurea is the main treatment for sickle cell disease [13], as it has been shown to reduce the frequency and severity of crises episodes and may even reduce the risk of stroke in children [18]. Treatment options focus on prevention and management of complications. Median life expectancy remains 30–40 years, but life expectancy data from India are limited [3].

Epidemiology

Prevalence of thalassemia/sickle cell carriers is the indicator used to determine the magnitude of thalassemia and sickle cell anaemia. Globally, the thalassemias are prevalent through the Mediterranean, sub-Saharan Africa, Middle East, Indian subcontinent, Southeast Asia, Melanesia and the Pacific islands (global carrier prevalence: 0.5–20%) (Fig. 11.1). Haemoglobin variants such as sickle cell anaemia (global carrier prevalence: 1–38%) and HbE disorders (global carrier prevalence: 1–70%) are also reported from these regions [19, 20]. Among the thalassemias, beta thalassemia, sickle cell anaemia and HbE disorders are more prevalent in India and constitute a public health problem. The reasons for the widespread and increasing prevalence of haemoglobin disorders has been attributed to natural selection for protection against malaria (the prevalence of haemoglobin disorders varies with that of malaria), consanguineous marriages (increasing the frequency of the Mendelian recessive disorder), epidemiological transition resulting in increasing visibility of the disorder due to reducing rates of other causes of mortality, and population migration from high prevalent regions to areas where the condition was not present [20–24]. The increasing detection of patients and lack of access to care have been brought to public attention by pronounced advocacy by patient organizations.

The prevalence rate of beta thalassemia in India has been estimated to be 0.2 [25]–37.9% [26] by various studies panning all over India. This huge variation in prevalence estimates can be attributed to different case definitions used, incorrect sampling methods and heterogeneous populations screened. On an average, the prevalence of beta thalassemia is between 3 and 4%, translating into 35–45 million carriers in India [27]. Using a systematic review and meta-analysis, the pooled prevalence of beta thalassemia carriers in India has been estimated to be 3.9% (95% CI 3.15–4.84) and estimated 8740 children with beta thalassemia are likely to be born each year with thalassemia major [Dharmarajan, unpublished data]. There may be as many as 100,000 patients with beta thalassemia in India [27]. Considering India's large population, there is a possibility that these may be underestimates, as there may be a significant number who could be beta thalassemia carriers, many of whom would be undiagnosed and unaware of their carrier status.

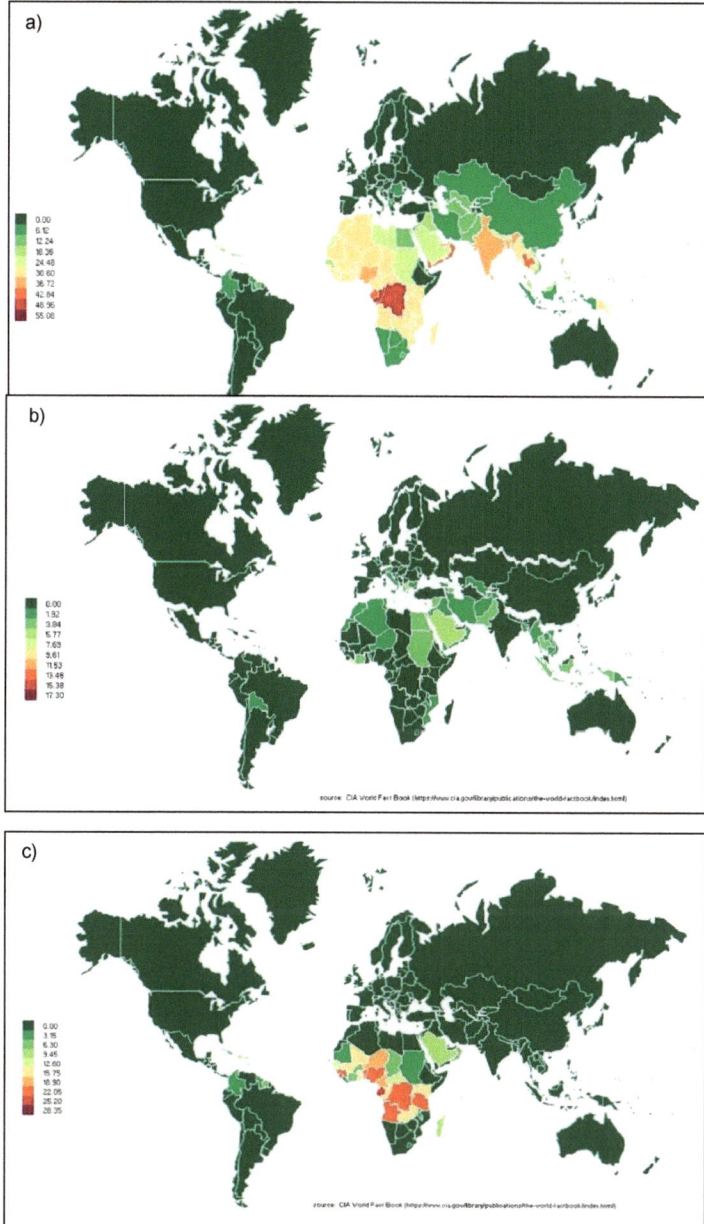

Fig. 11.1 Global distribution of haemoglobinopathies. **a** Heterozygous alpha thalassemia (prevalence range 5–60%), **b** heterozygous beta thalassemia (prevalence range 0.5–20%) and **c** sickle cell carriers (prevalence range 1–38%). Cyprus and Maldives which have the highest prevalence rates for beta thalassaemia carriers 15% and 17%, respectively, and are not visible in this map. Data from Ref. [29]

Due to the same methodological issues, the prevalence of sickle cell trait has to be interpreted with caution, with estimates varying from 5 to 35%. The condition is more prevalent in Scheduled Tribes and Scheduled Castes, but the distribution varies even within local areas. Sickle cell disease is frequently associated with beta thalassemia, but here again, the data vary. For example, 40% of patients in a hospital-based study had sickle cell-beta thalassemia, and 16% had sickle cell-alpha thalassemia. This was in contrast to studies conducted in Odisha and Gujarat that had reported 50 and 85% sickle cell-alpha thalassemia, leading to an erroneous assumption that sickle cell disease was more frequently associated with alpha thalassemia [30]. Non-randomly conducted small studies have also propagated the perception of sickle cell disease being restricted to tribal communities. These data have also been used to generate spatial maps of sickle cell disease in the country [31]. This is best reflected in a recently conducted study by the laboratory in Pune, India. The study screened a random sample of 360 pregnant women attending antenatal clinics in government hospitals in Pune, Maharashtra. Pune is the second largest city in Maharashtra, a western state in India. The study detected that 6.3% women were haemoglobinopathy carriers, of which 3.3% were beta thalassaemia carriers, 1.7% were sickle cell carriers and 1.4% were HbE carriers. Of these, sickle cell and HbE carriers have been considered to be prevalent in tribal areas and north-east India, respectively, but migration has resulted in persons carrying these mutations to be found in urban areas. The range of these mutations has diagnostic and treatment implications [28]. This highlights that multiple studies need to be conducted in different areas of India using appropriate study designs and correct diagnostic techniques. It is a lack of such studies resulting in data quality issues associated with measuring the magnitude of the disorder that makes it difficult to understand the true prevalence of haemoglobinopathies in India. Based on these best available data, it is estimated that there may be as many as 150,000 patients with sickle cell disease in India.

Impact of Thalassemia on Patients and Families in Low-Income Settings

Quality of Life (QoL)

Like parents of children with other genetic disorders, studies indicate that QoL of patients with thalassemia and their parents are affected. QoL studies report that the physical, psychological, social domains, as well as school functioning and home environment were affected. A study reported that patients were not satisfied with their body image and were depressed and anxious. They were more likely not to discuss the disorder with their friends and only depend on parents for their emotional support. This often resulted in patients dropping out from school and an adverse home environment, especially among older children [32, 33].

Parental quality of life was also affected with the earliest study in 1990 reporting that 86% of parents of patients with beta thalassemia reported emotional stress, anxiety and depression [8]. A smaller number of families reported experiencing social stigma. Majority of parents were diagnosed with psychological conditions such as depression and anxiety. Parents of newly diagnosed and older patients were most affected. The major cause for concern was expenses for treatment of thalassemia [9–12].

Out-of-Pocket Expenditure

India has the largest incidence of catastrophic health spending, wherein the largest proportion of the spending is on medications [34, 35]. There are only three cost-of-care studies conducted in India. Sangani et al. [8] estimated that the cost of blood transfusion services to be Rs. 900–3780 per year and reported that families spent 20–30% of income on management of the disorder [8]. Moirangthem and Phadke [36] reported that families spent 19,150–439,500 rupees per annum on management of the disorder with highest expenditure incurred on medications [36]. Families spent 29–67% of their family income on management with caregivers of older patients spending more on treatment. These data were reported from a government hospital where families from below the poverty line receive compensation for selected expenditures [33].

Emotional and financial costs of sickle cell disease have also been reported from across the world [37]. Indian studies report the economic burden on individuals and families. For example, a study on treatment of patients with sickle cell trait/disease observed that while 17% and 26% of patients availed free of charge services from the Primary Health Centres and the District Hospitals, respectively, majority (60%) made out-of-pocket payments for consulting private practitioners. Nearly 70% of transfusion was availed from government blood banks, where patients with haemoglobinopathies are not charged for the service. Among patients requiring hospital admissions, nearly 60% used a combination of government and private hospitals. For supporting treatment costs, 11% of families availed of loans [38].

Basic Components of a Genetic Service

The impact of thalassemia on patients and families in the absence of a genetic service is catastrophic. The number of patients, the widespread and increasing prevalence of haemoglobinopathies, and the severe consequences on those with the homozygous disorder were the main factors mobilizing the demand for recognition of the problem by the public health system in India and the development of a genetic service for these conditions in the country.

Genetic services should combine patient care and provide genetic counselling to families at risk (medical genetics services) and include population-based prevention strategies (public health genetics programmes). These services should be integrated and be an essential component of the maternal and child health services [39]. Development of such genetic services for haemoglobinopathies is important in LMICs because of (i) limited availability of management/treatment services which makes these disorders burdensome for patients and their families, (ii) prevention is a relatively low-cost tool that will reduce the prevalence of the disorder and consequently prevent the adverse impact on families, and (iii) diagnosis methods are inexpensive especially as the use of screening followed by diagnosis has reduced these costs. There is evidence from several countries on the reduction in birth prevalence of haemoglobinopathies due to introduction of genetic services as a component of public health services (discussed below). A study in Israel reported that the cost ratio of prevention to treatment was 1:4, illustrating the benefits of prevention [40].

A population-based genetic service can be organized only if the disorder to be screened is common, a clear diagnosis is possible, natural history of the condition is well understood, management strategy is available and acceptable and the programme is cost-effective [41]. Such a service is organized into primary, secondary and tertiary levels of prevention (Table 11.1). Primary prevention aims at reducing the incidence of genetic disorders. Secondary prevention aims at avoiding the birth of an affected child and minimizing the severity of clinical manifestations by early diagnosis. Tertiary prevention aims at proper management of the condition, thereby

Table 11.1 Primary, secondary and tertiary prevention of genetic disorders

Level of prevention	Interventions (with examples of thalassemia)
Primary	
Health promotion	Community education and awareness through Information, Education and Communication (IEC) on signs and symptoms of the disorder, mode of transmission, prevention and health facilities offering preventive services
Specific protection	Assessment of family history risk for genetic disorders among all women registering for antenatal care
	Voluntary premarital, preconception and antenatal screening and counselling: e.g. screening for beta thalassemia carriers in adolescents, in couples before marriage, after marriage before conception, pregnant women; cascade screening of extended family members
Secondary	Newborn screening
	Monitoring of child growth and development to identify sick children or children with developmental disabilities
Tertiary	Averting complications of diseases, e.g. transfusion, chelation or bone marrow transplant for patients with thalassemia
	Routine monitoring of patients, e.g. monthly monitoring for haemoglobin level and amount of blood transfused; periodic anthropometric assessment, liver biochemistry, kidney functioning, testing for transfusion transmitted infections and hemosiderosis

reducing complications including disability and providing psychosocial support to patients and their family members.

History of Development of Thalassemia Services in India

Perceiving the consequences of the thalassemia, other haemoglobinopathies and other genetic disorders, the World Health Organization (WHO) released a report in 1966, describing the pathology and clinical manifestations of these disorders [42]. The WHO emphasized the need for further research to investigate the distribution of these disorders, particularly in Africa, Asia, Oceania and some parts of Europe [43]. Subsequently, a Memorandum was released by the WHO in 1983, reporting expert recommendations to initiate national haemoglobinopathy control programmes, with prevention as the main component. The public health strategies recommended to be adopted were community awareness, foetal diagnosis and genetic counselling [44]. An updated set of guidelines were published in 1989 [45]. The guidelines included methods for determining the epidemiology of the condition, treatment guidelines, carrier screening methods, education and awareness, genetic counselling and prenatal diagnosis, service requirements for a control programme detailing personnel and equipments, and encouraging patient support groups in providing care and support to patients and families. The costing of such a programme and the suggested evaluation indicators are valid and relevant for India and other developing countries even today. The WHO in 2006 passed a resolution urging member states to regard genetic disorders such as haemoglobinopathies as a public health problem and introduce public health programmes for prevention and care for these disorders [46].

In India, thalassemia was not given a public health priority and development of services was on an ad hoc basis across different states. Early on, thalassemia services in India were spearheaded by a collaborative effort of parent–patient organizations, supported by the International Thalassemia Foundation, clinicians and the Indian Red Cross Society. Prolonged advocacy, together with the burgeoning HIV epidemic, was instrumental in ensuring free of charge transfusion from government blood banks. In the meantime, studies identified the high prevalence of thalassemia and its consequences on patients and their families [8]. Recognizing the chronicity of the disorder, the states of Odisha, Madhya Pradesh and Rajasthan started providing blood, free of charge, to patients with thalassemia and iron chelation, either free or at subsidized costs [47–49]. Other Indian studies identified the high prevalence of sickle cell disease among tribal populations. Certain states with high densities of these vulnerable populations initiated services for sickle cell anaemia. A programme for screening, followed by counselling and provision of care for patients with sickle cell disease, was initiated in 2006 in Gujarat and 2008 in Maharashtra.

Since the early 2000, studies reported not only a high prevalence of haemoglobinopathies in India, but the challenge of diagnosing the disease in remote and far-flung areas of the country (inhabited by tribal populations). A number of research studies implemented under a special funding scheme, the Jai Vigyan project, identified that a package of low-cost diagnostics (Naked Eye Single Tube Red Cell Osmotic Fragility Test (NESTROFT) test, red blood cell indices, solubility tests and di-chloro phenol indophenol (DCIP) test) could identify majority of beta thalassemia, sickle cell and HbE cases [7]. This package of diagnostics could make it feasible for screening among populations living in rural and remote communities. In 2014, the National Blood Transfusion Council declared blood and its components to be provided free of cost for patients with thalassemia and sickle cell anaemia, across the country [50].

It is important to note that while programmes for providing services for thalassemia patients were spearheaded by the non-governmental organization, Thalassemics India, programmes to provide services for sickle cell anaemia patients were initiated by the state governments. Parents of patients with thalassemia have needed to approach state governments to provide free of cost transfusion services and iron chelation, petition courts and approach the Prime Minister's Office in order to treat their child by bone marrow transplantation. In contrast, as sickle cell disease primarily affected vulnerable populations, prevention and care programmes were primarily initiated by state governments. Development of care and prevention services for the haemoglobinopathies and other genetic disorders therefore originated from piecemeal services, implemented in response to demand from patient support organizations. Unlike a public health service with defined goals and objectives, there are no achievable targets and no direction in terms of what these services were intended to achieve, other than providing some relief to patients.

As of now, thalassemia patients receive free blood and iron chelators, but leucocyte filters and other consumables have to be procured by patients. For patients with sickle cell disease living in remote areas, Primary Health Centres provide pain killers and antibiotics, and folic acid and hydroxyurea prophylaxis have been initiated at a few centres. In 2013, a programme to address birth defects was introduced through the launch of the Rashtriya Bal Swasthya Karkyakram (RBSK) (Chap. 13). It advocated screening of thalassemia carriers where the prevalence of the disorder was high [51]. In December 2016, there were specific guidelines launched for diagnosis, management and prevention of haemoglobinopathies in India. These guidelines closely follow the WHO guidelines [45], according to an adequate and necessary importance to prevention. The national guidelines form the first steps towards the development of genetic services in the country. The availability of these guidelines provides a basis for states to put in place services for providing care and offer services for prevention of haemoglobinopathies.

Guidelines on Haemoglobinopathies in India

The stated mission of the guidelines for the prevention and control of haemoglobinopathies in India is to improve care for patients with thalassemia and sickle cell disease, and lower the prevalence of these conditions through screening and awareness programmes [7].

Prevention Strategies

Prevention of haemoglobinopathies is to be implemented through screening at various stages of the life cycle. Newborn screening that is screening of newborns at birth for a panel of genetic disorders including haemoglobinopathies is mandated in order to improve the quality of life of the newborn. The guidelines mandate that children with anaemia (haemoglobin levels <7 g/dl) should be referred from community settings (i.e. playschools (*anganwadis*) and schools where screening for haemoglobin levels is routinely conducted) to district health facilities for further investigation. Adolescent screening implies intervention in the school age group. The guidelines suggest that school children should be provided information on haemoglobinopathies, their pathology, inheritance and the necessity of prevention, followed by screening after informed consent. Carriers are to be 'counselled' regarding avoiding marriage to another carrier and also informed about the availability of prenatal diagnosis.

The other stages when screening can be offered are during the preconceptional or the antenatal periods. Preconceptional screening involves screening of couples who are planning their pregnancy, and if both are carriers the option of prenatal diagnosis with pregnancy termination is provided. Antenatal screening involves universal screening of all pregnant women during the first trimester, screening husbands of those testing positive and provision of prenatal diagnostic services when both partners are positive. The guidelines suggest screening of siblings and cascade screening of other family members. Cascade screening involves screening of extended family members of patients and carriers, but the guidelines recommend caution as screening may not be accepted by extended family members [7, 52].

Key to identification of patients is the choice of screening and diagnostic tools. There are multiple methods available for screening and diagnosis of beta thalassemia. Choice of screening method is primarily governed by the specificity and sensitivity of the method, the range of mutations covered, ease of set-up, availability of technical knowledge and infrastructure, and its ease of use in remote and rural settings. As iron deficiency anaemia is widely prevalent in India, and its manifestations overlap with that of beta thalassemia carriers, it is important for the tests to provide a differential diagnosis between these two conditions.

The national guidelines recommend low-cost tests with high negative predictive value, such as NESTROFT, solubility test (for sickle cell trait/disease), DCIP, complete blood count or a peripheral blood smear as the first line of screening in community settings. Dried blood spots can be collected from newborns after informed consent. Diagnosis can be done using cellulose acetate electrophoresis (CAE), isoelectric focusing (IEF), capillary electrophoresis, cation-exchange high-performance liquid chromatography (CE-HPLC) using HbA_2 level indicator cut-offs at 3.5% (haemoglobin A_2 is a normal variant of haemoglobin, which is elevated in beta thalassemia carriers and patients), and molecular methods such as allele-specific oligonucleotide (ASO) hybridization, amplification-refractory mutation system polymerase chain reaction (ARMS PCR) and gap PCR [53].

It is worth noting that the national guidelines recommend cut-offs for diagnosis of carriers at HbA_2 values of 4% and further investigations for individuals with values between 3.5 and 3.9% [7]. International guidelines however recommend HbA_2 <

3.5% as diagnostic for beta thalassemia carriers and HbA_2 levels between 3.5 and 4.0% for further investigation [54]. The use of such cut-offs in the national guidelines will reduce the comparability of the data for making global estimates.

Pretest and Post-test Counselling in Community-Based Screening Programmes

Genetic counselling is the process through which information about the genetic aspects of illnesses is shared by trained professionals with those who are at an increased risk of having a heritable disorder or of passing it on to their unborn offspring [55]. Genetic counselling is composed of pretest and post-test counselling components. Pretest genetic counselling is offered before individuals undergo the genetic test. It is used to inform the utility, sensitivity and specificity of the test, what the test results will convey. In India's socio-cultural milieu, it can be used to sensitize the recipient and the partner, the necessity of the test and the impact of a positive diagnosis [56, 57].

Post-test counselling is offered to all individuals irrespective of the result of the test. In case of a negative test, the chances of an affected pregnancy in case of thalassemia disorders are low, but not zero due to prevalence of less severe and silent mutations. In case of a positive test, the partner needs to be screened. If both partners are carriers, the diagnosis and prognosis of the disorder for the child have to be explained. Availability of health services for the child has to be explained, and the alternative provision of termination of pregnancy services, if preferred, has to be explained. Appropriate referral linkages with the obstetrician are essential. All this counselling has to be non-directive, enabling an autonomous decision to be taken by the couple [56, 57].

Genetic Counselling Education, Testing Laboratories in India

There is no structured teaching programme in medical genetics in the undergraduate and postgraduate medical curriculum in the country [58]. The Indian Academy of Medical Genetics which has the objective of promoting the science and practice of medical genetics lists a single super-speciality course (DM Medical Genetics). Diplomate of National Board (DNB) Medical Genetics is offered/proposed to be offered from a few institutes. Master courses in Biomedical Genetics and Genetic Counselling, and a certificate course in Genetic Counselling are also being offered. There is a Board of Genetic Counselling registered in the state of Telangana. Prenatal diagnosis using invasive techniques is available at several centres, with the majority being at private medical facilities [58].

Since the establishment of the first genetic diagnostic laboratories in India, genetic testing capabilities have proliferated in the country [59, 60]. Table 11.2 shows the number of genetic diagnostic clinics in India [61]. This list is incomplete as there is no centralized data on the numbers of genetic clinics in India. The list does not include over ten large private genetic diagnostic laboratories, many of which are Indian offices of international diagnostic companies. The private diagnostic services also have a network of sample collection centres. Table 11.2 shows the variation in the number of genetic diagnostic companies by states of India. The socio-economically developed states, which also have several non-medical and medical universities and research institutions, tend to have greater numbers of genetic diagnostic testing facilities. It is notable that while the UK has 25 genetic testing centres catering to a population of 56 million [62], India has already established over one hundred such laboratories.

Table 11.2 Genetic clinics in India

State/Union	Genetic clinics	Area km^2	Population (census 2011)	Genetic clinics/100,000 population
Andhra Pradesh and Telangana	19	275,045	84,580,777	0.22
Chandigarh	3	114	1,055,450	0.28
Chhattisgarh	2	135,192	25,545,198	0.007
Gujarat	8	196,244	60,439,692	0.013
Haryana	8	44,212	25,351,462	0.031
Jammu & Kashmir*	2	222,236	12,267,032	0.016
Karnataka	14	191,791	61,095,297	0.022
Kerala	6	38,852	33,406,061	0.017
Madhya Pradesh	3	308,252	72,626,809	0.004
Maharashtra	23	307,713	112,374,333	0.02
Odisha	1	155,707	41,974,219	0.002
Punjab	2	50,362	27,743,338	0.007
Rajasthan	1	342,239	68,548,437	0.001
Tamil Nadu	17	130,060	72,147,030	0.02
Uttar Pradesh	5	240,928	199,812,341	0.002
West Bengal	3	88,752	91,276,115	0.003
Delhi	11	1483	16,787,941	0.065

*- The states in this table have been listed as in [61]

Medical Care

The national guidelines have proposed free of cost blood transfusion therapy with leuco-depleted, packed red blood cells (pRBCs). Chelation for iron overload is recommended to be initiated when the iron ferritin values are higher than 1000 μg/l. The iron chelators to be provided free of charge are desferrioxamine, deferiprone and deferasirox. Continuous monitoring for endocrine, cardiac, skeletal and other complications due to the disease and treatment of these complications is recommended. Monthly monitoring for haemoglobin level and amount of blood transfused and six monthly anthropometric assessments, liver biochemistry, kidney functioning, testing for transfusion transmitted infections and hemosiderosis are to be undertaken. Bone marrow transplantation (BMT)/hematopoietic stem cell transplant (HSCT) as the ultimate curative therapy is recommended, although these services are limited. For patients with sickle cell disease, prompt treatment of fever, penicillin prophylaxis and compulsory pneumococcal vaccination are recommended. Pain relief with non-steroidal anti-inflammatory drugs is recommended. Hydroxyurea therapy for decreasing pain and improving HbF levels is recommended. The guidelines recommend psychological support services for parents. Day care centres have been established at district hospitals providing transfusion services and are to be set up at District Early Intervention Centres [7].

Few state governments have put in place services for patients with haemoglobinopathies in India. Leuco-depleted, packed red blood cells are not available at all blood banks, so that patients have to travel long distances for transfusion. Iron chelators, if supplied, are predominantly desferioxamine. Due to the high reactions and the visible distress of patients, it is necessary to provide oral chelators. If supplied at some centres, the services are erratic. At present, treatment from private practitioners, through personal expenditure, remains the main source of treatment for patients. For patients with sickle cell disease, accessing treatment is a greater challenge due to the difficult terrain of tribal areas where the disease is most prevalent. Non-governmental organizations have been instrumental in providing care.

Global Strategies for Prevention of Haemoglobinopathies

A genetic service with prevention at its core can only be successful if it is acceptable to the community at large. Prevention and control is largely dependant on availability of services to terminate a pregnancy, if required, after a positive genetic test. The type of carrier screening (premarital, prenatal or antenatal) to be introduced, depends on acceptance of this ethically difficult decision by the cultural leaders and subsequently the larger community. In certain high burden countries such as Cyprus, Greece and Italy, thalassemia screening was voluntary and premarital and antenatal screening services, and prenatal diagnostic services were available [63]. Screening was mandated in most Islamic countries such as Iran, Turkey, Jordan and Saudi

Arabia, due to high prevalence of haemoglobinopathies and high consanguinity rates. Only premarital screening was provided in these countries resulting in a high marriage cancellation rate among carrier couples. Due to limited abortion services and strict laws, carrier couples in these countries were *counselled* to and opted to cancel their marriage [64].

Such strategies raise the public health debate of ethics of individual choice versus reducing the magnitude of a chronic debilitating disorder. Cyprus has the highest carrier prevalence rates in the world, and thalassemia screening is mandatory [65]. Despite debates surrounding ethics, mandatory screening had yielded effects, as all countries with thalassemia screening programmes had reported a reduction. Maldives and Iran showed 55–80% reduction in the number of affected births after implementation of the programme [64, 66]. In all these countries, screening programmes were accompanied by extensive media campaigns for increasing the awareness about the thalassemias, which has been attributed to the successful uptake of the thalassemia screening services. Cost-benefit analysis supports preventive services to treatment programmes [8, 63, 64, 66–68].

Ethical Considerations for Indian Genetic Screening Services

One of the key issues of the genetic service is that interventions should be ethical, upholding the four key principles of ethics that is autonomy, justice, beneficence and non-maleficence. A genetic service only focused on prevention and not on treatment is unethical. Hence, a comprehensive programme encompassing preventive and treatment services is a must. This has been highlighted in each of the statements and guidelines released by the WHO. Following these recommendations, the Indian guidelines on haemoglobinopathies have also given appropriate weightage to treatment. Newborn screening, for example, has little utility if it is not followed by access to appropriate treatment.

One of the major issues of concern is the marriage counselling strategies being used by different state agencies and some NGOs. The Gujarat sickle cell anaemia programme, for example, provides colour-coded cards, with patients given a yellow card, carriers given a yellow and white card and non-carriers a white card (Fig. 11.2) [69]. Providing such cards may appear an effective public health tool to increase awareness about carrier status among populations with low literacy levels. This method however is highly unethical, as it violates the principle of autonomy of choosing a partner and causes psychosocial harm to the individual. In the social–cultural milieu of India, adolescent screening and premarital screening strategies suggested in the guidelines for detecting both thalassaemia and sickle cell anaemia carriers place adolescents and young adults at risk, with the risk being higher in girls. The potential of screening strategies in stigmatizing individuals might impact the acceptance of screening by the community [70]. Lack of community acceptance will lead to failure of an otherwise well-structured genetic services programme.

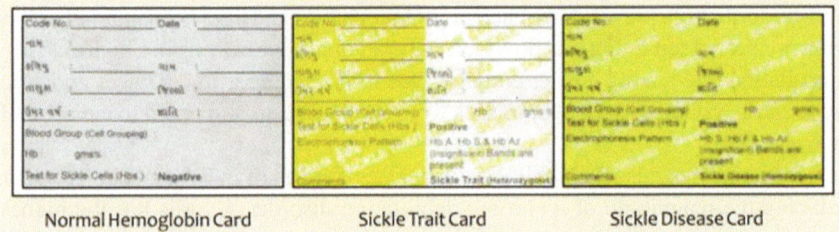

| Normal Hemoglobin Card | Sickle Trait Card | Sickle Disease Card |

Fig. 11.2 Sickle cell anaemia counselling cards. Counselling cards issued to persons screened for sickle cell anaemia. The white shaded portion indicates a normal beta globin allele, and yellow shaded portion indicates a mutant allele. A non-carrier will be issued a white card, a sickle cell carrier will be issued a card shaded white and yellow, whereas a patient with sickle cell anaemia will be issued a yellow card. From [69]

The methods that will be most acceptable to the community without violating the principles of ethics are the antenatal and cascade screening strategies, wherein carriers can be detected through screening, and prenatal diagnosis can be provided on time. This indicates that the timing of provision of carrier screening strategies is vital for ensuring success of a genetic services programme in a community which is governed by strict religious and cultural norms. Integrating genetic services with maternal health services and ensuring that such a programme is governed by strict guidelines safeguarding the interests of patients, parents and persons screened for genetic disorders are imperative.

Monitoring

Important for a genetic service is monitoring, particularly when there are multiple programme components. Periodic monitoring will ensure correction of programmatic approaches and modification of interventions as and when needed. Modell and Darlison have prepared a list of such indicators that will be useful for determining the progress of the service towards achieving its stated goals [71] (Box 11.2).

Box 11.2 Programme indicators for a genetic service [71]

- **Indicator for patient care**: Annual conceptions with a haemoglobin disorder in the absence of prevention.
- **Indicator for carrier screening**: Annual number of carrier tests required.
- **Indicator for carrier information and offer of partner testing**: Annual number of carriers detected using the chosen strategy.

- **Indicator for expert risk assessment and genetic counselling**: Annual at-risk pregnancies of carrier couples detected by screening using the chosen strategy.
- **Indicator for the offer of prenatal diagnosis**: Annual pregnancies at risk detected.

Research

There are multiple institutes across India researching thalassemia and sickle cell anaemia. The major publications from these institutes involve molecular characterization of *HBB* mutations. The research has determined the most prevalent mutations in India, thereby reducing costs of diagnostic testing. Other research studies have yielded data on the usefulness of low-cost diagnostics that have come into routine use in rural and remote settings. Description of clinical manifestations and effectiveness of medications have informed which treatment modalities are acceptable and effective. The clinical and large technical competency in India to offer genetic testing is apparent.

There are however a number of research gaps. First and foremost, systematic studies are required to understand the true prevalence of haemoglobinopathies in India, as existing studies have methodological limitations described earlier in the article. Appropriately powered studies, using globally relevant case definitions and cut-offs, are essential. Such studies are important not only to identify the true magnitude of the problem, but also to understand whether the problem extends beyond the selected communities that have been repeatedly studied. Clinical research also needs to be systematic [72]. The natural history of haemoglobinopathies needs to be studied, preferably through cohort studies, so that the clinical course of disease and the effectiveness of health interventions can appropriately inform public health service providers. There is limited research on the quality of life of patients with genetic disorders and their parents, limited needs assessment studies, cost of care studies and investigations on genetic counselling in context to India's cultural milieu. Such studies, conducted across the country, are urgently required to provide evidence for further developing the proposed programme for the prevention and control of haemoglobinopathies in India, which can be generally applicable at the national level and streamlined to meet the local needs.

References

1. Christianson A, Modell B (2004) Medical genetics in developing countries. Annu Rev Genomics Hum Genet 5:219–265

2. Galanello R, Origa R (2010) Beta-thalassemia. Orphanet J Rare Dis 5(11). https://doi.org/10.1186/1750-1172-5-11
3. Rees DC, Brousse VA (2016) Sickle cell disease: status with particular reference to India. Indian J Med Res 143(6):675–677
4. Weatherall DJ, Clegg JB (2008) Distribution and population genetics of the thalassemias. In: Weatherall DJ, Clegg JB (eds) The thalassemia syndromes, 4th edn. Blackwell, London, pp 237–284
5. Sinha S, Black ML, Agarwal S, Colah R, Das R, Ryan K, Bellgard M, Bittles AH (2009) Profiling β-thalassemia mutations in India at state and regional levels: implications for genetic education, screening and counselling programmes. HUGO J 3:51–62
6. Origa R (2000) Beta-thalassemia. In: GeneReviews®(Internet). University of Washington, Seattle. Available at https://www.ncbi.nlm.nih.gov/books/NBK1426/?report=printable. Accessed 28 July 2020
7. National Health Mission, Ministry of Health and Family Welfare, Govt. of India (2016) Prevention and control of hemoglobinopathies in India—thalassemias, sickle cell disease and other variant hemoglobins. https://nhm.gov.in/images/pdf/programmes/RBSK/Resource_Documents/Guidelines_on_Hemoglobinopathies_in%20India.pdf. Accessed 28 July 2020
8. Sangani B, Sukumaran PK, Mahadik C et al (1990) Thalassemia in Bombay: the role of medical genetics in developing countries. Bull World Health Organ 68(5048):75–81
9. Ajij M, Pemde HK, Chandra J (2015) Quality of life of adolescents with transfusion-dependent thalassemia and their siblings: a cross-sectional study. J Pediatr Hematol Oncol 37(3):200–203
10. Khurana A, Katyal S, Marwaha RK (2006) Psychosocial burden in thalassemia. Indian J Pediatr 73(10):877–880
11. Saha R, Misra R, Saha I (2015) Health related quality of life and its predictors among Bengali thalassemic children admitted to a tertiary care hospital. Indian J Pediatr 82:909–916
12. Shaligram D, Girimaji SC, Chaturvedi SK (2007) Psychological problems and quality of life in children with thalassemia. Indian J Pediatr 74(8):727–730
13. Ware RE, de Montalembert M, Tshilolo L, Abboud MR (2017) Sickle cell disease. Lancet 390(10091):311–323
14. Jain D, Warthe V, Dayama P et al (2016) Sickle cell disease in Central India: a potentially severe syndrome. Indian J Pediatr 83:1071–1076
15. Modell B, Khan M, Darlison M (2000) Survival in β-thalassemia major in the UK: data from the UK Thalassemia Register. Lancet 355(9220):2051–2052
16. Modell B, Khan M, Darlison M, Westwood MA, Ingram D, Pennell DJ (2008) Improved survival of thalassemia major in the UK and relation to T2* cardiovascular magnetic resonance. J Cardiovasc Magn Reson 10(1):42. https://doi.org/10.1186/1532-429X-10-42
17. Ladis V, Chouliaras G, Berdoukas V et al (2011) Survival in a large cohort of Greek patients with transfusion-dependent beta thalassemia and mortality ratios compared to the general population. Eur J Haematol 86:332–338. https://doi.org/10.1111/j.1600-0609.2011.01582.x
18. Thornburg CD, Files BA, Luo Z (2012) Impact of hydroxyurea on clinical events in the BABY HUG trial. Blood 120(22):4304–4310. https://doi.org/10.1182/blood-2012-03-419879
19. Weatherall D, Akinyanju O, Fucharoen S et al (2006) Inherited disorders of hemoglobin. In: Jamison DT, Breman JG, Measham AR et al (eds) Disease control priorities in developing countries, 2nd edn, Chap 34. The International Bank for Reconstruction and Development/The World Bank, Washington. https://www.ncbi.nlm.nih.gov/books/NBK11727/. Accessed 28 July 2020
20. Weatherall DJ, Clegg JB (2001) Inherited haemoglobin disorders: an increasing global health problem. Bull World Health Organ 79(8):704–712
21. Weatherall D (2011) The inherited disorders of haemoglobin: an increasingly neglected global health burden. Indian J Med Res 134:493–497
22. Piel FB, Tatem AJ, Huang Z et al (2014) Global migration and the changing distribution of sickle haemoglobin: a quantitative study of temporal trends between 1960 and 2000. Lancet Glob Health 2(2):e80–e89. https://doi.org/10.1016/S2214-109X(13)70150-5

23. Piel FB (2016) The present and future global burden of the inherited disorders of hemoglobin. Hematol Oncol Clin North Am 30(2):327–341
24. Weatherall DJ (2011) The challenge of haemoglobinopathies in resource-poor countries. Br J Haematol 154(6):736–744. https://doi.org/10.1111/j.1365-2141.2011.08742.x
25. Achoubi N, Asghar M, Saraswathy KN et al (2012) Prevalence of β-thalassemia and hemoglobin E in two migrant populations of Manipur, North East India. Genet Test Mol Biomarkers 16(10):1195–1200. https://doi.org/10.1089/gtmb.2011.0373
26. Chandrashekar V, Soni M (2011) Hemoglobin disorders in South India. ISRN Hematol. https://doi.org/10.5402/2011/748939
27. Colah R, Italia K, Gorakshakar A (2017) Burden of thalassemia in India: the road map for control. Pediatr Hematol Oncol J 2(4):79–84. https://doi.org/10.1016/j.phoj.2017.10.002
28. Dharmarajan S, Pawar A, Bhide P, Kar A (2021) Undiagnosed haemoglobinopathies among pregnant women attending antenatal care clinics in Pune, India. J Community Genet. https://doi.org/10.1007/s12687-021-00505-8. Epub ahead of print. PMID: 33486692
29. Modell B, Darlison MW, Moorthie S et al (2016) Epidemiological Methods in Community Genetics and the Modell Global Database of Congenital Disorders (MGDb). Downloaded from UCL Discovery: http://discovery.ucl.ac.uk/1532179/
30. Serjeant GR, Ghosh K, Patel J (2016) Sickle cell disease in India: a perspective. Indian J Med Res 143(1):21–24. https://doi.org/10.4103/0971-5916.178582
31. Hockham C, Bhatt S, Colah R et al (2018) The spatial epidemiology of sickle-cell anaemia in India. Sci Rep 8(1):17685. https://doi.org/10.1038/s41598-018-36077-w
32. Khairkar P, Malhotra S, Marwaha R (2010) Growing up with the families of β-thalassemia major using an accelerated longitudinal design. Indian J Med Res 132:428–437
33. Shaligram D, Girimaji SC, Chaturvedi SK (2007) Quality of life issues in caregivers of youngsters with thalassemia. Indian J Pediatr 74(3):275–278
34. Wagstaff A, Flores G, Hsu J et al (2017) Progress on catastrophic health spending in 133 countries: a retrospective observational study. Lancet Glob Health 6(2):e169–e179. https://doi.org/10.1016/S2214-109X(17)30429-1
35. Selvaraj S, Farooqui HH, Karan A (2018) Quantifying the financial burden of households' out-of-pocket payments on medicines in India: a repeated cross-sectional analysis of National Sample Survey data, 1994–2014. BMJ Open 8(5):e018020. https://doi.org/10.1136/bmjopen-2017-018020
36. Moirangthem A, Phadke SR (2018) Socio-demographic profile and economic burden of treatment of transfusion dependent thalassemia. Indian J Pediatr 85:102–107. https://doi.org/10.1007/s12098-017-2478-y
37. Kuerten BG, Brotkin S, Bonner MJ et al (2020) Psychosocial burden of childhood sickle cell disease on caregivers in Kenya. J Pediatr Psychol 45(5):561–572
38. Kadam DD, Bose AK, Chandekar PL (2019) A study to assess the socio-epidemiological profile of sickle cell disease among affected tribal population. Int J Med Sci Public Health 8(12):1029–1033
39. World Health Organization (1999) Services for the prevention and management of genetic disorders and birth defects in developing countries. https://apps.who.int/iris/handle/10665/66501. Accessed 29 July 2020
40. Koren A, Profeta L, Zalman L et al (2014) Prevention of β thalassemia in Northern Israel—a cost-benefit analysis. Mediterr J Hematol Infect Dis 6(1):e2014012. https://doi.org/10.4084/MJHID.2014.012
41. Donnai D (2002) Genetic services. Clin Genet 61(1):1–6. https://doi.org/10.1034/j.1399-0004.2002.610101.x
42. World Health Organization (1966) Haemoglobinopathies and allied disorders. Report of a WHO scientific group. Technical report series no. 338. https://apps.who.int/iris/handle/10665/39852. Accessed 29 July 2020
43. Howson CP, Christianson AC, Modell B (2008) Controlling birth defects: reducing the hidden toll of dying and disabled children in lower-income countries. Disease control priorities project. Washington (District of Columbia). https://www.marchofdimes.org/materials/partner-

controlling-birth-defects-reducing-hidden-toll-of-dying-children-low-income-countries.pdf. Accessed 29 July 2020

44. Community control of hereditary anaemias: memorandum from a WHO meeting (1983) Bull World Health Organ 61(1):63–80
45. World Health Organization (1989) Guidelines for the control of haemoglobin disorders. Report of the VIth annual meeting of the WHO working group on haemoglobinopathies. In: Hereditary diseases programme. World Health Organization, Sardinia. https://www.who.int/iris/handle/10665/66665. Accessed 29 July 2020
46. World Health Organization (2006) Report by the secretariat. Thalassemia and other haemoglobinopathies. EB118/5. https://apps.who.int/gb/archive/pdf_files/EB118/B118_5-en.pdf. Accessed 29 July 2020
47. Mission Director, National Rural Health Mission Rajasthan (2011) Regarding provision of free diagnosis and treatment for patients with thalassemia and haemophilia. https://www.thalassemicsindia.org/images/pdf/rajasthan1.pdf. Published 2011. Accessed 5 June 2018
48. Directorate of Health Services, Madhya Pradesh (2013) Provision of free blood and day care services to patients of thalassemia and sickle cell anaemia. https://www.thalassemicsindia.org/images/pdf/M.P.1.pdf. Accessed 5 June 2018
49. Governor, Government of Odisha (2013) Free medicines for thalassemia, sickle cell disease and haemophilia. https://www.thalassemicsindia.org/images/pdf/orrisa.jpg. Accessed 5 July 2018
50. Department of AIDS Control (2014) Guidelines for recovery of processing charges for blood and blood components-reg. https://nbtc.naco.gov.in/assets/resources/policy/Guidelines_on_processing_charges_for_blood_and_blood_components.pdf. Accessed 29 July 2020
51. Ministry of Health and Family Welfare, Government of India (2013) Operational guidelines. Rashtriya Bal Swasthya Karyakram (RBSK) child health screening and early intervention services under NRHM. https://rbsk.gov.in/RBSKLive/. Accessed 23 June 2018
52. Verma IC, Saxena R, Kohli S (2011) Past, present & future scenario of thalassaemic care & control in India. Indian J Med Res 134:507–521
53. Weatherall DJ, Clegg JB (2008) The laboratory diagnosis of the thalassemias. In: Weatherall DJ, Clegg JB (eds) The thalassemia syndromes, 4th edn. Blackwell, London, pp 686–723
54. Stephens AD, Angastiniotis M, Baysal E et al (2012) ICSH recommendations for the measurement of haemoglobin A2. Int J Lab Hematol 34(1):1–13. https://doi.org/10.1111/j.1751-553X.2011.01368.x
55. World Health Organization (2019) Genetic counselling services. https://www.who.int/genomics/professionals/counselling/en/. Accessed 9 Sept 2019
56. Rink BD, Kuller JA (2018) What are the required components of pre- and post-test counseling? Semin Perinatol 42(5):287–289. https://doi.org/10.1053/j.semperi.2018.07.005
57. Fonda Allen J, Stoll K, Bernhardt BA (2016) Pre- and post-test genetic counseling for chromosomal and Mendelian disorders. Semin Perinatol 40(1):44–55. https://doi.org/10.1053/j.semperi.2015.11.007
58. Gupta N, Kabra M (2012) The current status of medical genetics in India. In: Kumar D (ed) Genomics and health in the developing world. Oxford University Press
59. Madon P (2012) Conventional and molecular cytogenetics in India. In: Kumar D (ed) Genomics and health in the developing world. Oxford University Press
60. World Health Organization (2013) Birth defects in South-East Asia: a public health challenge: situation analysis (No. SEA-CAH-13). WHO Regional Office for South-East Asia. https://apps.who.int/iris/handle/10665/204821. Accessed 28 July 2020
61. Kar B, Sivamani S (2016) Directory of genetic test services and counselling centres in India. Int J Hum Genet 16(3–4):148–157
62. Raeburn S, Kent A, Gillott J (1997) Genetic services in the United Kingdom. Eur J Hum Genet 5(2):188–195
63. Cousens NE, Gaff CL, Metcalfe SA, Delatycki MB (2010) Carrier screening for beta-thalassemia: a review of international practice. Eur J Hum Genet 18(10):1077–1083. https://doi.org/10.1038/ejhg.2010.90

64. Saffi M, Howard N (2015) Exploring the effectiveness of mandatory premarital screening and genetic counselling programmes for β-thalassemia in the Middle East: a scoping review. Public Health Genomics 18(4):193–203. https://doi.org/10.1159/000430837
65. Maharashtra Arogya Mandal. Sickle cell anaemia. https://www.mam.org.in/Living_With_Sickle_Cell.html. Accessed 29 July 2020
66. Firdous N, Gibbons S, Modell B (2011) Falling prevalence of beta-thalassemia and eradication of malaria in the Maldives. J Community Genet 2(3):173–189. https://doi.org/10.1007/s12687-011-0054-0
67. Ostrowsky JT, Lippman A, Scriver CR (1985) Cost-benefit analysis of a thalassemia disease prevention program. Am J Public Health 75(7):732–736. https://doi.org/10.2105/AJPH.75.7.732
68. Modell B, Kuliev AM (1991) Services for thalassemia as a model for cost-benefit analysis of genetics services. J Inherit Metab Dis 14(4):640–651. https://doi.org/10.1007/BF01797934
69. Commissionerate of Health Family Welfare and Medical Services Gujarat. Sickle cell anaemia program manual. https://nrhm.gujarat.gov.in/sickle-cell.htm. Accessed 28 July 2020
70. Phadke SR, Puri RD, Ranganath P (2017) Prenatal screening for genetic disorders: suggested guidelines for the Indian scenario. Indian J Med Res 146:689–699. https://doi.org/10.4103/ijmr.IJMR_1788_15
71. Modell B, Darlison M (2008) Global epidemiology of haemoglobin disorders and derived service indicators. Bull World Health Organ 86(6):480–487. https://doi.org/10.2471/BLT.06.036673
72. Serjeant G (2017) World sickle cell day: lessons for India. Indian J Med Res 145(6):705

Part IV
Services

The two articles in this part (12. "Medical, Rehabilitation and Social Welfare Services for Children with Birth Defects and Developmental Disabilities in India" and 13. "Early Childhood Intervention Services in India") discuss the services that are available for children with birth defects and developmental disabilities in the country. Both articles identify the lack of coherence between policies, and lack of referral pathways that can direct caregivers to needed medical, physical and social welfare services.

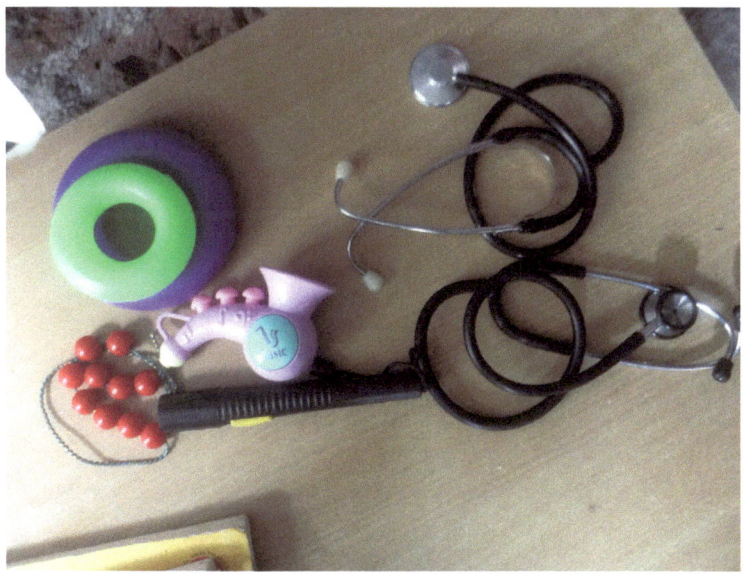

Chapter 12
Medical, Rehabilitation and Social Welfare Services for Children with Birth Defects and Developmental Disabilities in India

Anita Kar

Abstract Birth defects and developmental disabilities are highly disabling, life-long conditions. Children have special medical and rehabilitation needs. Families require additional social support. The essence of an efficient birth defects service is to ensure that all these activities are integrated through an organized referral system, so that caregivers can access needed services. India forms an example of an emergent birth defects service. Several programme components of the maternal and child health services and social welfare services for persons with disabilities can cater to the needs of children and their families. This article describes these programme components which span through the life course. There are services for preterm and sick newborns, screening and medical care for specific types of birth defects and developmental disabilities, and early intervention services. Several social welfare programmes are available for children with disabilities and their care-givers. The article concludes that India already has in place medical and rehabilitation services for birth defects. These services need to be strengthened and referral linkages between medical and welfare services have to be created to ensure that caregivers are not confronted by a poorly functioning, fragmented set of services.

Keywords Congenital · Disability · Children · Rehabilitation · Social welfare

Birth defects, exemplified by a diverse group of congenital disorders like spina bifida, congenital heart defects and clubfoot, and developmental disabilities like Down syndrome, autism or intellectual disability, are major causes of childhood disabilities and chronic medical conditions. Children have special healthcare needs that include specialized medical care and physical and social rehabilitation services (Fig. 12.1). These services are not usual components of child health services, especially in low- and middle-income countries (LMICs). Routine child health services provide immunization, supplemental nutrition and care for common childhood diseases and deficiencies. Several birth defects require additional services, such as life-saving surgeries for newborns with spina bifida, imperforate anus, critical

A. Kar (✉)
Birth Defects and Childhood Disability Research Centre, Pune 411020, India

© Springer Nature Singapore Pte Ltd. 2021　　　　　　　　　　　　　　275
A. Kar (ed.), *Birth Defects in India*,
https://doi.org/10.1007/978-981-16-1554-2_12

congenital heart defects, omphalocele and other malformations. Paediatric surgical services (including specialists) are limited in LMICs [1], so that there are missed opportunities for not only reducing preventable mortality among newborns, but also limiting complications and disabilities [2]. Although correctable, several birth defects cannot be fully cured, so that complications and comorbidities persist, requiring care throughout life.

The need for physical rehabilitation services is frequent. Surgically corrected spina bifida, for example, is associated with physical impairment and loss of lower body mobility. Routine physical rehabilitation therapies are required. Pressure sores and incontinence due to impaired bowel and bladder control hamper routine life. Hydro-cephalus, often but not always associated with spina bifida, occurs from accumulation of fluid in the ventricles of the brain. Despite surgery, hydrocephalus is associated with impairment in motor and executive skills, epilepsy and difficulties in learning, attention and behaviour. Physiotherapy and occupational therapy are required for rehabilitation. Down syndrome is a common birth defect where children are born with mild to severe intellectual impairment. Rehabilitation needs may encompass physiotherapy, speech and auditory therapy, occupational and behavioural therapy. Some children with Down syndrome may require surgery for congenital heart defects. Cerebral palsy is yet another example where children require extensive physical reha-bilitation. For young children under three years of age, these therapies are offered through early intervention services, whose goal is to assist in achieving basic skills of daily living (Chap. 13). Rehabilitation services are underdeveloped in LMICs [3], enhancing the complications resulting in disability, morbidity and premature mortality.

Other congenital disorders such as epilepsy, or those caused by single-gene disor-ders require specialized medical management, often with high cost therapeutic prod-ucts. Patients with haemophilia, a bleeding disorder, require prophylactic manage-ment with clotting factor concentrate (Chaps. 5 and 16). The cost of this replacement therapy limits access to optimal treatment [4], and parents incur extensive out-of-pocket expenditure in countries where haemophilia care is not subsidized through public services or insurance [5]. Children with transfusion-dependent thalassemia require frequent transfusions, but the cost of iron chelators makes excessive finan-cial demands on parents [6]. Children may require splenectomy and other treat-ments such as those required to address endocrine complications. Routine monitoring involves high-cost diagnostics. Without access to services, frequent morbidity and comorbidities result in premature mortality and progressive disability.

In addition to medical and physical rehabilitation services, children require social rehabilitation, which include services such as regular or special schooling, vocational training and skilling, and opportunities for employment. Recreational facilities for children and adults with disabilities are important, as they may not be able to share common facilities with others. Of prime importance are regulations to safeguard the rights of children and adults with disabilities. Thus, while a birth defects service needs to address specific medical needs and physical rehabilitation, appropriate social welfare measures to ensure social rehabilitation are of equal importance.

Congenital disorders affect the quality of life of parents [7]. Parents receive information from multiple sources, that is, from medical professionals, therapists and educators, but are the main decision-makers and caregivers. The efficiency of caregiving depends not only on available facilities, their physical location, ease of access to these services and costs incurred, but also on parent comprehension of how these services are to be used and their ability to organize and manage multiple therapies, medical appointments, education and daily living activities [8]. Birth defects affect family life and alter family routine from the time of diagnosis, with new challenges emerging as the child grows up. Psychosocial support for parents is essential. Knowledge on where to access medical and rehabilitation services including assistive devices, medical equipment and therapeutics, information on schooling, recreation and skilling are required. Subsidized and accessible transportation for moving the child is required, as many types of birth defects cause movement difficulty. Home assistance in caring for the child, day care and even respite care, where others care for the child while parents take a break from the challenging task of caregiving, is needed. Legal guardianship is also of prime importance, where a family member is granted the legal right to decide for a person with a disability who is unable to take decisions. Disability allowance is a valuable social welfare measure, as children with birth defects have higher utilization of healthcare services, and medical visits and hospitalizations may be more frequent, complicated and expensive. Physical therapy is expensive and required for a significant duration of time. Thus, children as well as their parents are important stakeholders and beneficiaries of a birth defects service (Fig. 12.1).

Without access to these services, the quality of life of children is affected, and attainment of possible skills and social participation are delayed. Social isolation, higher rates of depression, suicidal ideation and poorer quality of life are common. Access to appropriate resources is important, as quality of life is impaired when parents have unmet needs [9, 10]. Social isolation among families is common, and the medical fragility of the child or the physical disability may cause families to withdraw from social activities [7]. These needs, shown in Fig. 12.1, underline the importance of a comprehensive birth defects service.

Organization of Health, Rehabilitation and Social Welfare Services in India

Birth defects services are emerging in LMICs. India represents a typical example. In India, medical and rehabilitation care for birth defects is available through multiple ministries (Fig. 12.2). Medical and physical rehabilitation services are provided through the Ministry of Health and Family Welfare (MoHFW) [11]. The Department of Empowerment of Persons with Disabilities (DEPwD) (a department in the Ministry of Social Justice and Empowerment, MSJ&E) is entrusted with "the task of physical, educational, economic and social rehabilitation and empowerment of

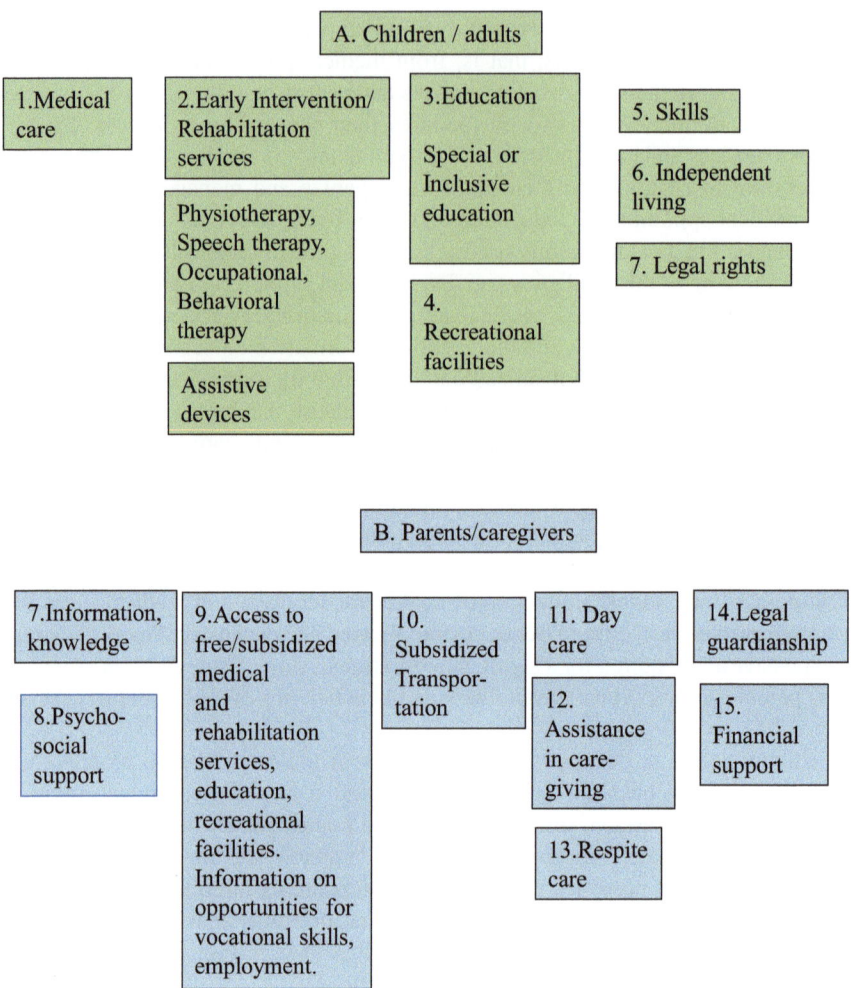

Fig. 12.1 Essential service needs for children and caregivers

people with disabilities" [12]. Other ministries are involved in education, sports, skill development and employment generation activities (Fig. 12.2).

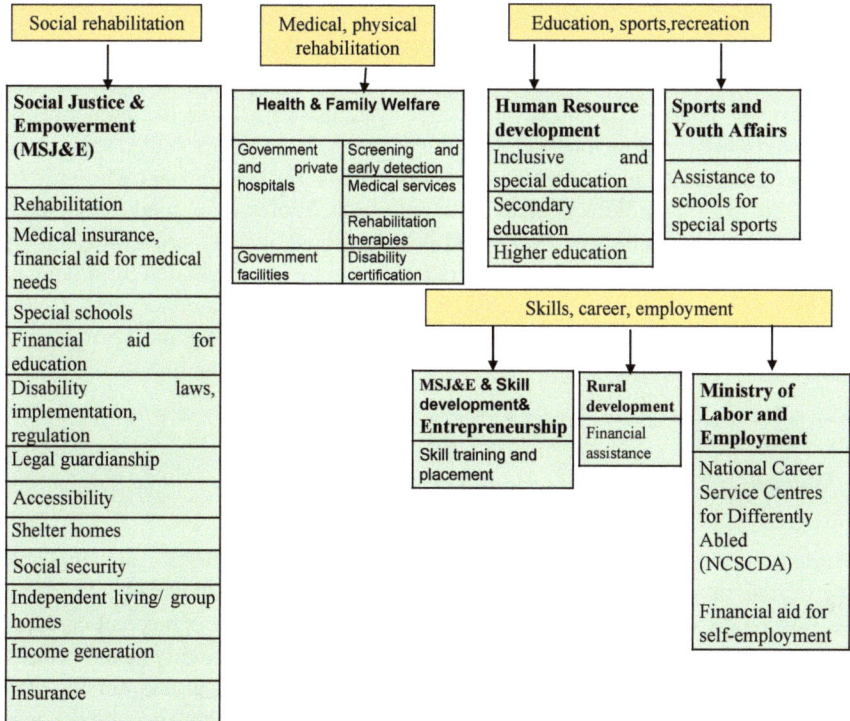

Fig. 12.2 Ministries offering services for children with disabilities

Medical Services

Organization of Government and Private Medical Services in India

Children with birth defects in India can access medical care from government (public) [13] or private hospitals [14]. Public hospitals provide free of cost or highly subsidized services. Private hospitals and clinics provide services on payment basis. Public sector services are limited. Private hospitals often offer the state-of-the-art medical care. In 2017–18, a nationwide survey reported that 65% of outpatient and 55% of inpatient users sourced care from private hospitals and private doctors [15]. Public services were the mainstay of families belonging to the lowest socioeconomic strata. Personal insurance was low, with 81% of urban and 86% of rural population reporting not having any personal insurance [15]. A national insurance scheme for families below the poverty line (Pradhan Mantri Jan Arogya Yojana) and schemes offered by state governments can be approached for financial assistance to support hospitalization costs, including surgeries.

The Indian private healthcare sector is very large and is made up of small and large hospitals and medical practitioners in individual practice. In addition to modern medicine, there are practitioners from the Indian systems of medicine such as Ayurveda, unani or siddha, or alternative systems of medicine like homeopathy. The quality of private medical sector services ranges from highly skilled specialists, practising in well-equipped medical centres, to general practitioners who may be unfamiliar with birth defects or their management. Professional medical societies are involved in continuing education and skill development, but there is little activity in terms of skilling general practitioners in management of common birth defects. Furthermore, there is incomplete regulation of private healthcare providers, so that there is no monitoring of types, quality, efficacy or cost of treatments provided to patients.

Government General Medical Services in India

Medical services for children with birth defects and disabilities are available at government medical college hospitals or district hospitals. Rural medical services are offered through sub-district hospitals which provide routine medical care and minor surgeries. Basic health services are available from community health centres, Primary Health Centres and sub-centres. In addition to medical and nursing staff at health facilities, community health workers (CHWs), like the Accredited Social Health Activist (ASHA), health educators and other workers, are in routine contact with communities. The CHWs are supervised by the medical officer at the Primary Health Centre. Birth defects, genetic disorders and developmental disabilities may be suspected at these facilities, and parents may be referred to tertiary care facilities for further investigation and treatment.

Specific services are available for birth defects. These include care for small and sick newborns, screening and early intervention for children with birth defects and developmental disabilities (Rashtriya Bal Swasthya Karyakram, RBSK) [16], early intervention for deafness including provision of hearing aids [17], a programme for addressing blindness and vision impairment through cataract surgeries and distribution of free eyeglasses for children [18] and a national mental health programme with provision of day care and continuing care facilities in addition to therapeutic care [19].

Birth Defects Service Components in the Maternal and Child Health Programme

Birth defects services are available through different components of the maternal and child health (MCH) programme (Reproductive, Maternal, Newborn, Child and

Adolescent Health (RMNCH + A) programme) [20]. Under this programme, early registration, antenatal check-up and deliveries are conducted at sub-centres and Primary Health Centres, sub-district, district and medical college hospitals. In 2017–18, 29 million women registered for antenatal services [21]. In 2017–18, 69% of rural and 48% of urban births occurred at government hospitals [15].

Prenatal Detection of Foetal Anomalies

Women are offered ultrasound scans, mostly during the first trimester (dating scans), and anomaly scans during the second trimester. Data on the numbers of congenital anomalies detected are unavailable. A national survey identified that the usage of ultrasonography during pregnancy was 61% in 2015–16 [22]. However, details in terms of gestational age when ultrasound investigation was conducted are unavailable. Ultrasound technology is widely used in India, but it is not strongly promoted, due to its use in sex-selective abortion. A cohort study from Pune identified that among 791 pregnant women, the average number of ultrasound scans received by the women was 3.6 ± 1.3 (range 0–7). Majority were dating scans, and only one-third of the women (266, 34%) reported a scan during 18–20 weeks of gestation [23]. Although anomaly scans appeared not to be frequently used in this cohort, nearly half of the 42 major anomalies (20, 48%) were detected prenatally, giving a congenital anomaly prenatal diagnosis prevalence of 10.98 per 1000 births [23].

In India, medical termination of pregnancy (MTP) is legal [24]. A recent amendment (The Medical Termination of Pregnancy (Amendment) Bill, 2020) permits pregnancy termination up to 24 weeks of gestation if there is risk of "physical or mental abnormalities in the child (if born) so as to be seriously handicapped" [24]. This upper limit is not applicable in case of detection of substantial foetal anomalies. Pregnancy termination beyond 24 weeks of gestation requires certification by a Medical Board. Although all causes of pregnancy termination are documented, these data are not public. The cohort study described above identified that out of 20 malformations detected in the prenatal period, eight pregnancies were terminated after detection of congenital anomalies. This gave an elective pregnancy termination rate of 4.39 per 1000 births. In this cohort study, one-fifth of severe malformations were avoided through prenatal detection followed by elective termination of the pregnancy [23].

Care of Small and Sick Newborns

Newborns with birth defects may be born before term, have low birth weight or require urgent medical or surgical services. The RMNCH + A service in India offers essential newborn care as well as facility-based and home-based care. Facility-based newborn care refers to clinical care delivered by trained personnel [25]. Services consist of resuscitation care including provision of warmth, early initiation of breastfeeding and initial care for sick newborns. Newborn Stabilization Units (NBSUs) have been established at sub-district-level centres. Special Newborn Care Units (SNCUs) are available at district level, and Neonatal Intensive Care Units with capabilities for offering assisted ventilation, surfactants and surgery are available at selected district hospitals. Till middle of 2017, 712 SNCUs and 2321 NBSUs have been established across India.

Financial provisions ensure that parents are protected from financial risk in case a baby needs to be shifted to a higher-level facility. The Janani Shishu Suraksha Karyakram (JSSK) provides free of cost hospital and diagnostic services and free transportation from and to the mothers' residence [24]. With specific reference to birth defects, this service provides transportation facilities, so that a newborn requiring emergency care can be transported to a higher-level facility from community centres. In addition to transportation, high-risk infants discharged from SNCUs are followed up by CHWs till one year of age. Home-based newborn care involves 6 visits within 42 days by healthcare workers for all births occurring at public institutions. This ensures that any problems with the newborn, including birth defects, can be identified, and the baby transferred from the community to medical facilities.

Medical Care for Selected Birth Defects and Developmental Disabilities: RBSK

After a newborn is discharged from the health facility, contact with the health system is restricted to immunization visits, or if the child is ill and is brought to the health centre by parents. Immunization is usually conducted by nursing staff, so that contact with a paediatrician is limited. Growth monitoring is done during immunization visits and subsequently through public play schools (anganwadis). There is, however, no monitoring of child development, so that developmental delays or growth faltering due to an underlying morbidity may not be detected till the parents present to a medical provider. Studies show that even though parents are aware of problems in the child, advice may be sought from multiple lay people (relatives, friends, neighbours, teachers), and children may be taken to traditional healers before the appropriate professional is reached [26].

In 2013, services for birth defects and developmental delays and disabilities were introduced in a child screening and intervention service, the Rashtriya Bal Swasthya Karyakram (RBSK) [27]. The RBSK service has three key components (Fig. 12.3) [9]. The first component is community-based screening for common childhood diseases, nutritional deficiencies, selected birth defects and developmental disabilities. The conditions included in the latter two groups are congenital heart defects, Down syndrome, cleft lip and palate, congenital talipes equinovarus, neural tube defects, developmental dysplasia of hip, retinopathy of prematurity, congenital deafness, congenital cataract, vision, hearing, neuromotor impairments, motor, cognitive

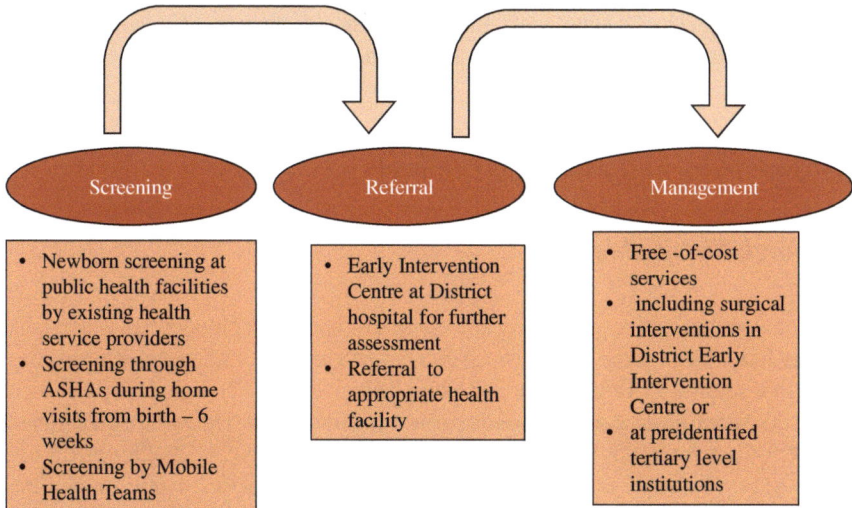

Fig. 12.3 RBSK model

and language delay, behaviour disorder and attention deficit hyperactivity disorder. Screening for congenital hypothyroidism, sickle cell anaemia and thalassemia is recommended in areas where these conditions are prevalent. As per the plan, the service is to initiate with newborn examination at birth. Resource materials have been created for nurse midwives for physical examination of newborns for birth defects [28]. However, this activity is yet to be implemented across the country. ASHAs are entrusted with the task of identifying children with birth defects and disabilities from communities. These community health workers have been provided with a pictorial manual for identifying birth defects and trained in the process of referral to nearest health facility [27, 28].

The most active component of the RBSK is screening of children between 2 and 18 years of age at government preschools (anganwadis) and public schools in rural and urban areas. Screening is conducted by mobile teams, which include appropriately trained ayurvedic doctors, nurse and pharmacist. Screening is conducted using standardized tools [29].

The second component of the model is referral to District Early Intervention Centres (DEIC) [30], which forms the central hub from where children are referred for either medical services (such as surgery for congenital heart defects, repair of orofacial clefts, orthopaedic care for developmental dysplasia of the hip and other conditions) and/or rehabilitation services. The functions of the DEIC are depicted in Fig. 12.2. As specialized skills (such as paediatric cardiac surgery) and infrastructure are of limited availability at government hospitals, the RBSK has partnered with private hospitals where specialist services and necessary infrastructure are available. International non-government organization like the Smile Train [31] is also a partner.

Services are provided at negotiated costs with the health ministry [32], so that parents do not incur any costs, even for complex surgeries.

Medical Services for Genetic Disorders

Haemoglobinopathies

A national guideline for prevention and control of haemoglobinopathies is available [33], as India has a large number of patients and carriers of haemoglobinopathies. Beta thalassemia carrier prevalence is estimated around 3–4%, indicating that there may be 35–40 million beta thalassemia carriers in the country. The estimated numbers of patients (between 100,000 and 150,000) are also very high [34]. Sickle cell disease is also of considerable magnitude in India, with an estimated 1.4 million children with sickle cell disease and 18 million with sickle cell trait [35]. Taken together, at an estimated prevalence of 1.2 per 1000 live births, there are likely to be 324,000 births with serious haemoglobin disorders each year in India [34]. A spatial map of distribution of sickle cell anaemia in India identifies that the disorder is prevalent in specific geographical regions of the country [36]. Prevalence estimates of haemoglobinopathies in India are however compromised by non-random samples and poor study designs (Chap. 12).

Providing treatment for such a magnitude of patients exerts considerable challenge for public health planners. In India, 50 and 67% of patients with beta thalassaemia major were being treated with iron chelation therapy [37]. Treatment for thalassemia is partially subsidized and available free of charge from government hospitals and health centres in areas where the condition is prevalent. Leucodepleted, packed red blood cell is recommended for transfusion of patients with thalassemia. As many blood banks lack component separation facilities, transfusion often requires travel to distant centres where leucodepleted, packed red blood cell is available. Where leucodepletion facilities are unavailable, bedsides filtration is the usual method that is adopted. Iron overload is one of the major consequences of repeated transfusions. Desferrioxamine, provided subcutaneously via an infusion pump over a period of 8–12 h, is still used at government hospitals, even though two oral chelators, deferiprone and deferasirox are being used by patients who are able to afford these drugs.

Transfusion services include compulsory screening of blood for HIV, hepatitis B, hepatitis C, malaria and syphilis, but the more sensitive nucleic acid amplification tests are not conducted. Iron overload is a complication arising from repeated transfusions. Serum ferritin levels are monitored, and organ-specific iron overload is monitored using MRI T2* (Magnetic Resonance Imaging T2 star) which permits the visualization of hemosiderin deposits. Hematopoietic stem cell transplant (bone marrow transplant) is not supported by government funds. Day care centres are supposed to be created at DEICs, but currently, transfusion services are associated with haematology departments of government hospitals. The state-of-the-art thalassemia management services are available at private hospitals, but limited to those who can afford these services. Despite availability of government services, a recent study showed that the expenditure on thalassemia treatment in India ranges from US$629 to US$2300, with nearly half of expenditure being spent for iron chelation and other medications [6].

In India, sickle cell programmes have been set up in states like Gujarat, Maharashtra, Madhya Pradesh and Tamil Nadu, where the condition is prevalent among tribal populations [38]. Due to the remoteness of these populations, service usage is low, and mortality during childhood is high. The most visible component of these programmes is carrier screening, followed by issuance of coloured cards to carriers, with counselling on marriage. Antenatal diagnosis at national centres is another service, but is challenging due to community hesitation, difficulty in transportation and other issues. In terms of medical care, supportive treatment is offered. As children with sickle cell disease are at a higher risk of infections due to compromised spleen function, pneumococcal vaccination and daily folic acid are provided. Hydroxyurea treatment is provided free of charge to patients below the poverty line. Patients with sickle cell disease experience repeated bouts of crisis caused by vaso-occlusion that is accompanied by severe pain. Non-steroidal anti-inflammatory drugs (NSAIDs) are provided free of cost from health centres. Children with any signs of infection are treated with broad spectrum antibiotics, and in few areas, penicillin prophylaxis is being considered [38].

Haemophilia

Haemophilia is a highly prevalent bleeding disorder, affecting 1 in 5000 males [39]. Patients lack or have low levels of clotting factor, predisposing them to frequent bleeding episodes. The two most prevalent types of bleeding disorders are haemophilia A and haemophilia B, caused by deficiency of clotting factor VIII and IX, respectively. Despite under-diagnosis, India reports the highest numbers of people living with haemophilia globally (20,778), exceeding that of the USA, which reported 17,757 patients in 2018. In 2018, there were 17,606 patients with haemophilia A (US 13,616). There were 2715 patients with haemophilia B (US 4141) [40]. There were 769 patients with clinically significant inhibitors, slightly less than that in the USA (893) [40].There were 114 HIV positive patients with haemophilia.

Haemophilia A and B are expensive to treat. Plasma-derived clotting factor concentrate which is used by 60% of patients is imported [40]. Although there is no haemophilia programme, currently, clotting factor concentrate is being provided to patients, free of cost, in some states of the country. This programme was initiated after litigation by patient organizations across the states, which led to judicial directives to ensure free clotting factor concentrate for patients [41]. Prior to this, out-of-pocket expenditure for treatment was unaffordable, leading to suboptimal usage [4]. Infrequent usage is not only associated with painful and prolonged incapacitation due to bleeding, but associated with joint damage, and progressive, crippling disability. Lack of access to clotting factor concentrate can also be life-threatening.

Without government subsidy, cost of treatment was estimated to be 21–314 times the monthly family income, with nearly 70% of families experiencing catastrophic expenditure [5]. Expectedly, provision of free clotting factor concentrate has significantly reduced the economic burden on parents in India [42]. Despite an excellent model of genetic counselling through a partnership between the haemophilia non-government organization and a national institute [43], there is no genetic counselling service. The ad hoc nature of the haemophilia programme is illustrated by the fact that it is not associated with the RBSK, but maintained under the State Blood Transfusion Centre, and is a stand-alone programme.

Rehabilitation Services

Rehabilitation services are available through DEICs under the RBSK programme and through national institutes of rehabilitation and their satellite centres. These services are free of charge, but are limited in availability. Rehabilitation services are also available at private hospitals. There are large numbers of physiotherapists, speech and occupational therapists, special educators and remedial teachers in individual practice. Like medical services, there is no regulation of charges or monitoring of quality of services, so that parents encounter considerable financial expenditure. One of the major issues is that there is no coordination between medical and rehabilitation services, so that the duration and type of treatment are decided by parents and most frequently guided by economic considerations.

Rehabilitation Services Through DEICs

Rehabilitation services are offered for children below the age of six years through the DEIC. Figure 12.4 summarizes the DEIC activities [30]. Occupational and physical therapy services are offered in order to nurture self-help skills, adaptive behaviour, play, sensory and motor functions and to prevent and reduce movement difficulties. Psychological services include evaluation of developmental milestones, learning and mental health status. Counselling to familiarize parents in caregiving are provided by clinical psychologists. Identifying cognitive delays and providing occupational therapy to enhance cognitive development, adaptive and learning behaviours are another service offered from the DEIC. Audiology services are available at the DEIC or at the hospital department for children with hearing impairments including those with congenital deafness. Vision services are available for children with visual disorders [30].

The early intervention activities are backed up with laboratory services for routine blood investigations. It is envisioned that these services will subsequently be scaled up to offer services for confirming congenital hypothyroidism, thalassemia and sickle cell anaemia or other inborn errors of metabolism. Nutrition services are provided by

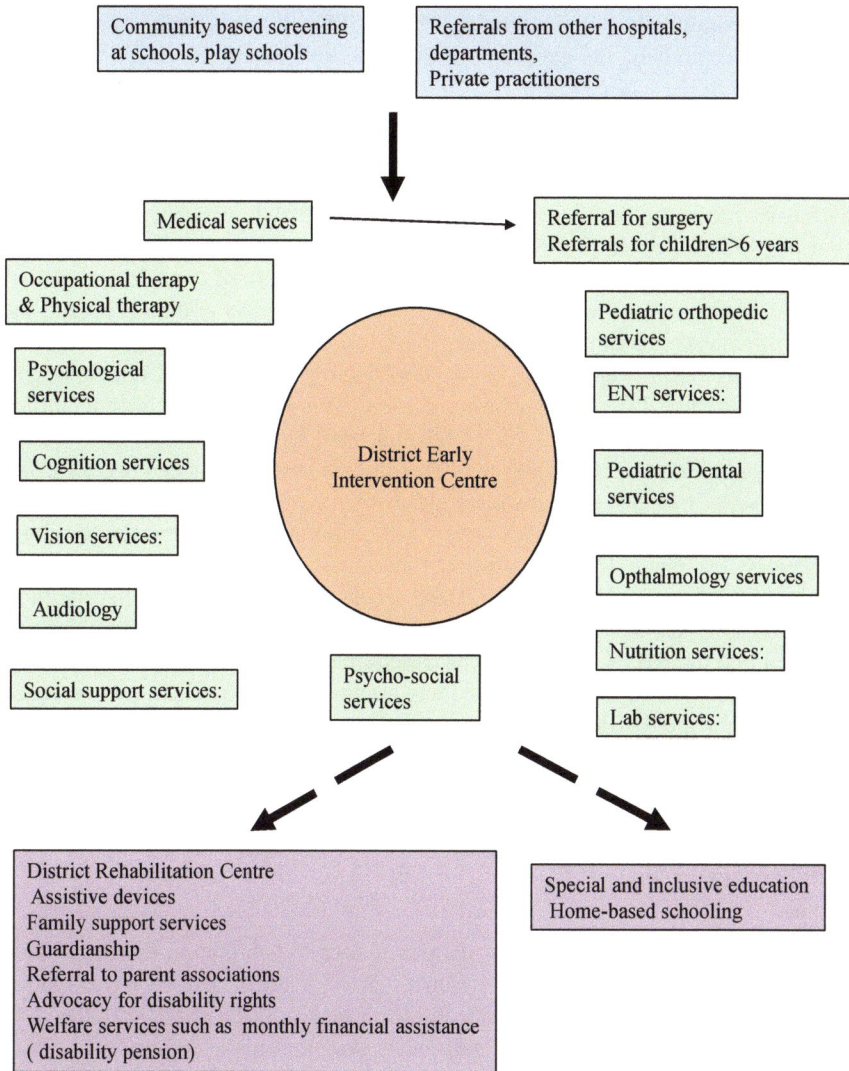

Fig. 12.4 DEIC services

nutritionists/dietician at the DEIC. Social support services at the DEIC link families with needed social services. Special educators are available for designing learning environments and activities that promote the child's development. The DEIC has the responsibility for providing families with information, skills and support to enhance the child's development [30]. Parents are guided to a disability board to facilitate disability certification process.

As of September 2017, there were only 92 functional DEICs (India has 723 districts) [44]. There were 11,020 mobile teams in 36 state/union territories conducting community-based screening. Nearly 825 million children were screened since the inception of the programme in 2013, and nearly 20 million children benefitted from the service [32].

Despite this impressive plan, DEIC services, if available, are plagued by lack of staff and equipment. A study from Chhattisgarh reported that only 30–40% of children reached DEICs [45]. Poor uptake of DEIC services was ascribed to unwillingness on parents to continue treatment at the DEIC and oversight on part of the team providing referrals [7, 45, 46]. Uptake and compliance were especially low for children with developmental disabilities. The necessity of considerable paperwork for availing of these services, lack of specialized tests at the DEIC and the need for personal expenditure to access these tests, lack of specialists, distance to the DEIC and loss of wages were factors contributing to suboptimal functioning of services. A study of 115 parents who were users of the RBSK programme identified high usage of medical services for congenital heart defects and other correctable conditions [7]. However, the distance to district early intervention centres, difficulty and expense of transporting a disabled child, comprehension of the complexity of treatment, social support, attitudes towards disability and competing family issues were factors leading to discontinuation of rehabilitation services. Another issue was that there was no referral between medical services and social welfare services so that most caregivers were unaware of these services [7]. Overall however, the study identified that the RBSK programme provided treatment opportunities for children with birth defects who had not received any treatment until they had been screened and referred by the RBSK mobile teams.

Rehabilitation Services Through National Institutes and Centres

There are eight national institutions that provide free rehabilitation services. These institutes are also involved in rehabilitation research for guidance of the rehabilitation policy in India [47]. Several of these institutes have satellite centres for outreach of services. The services are limited to selected major cities in the country.

Social Welfare Services

Disability Certification

In India, disability certification is necessary for availing free of cost medical and social welfare services for people with disabilities. A medical model of disability is adopted in India. There are standard guidelines and medical tests for evaluation of disabilities arising from 21 different medical conditions, which are recognized under the Rights of Persons with Disabilities Act, 2016 (RPwDA) [48]. Physical disabilities that are recognized under the Act are locomotor disabilities arising from different causes. These may be medical (cerebral palsy, leprosy, extreme physical disability, dwarfism) or arise from social issues (acid attack victim). Other categories of physical disability include visual impairment (blindness and low vision), hearing impairment (deaf and hard of hearing), and speech and language disability. The second category is intellectual disability that includes specific learning disabilities (perceptual disabilities, dyslexia, dysgraphia, dyscalculia, dyspraxia and developmental aphasia) and autism. A third category is mental behaviour (mental health conditions). A fourth category includes disability caused by chronic neurological conditions (multiple sclerosis and Parkinson's disease) and blood disorders (haemophilia, thalassemia, sickle cell disease). A fifth category is multiple disabilities. Certification of visible disabilities can be done by the Medical Officer in community settings at Primary Health Centres, or a specialist can be invited for the certification. Multiple disabilities require certification by a Medical Board.

Welfare services for people with disabilities in India are guided by a national policy for persons with disabilities [49]. There is an elaborate system in place, with several ministries offering services. There is however no centralized information system, so that users are unaware about these services and how to access them.

Disability Policy and Legislations

Disability services in India are informed by several acts and policies to ensure the rights of persons with disabilities in the country. In 2006, the United Nations General Assembly adopted the Convention on the Rights of Persons with Disabilities (CRPD) adopting the principles of respect, non-discrimination, inclusion, diversity, equality, accessibility and rights of persons with disabilities [50]. The Act laid down the ten principles for the empowerment of people with disabilities (Box 12.1). India ratified the Convention in 2008. The two important obligations that arose out of the Convention were the need to implement the provisions of the CRPD and harmonization of Indian laws with the CRPD principles.

Box 12.1 Key points of the UN Convention on the Rights of Persons with Disabilities (from 50)

- respect for inherent dignity, individual autonomy including the freedom to make one's own choices;
- independence of persons;
- non-discrimination; full and effective participation and inclusion in society; respect for difference and acceptance of persons with disabilities as part of human diversity and humanity;

- equality of opportunity;
- accessibility;
- equality between men and women;
- respect for the evolving capacities of children with disabilities and respect for the right of children with disabilities to preserve their identities.

In 2006, India's National Policy for Persons with Disabilities was enacted which recognized persons with disabilities as "… valuable human resources for the country …" and sought to "… create an environment … that would provide … equal opportunities, protection of rights and full participation in society …" [49]. The National Policy recognized that most persons with disabilities can lead a better quality of life if equal opportunities and effective access to rehabilitation measures are provided. The National Policy flagged certain key components including prevention of disabilities, rehabilitation measures, economic rehabilitation, opportunities for education and vocational training, disability data collection, disability research, opportunities for sports, recreation and barrier-free environment, cultural life, social security, a special focus on women and children and promotion of non-governmental organizations supporting persons with disabilities. In 2016, the Rights of Persons with Disabilities Act, 2016 (RPwD), gave effect to the United Nations Convention on the Rights of Persons with Disabilities [48].

Social Rehabilitation Services

There is a large repertoire of welfare policies and services in India (Fig. 12.2). As mentioned above, the services are fragmented and disconnected, and getting information about and access to these services is challenging.

The National Trust

The National Trust for Welfare of Persons with Autism, Cerebral Palsy, Mental Retardation, and Multiple Disabilities is a statutory organization that provides care for children with intellectual impairment [51]. The National Trust provides early intervention and day care centres. Residential care is available for persons from below the poverty line. The National Trust has a scheme of training skilled disability assistants. There is provision for respite care for parents. The Trust provides funds for students with these disabilities for enrolling into professional courses and for pursuing higher studies.

Education, Skills, Sports, Employment

Inclusive or special education is available for children with disabilities between the ages of six to sixteen, followed by pre-vocational training (16–18 years) and vocational training (18 years) opportunities. The Sarva Shiksha Abhiyan (SSA), an initiative to promote universal elementary education, deems the education of children with special needs as an important component of universal elementary education [52]. It provides fee waiver and scholarships to support the education of children with disabilities [53]. Physical participation in sports is encouraged by providing assistance to schools for training programmes and sports competitions for children with disabilities.

National Career Service Centres for Differently Abled have been developed for skilling and employment. A National Action Plan for skill training and placement of skilled workforce has been developed [47]. The National Handicapped Finance and Development Corporation promotes economic independence of people with disabilities through skill training and self-employment ventures, and by providing loans on flexible conditions to people with disabilities. There is reservation for people with disabilities in public employment. Non-governmental organizations are also supported for marketing products made by people with disabilities.

Children with disabilities receive a nominal monthly allowance, and in some places, families also receive financial support. There is income tax exemption for families with a person with a disability. Travel concession is available for people with disabilities and an accompanying person.

Medical Insurance

Medical insurance up to 100,000 rupees, for expenditure on medicines, diagnostic tests, regular medical evaluation, dental treatment and surgery to prevent complications of disability, is available [47]. Non-surgical treatments, hospital costs for surgeries for children with cerebral palsy, autism, intellectual impairment and multiple disabilities are available.

Assistive Devices

Indigenous prosthetic and assistive devices are manufactured by the Artificial Limbs Manufacturing Corporation which was established in 1970. Assistive aids are distributed free of cost through camps organized by District Disability Rehabilitation Centres. The functioning of these centres is entrusted with NGOs. NGOs provide information on aids and assistive devices, and spread awareness about disability and the various available schemes in the community.

Certification of Rehabilitation Professionals

All rehabilitation professionals and organizations offering rehabilitation services in India need to be registered with the Rehabilitation Council of India, the body that is entrusted with the development of human resources in rehabilitation and special education in the country. This organization is also responsible for maintaining a database of rehabilitation professionals.

Legal Services

Ensuring legal rights and provisions such as legal guardianship are entrusted with the Chief Commissioner for Persons with Disabilities. Legal duties are discharged through Local Level Committees in each district. These Committees oversee the process of providing legal guardianship that is handing over decision-making powers to legal guardians when decision-making abilities are impaired. All these welfare activities are available free of charge [53].

Conclusion

Medical, physical rehabilitation and social welfare services for birth defects are available in India, but referral pathways have not been developed so as to ensure that caregivers can access a holistic platform of needed services. The RBSK is a major development that identifies children with birth defects from community settings and provides opportunities for care [7]. The service however needs to be fully implemented across the country, and the scope and quality of services need to be strengthened. Most importantly, evaluation of newborns prior to discharge has to be implemented, as birth defects are most likely to be identified at this age. One of the important components of the RBSK is the establishment of early intervention centres. This is the first instance of providing rehabilitation for children with disabilities in India through the public health service. However, the urban, hospital-based nature of rehabilitation services can be considered a deterrent, as caregivers may not have the means for transporting the child for therapy sessions [7]. Social welfare services are available, but these services are implemented through NGOs, without any public documentation on the quality of their functioning.

It is well known that people with disabilities encounter several barriers in accessing medical and rehabilitation services [3, 54]. Several of these barriers arise from poor functioning of services and lack of intersectoral coordination [55]. There is no centralized information system, so that parents usually learn about services through personal efforts. Disability magnifies equity issues, as in India, disability is more prevalent among socio-economically deprived [56]. Special education is most often limited to the affluent and obtained on payment from private rather than government special schools [57]. People with disabilities are less likely to be employed and are at a higher risk of non-communicable diseases and depression [58], experiencing significantly more barriers in accessing health services. Majority of these barriers arise from ignorance about availability of free government services, the poor organization of these services and the cost of paying for services and transportation [58]. Specifically for birth defects and developmental disabilities, well-planned studies are needed to explore caregivers knowledge and understanding about these conditions, care seeking, types of practitioners and treatments, cost and barriers encountered. Health service research for alternative models of services for these conditions also needs to be explored.

References

1. Wright NJ, Anderson JE, Ozgediz D, Farmer DL, Banu T (2018) Addressing paediatric surgical care on World Birth Defects day. Lancet 391(10125):1019
2. Higashi H, Barendregt JJ, Kassebaum NJ, Weiser TG, Bickler SW, Vos T (2015) The burden of selected congenital anomalies amenable to surgery in low and middle-income regions: cleft lip and palate, congenital heart anomalies and neural tube defects. Arch Dis Child 100(3):233–238
3. Bright T, Wallace S, Kuper H (2018) A systematic review of access to rehabilitation for people with disabilities in low-and middle-income countries. Int J Environ Res Public Health 15(10):2165
4. Dharmarajan S, Gund P, Phadnis S, Lohade S, Lalwani A, Kar A (2012) Treatment decisions and usage of clotting factor concentrate by a cohort of Indian haemophilia patients. Haemophilia 18(1):e27–e29
5. Dharmarajan S, Phadnis S, Gund P, Kar A (2014) Out-of-pocket and catastrophic expenditure on treatment of haemophilia by Indian families. Haemophilia 20(3):382–387
6. Moirangthem A, Phadke SR (2018) Socio-demographic profile and economic burden of treatment of transfusion dependent thalassemia. Indian J Pediatr 85(2):102–107
7. Lemacks J, Fowles K, Mateus A, Thomas K (2012) Insights from parents about caring for a child with birth defects. Int J Environ Res Public Health 10(8):3465–3482
8. Kar A, Radhakrishnan B, Girase T, Ujagare D, Patil A. Community based screening and early intervention for birth defects and developmental disabilities: lessons from the RBSK programme in India (in press, Disability, CBR & Inclusive Development)
9. Pit-ten Cate IM, Kennedy C, Stevenson J (2002) Disability and quality of life in spina bifida and hydrocephalus. Dev Med Child Neurol 44(5):317–322
10. Kirpalani HM, Parkin PC, Willan AR, Fehlings DL, Rosenbaum PL, King D, Van Nie AJ (2000) Quality of life in spina bifida: importance of parental hope. Arch Dis Child 83:293–297
11. Government of India, Ministry of Health and Family Welfare. Available at https://www.mohfw.gov.in/. Accessed June 2019
12. Government of India, Department of Empowerment of Persons with Disabilities (Divyangjan). https://disabilityaffairs.gov.in/content/. Accessed June 2019
13. Government of India, Ministry of Health and Family Welfare Rural Health Statistics 2018–19. Available at https://www.thehinducentre.com/resources/article31067514.ece/binary/Final%20RHS%202018-19_0-compressed.pdf. Accessed June 2019
14. Mackintosh M, Channon A, Karan A, Selvaraj S, Cavagnero E, Zhao H (2016) What is the private sector? Understanding private provision in the health systems of low-income and middle-income countries. Lancet 388(10044):596–605

15. Government of India, Ministry of Statistics and Programme Implementation. Key indicators of social consumption in India: health NSS 75th round (July 2017–June 2018). Available at https://www.mospi.gov.in/sites/default/files/publication_reports/KI_Health_75th_Final.pdf. Accessed June 2019
16. Singh AK, Kumar R, Mishra CK, Khera A, Srivastava A (2015) Moving from survival to healthy survival through child health screening and early intervention services under Rashtriya Bal Swasthya Karyakram (RBSK). Indian J Pediatr 82(11):1012–1018
17. Garg S, Chadha S, Malhotra S, Agarwal AK (2009) Deafness: burden, prevention and control in India. Natl Med J India 22(2):79–81
18. Verma R, Khanna P, Prinja S, Rajput M, Arora V (2011) The national programme for control of blindness in India. Australas Med J 4(1):1–3
19. Wig NN, Murthy SR (2015) The birth of national mental health program for India. Indian J Psychiatry 57(3):315–319
20. Ministry of Health and Family Welfare. National health mission RMNCH+A. https://nhm.gov.in/index1.php?lang=1&level=1&sublinkid=794&lid=168. Accessed June 2019
21. Ministry of Health and Family Welfare. Health management information system. Available at https://nrhm-mis.nic.in/SitePages/Home.aspx Accessed June 2019
22. International Institute for Population Sciences (IIPS) and ICF (2017) National family health survey (NFHS-4), 2015–16: India. IIPS, Mumbai. Available at https://rchiips.org/nfhs/pdf/NFHS4/India.pdf. Accessed Mar 2018
23. Bhide P, Gund P, Kar A (2016) Prevalence of congenital anomalies in an Indian maternal cohort: healthcare, prevention, and surveillance implications. PLoS ONE 11(11):e0166408. https://doi.org/10.1371/journal.pone.0166408
24. The Medical Termination of Pregnancy (Amendment) Bill (2020) https://www.prsindia.org/sites/default/files/bill_files/The%20Medical%20Termination%20of%20Pregnancy%20%28Amendment%29%20Bill%2C%202020.pdf. Accessed Jan 2020
25. Toolkit for setting up special care newborn units, stabilisation units and newborn care corners. Available from https://www.unicef.org/india/SCNU_book1_April_6.pdf. Accessed June 2019
26. Chakraborty S, Kommu JVS, Srinath S, Seshadri SP, Girimaji SC (2014) A comparative study of pathways to care for children with specific learning disability and mental retardation. Indian J Psychol Med 36(1):27–32
27. Government of India, Ministry of Health and Family Welfare, National Health Mission (2013a) Rashtriya Bal Swasthya Karyakram A child health screening and early intervention services under NRHM. Ministry of Health and Family Welfare. Available at https://nhm.gov.in/images/pdf/programmes/RBSK/Operational_Guidelines/Operational%20Guidelines_RBSK.pdf. Accessed 6 Aug 2019
28. Government of India, Ministry of Health and Family Welfare, National Health Mission (2013b) Rashtriya Bal Swasthya Karyakram resource documents. Available from https://nhm.gov.in/images/pdf/programmes/RBSK/Resource_Documents/participants_manual_full.pdf. Accessed 18 Feb 2020
29. Government of India, Ministry of Health and Family Welfare, National Health Mission (2014) Rashtriya Bal Swasthya Karyakram job aids child health screening and early intervention services. National Rural Health Mission. Available from https://nhm.gov.in/images/pdf/programmes/RBSK/Resource_Documents/RBSK_Job_Aids.pdf. Accessed 18 Feb 2020
30. Rashtriya Bal Swasthya Karyakram, Ministry of Health and Family Welfare (2014) Resource manual for equipment and infrastructure for DEIC (including model DEIC) under RBSK. Available at https://nhm.gov.in/images/pdf/programmes/RBSK/Resource_Documents/Resource_Manual_for_equpment.pdf. Accessed June 2019
31. SmileTrain India. https://www.smiletrainindia.org/. Accessed June 2019
32. Rashtriya Bal Swasthya Karyakram, Ministry of Health and Family Welfare (2014) Procedures and model costing for surgeries. Available at https://www.nrhmhp.gov.in/sites/default/files/files/PROCEDURES%20AND%20MODEL%20COSTING%20RBSK.pdf. Accessed June 2019

33. National health mission guidelines on hemoglobinopathies in India. Prevention and control of hemoglobinopathies in India. Thalassemias, sickle cell disease and other variant hemoglobins (2016). Available at https://nhm.gov.in/images/pdf/in-focus/NHM_Guidelines_on_Hemoglo binopathies_in_India.pdf. Accessed Jan 2019
34. Colah R, Italia K, Gorakshakar A (2017) Burden of thalassemia in India: the road map for control. Pediatr Hematol Oncol J 2(4):79–84
35. Serjeant GR, Ghosh K, Patel J (2016) Sickle cell disease in India: a perspective. Indian J Med Res 143(1):675–677
36. Hockham C, Bhatt S, Colah R, Mukherjee MB, Penman BS, Gupta S, Piel FB (2018) The spatial epidemiology of sickle-cell anaemia in India. Sci Rep 8(1):1–10
37. Shah N, Mishra A, Chauhan D, Vora C, Shah N (2010) Study on effectiveness of transfusion program in thalassemia major patients receiving multiple blood transfusions at a transfusion centre in Western India. Asian J Transfus Sci 4(2):94–98. https://doi.org/10.4103/0973
38. Epidemic Branch. Commissionerate of Health, Family Welfare and Medical Services. Sickle cell anemia control program manual. Gujarat State. Available at https://nhm.gujarat.gov.in/Ima ges/pdf/SickleCellAnemiaManual.pdf. Accessed June 2019
39. Kar A, Phadnis S, Dharmarajan S, Nakade J (2014) Epidemiology and social costs of haemophilia in India. Indian J Med Res 140(1):19–31
40. World Federation of Hemophilia (2018) Report on the annual global survey 2012. https://www1.wfh.org/publications/files/pdf-1731.pdf. Accessed June 2019
41. Ghosh K (2019) Evolution of hemophilia care in India. Indian J Hematol Blood Transfus 35:716–721
42. Singh P, Mukherjee K (2017) Cost-benefit analysis and assessment of quality of care in patients with hemophilia undergoing treatment at National Rural Health Mission in Maharashtra, India. Value Health Reg Issues 12:101–106
43. Nakade J, Potnis-Lele M, Kar A (2013) Impact of genetic counselling? The potential utility of haemophilia surveillance data in developing countries. Haemophilia 19(6):e388–e390
44. Ministry of Health and Family Welfare (2019) Annual report 2017–18. Chapter 04, Child health programme. Available at https://main.mohfw.gov.in/sites/default/files/04Chapter.pdf. Accessed June 2019
45. Prabhu SA, Shukla NK, Roshni MS (2021) Rapid assessment of Rashtriya Bal Swasthya Karyakram program implementation and beneficiary feedback at two district early intervention centers in Chhattisgarh State in India. Curr Med Issues 19(1):3
46. Rameshbabu B, Kumaravel KS, Balaji J, Sathya P, Shobia N (2019) Health conditions screened by the 4Ds approach in a District Early Intervention Centre (DEIC) under Rashtriya Bal Swasthya Karyakram (RBSK) program
47. Ministry of Social Justice and Empowerment, Department of Empowerment of Persons with Disabilities (Divyangjan) (2019) Annual report 2017–18. Available at https://disabilit yaffairs.gov.in/upload/5bd14e4020c5dAnnual%20Report%202017-18%20(E).pdf. Accessed June 2019
48. The Gazette of India, Ministry of Law and Justice (2016) The rights of persons with disabilities act 2016. https://www.disabilityaffairs.gov.in/upload/uploadfiles/files/RPWD%20ACT% 202016.pdf. Accessed June 2019
49. Government of India, Ministry of Social Justice and Empowerment (2006) National policy for persons with disabilities. Available at https://disabilityaffairs.gov.in/upload/uploadfiles/files/ National%20Policy.pdf. Accessed June 2019
50. UN General Assembly (2007) Convention on the rights of persons with disabilities: resolution/adopted by the General Assembly. Available at https://www.un.org/en/development/desa/ population/migration/generalassembly/docs/globalcompact/A_RES_61_106.pdf. Accessed June 2019
51. Government of India, National Trust (2019) Empowering abilities, creating trust. Available at https://thenationaltrust.gov.in/upload/uploadfiles/files/Annual_Report_2017-18_Eng lish.pdf. Accessed June 2019

52. Thirumurthy V, Thirumurthy V (2007) Special education in India at the crossroads. Child Educ 83(6):380–384
53. Government of India, Ministry of Social Justice and Empowerment, Department of Empowerment of Persons with Disabilities (Divyangjan) (2019) Compendium of schemes for the welfare of persons with disability 2018. Available at https://innovationclustersarchive.nic.in/upload/upl oadfiles/files/Compendium%20of%20Schemes-2018%20(Eng).pdf. Accessed June 2019
54. Gudlavallet VSM (2018) Challenges in accessing health care for people with disability in the South Asian context: a review. Int J Environ Res Public Health 15(11):2366
55. Kumar SG, Roy G, Kar SS (2012) Disability and rehabilitation services in India: issues and challenges. J Fam Med Primary Care 1(1):69–73
56. Saikia N, Bora JK, Jasilionis D, Shkolnikov VM (2016) Disability divides in India: evidence from the 2011 census. PLoS ONE 11(8):e0159809
57. Kalyanpur M (2008) The paradox of majority underrepresentation in special education in India: constructions of difference in a developing country. J Spec Educ 42(1):55–64
58. Gudlavalleti MVS, John N, Allagh K, Sagar J, Kamalakannan S, Ramachandra SS, South India Disability Evidence Study Group (2014) Access to health care and employment status of people with disabilities in South India, the SIDE (South India Disability Evidence) study. BMC Public Health 14(1):1125

Chapter 13
Early Childhood Intervention Services in India

Humaira Ansari and Supriya K. Nikam

Abstract Early childhood development is determined by intrinsic and environmental factors. Development is variable and rapid in early childhood and influenced by many factors. Delay in achievement of milestones beyond a certain age indicates underlying health complications. Early intervention is a set of coordinated services and support systems that are available to infants and young children with developmental delays and disabilities and their families that aim at maximizing the developmental and health outcomes of children. Early intervention may be in the form of physical and occupational therapy, speech therapy, behavioural education and play therapy. Early intervention is more beneficial and cost-effective if provided early than at a later stage in life. In India, policies for early intervention are disconnected, with services being offered through the District Early Intervention Centres of the Rashtriya Bal Swasthya Karyakram, and programmes of the National Trust. This article describes early intervention services in India. Considering the magnitude of childhood disability in the country, the article identifies that there is a need to focus on developing integrated services for children needing these services.

Keywords Early intervention · Developmental disabilities · Child development · India

Early Child Growth and Development

Child growth and development are a result of constant interplay of genetics and environmental influences [54]. Growth is increase in size of the body, body organs, and increase in height and weight that are visible physically. Overall, growth is compatible with the established standards of a given population, although there are individual, ethnic and geographic variations among the human population. For example, European children have different growth parameters than children of Indian

H. Ansari (✉)
Symbiosis International (Deemed) University, Pune, Maharashtra, India

S. K. Nikam
Birth Defects and Childhood Disability Research Centre, Pune 411020, India

© Springer Nature Singapore Pte Ltd. 2021　　　　　　　　　　　　　　　　　297
A. Kar (ed.), *Birth Defects in India*,
https://doi.org/10.1007/978-981-16-1554-2_13

or Asian origin. The World Health Organization (WHO) has developed the WHO Child Growth Standards that provides an international standard that gives description of physiological growth for all children from birth to five years of age [55]. The WHO Multicentre Growth Reference Study (MGRS) conducted across varied ethnic and cultural backgrounds across the world is the foundation of these growth curves [34].

Human growth and development go hand in hand. Development is different than growth, being the physical, language, cognitive, social and emotional development that initiates from birth [49]. Physical development refers to the development of the capability to skillfully conduct complicated activities. Motor development refers to muscular coordination that enables the infant to hold the neck, roll over on the side, crawl and walk. Gross motor development refers to co-ordination of major muscles of the body such as those of the arms, forearms, thighs and legs, so that the child can perform functions like throwing a ball, riding a tricycle or climbing stairs. Fine motor development refers to co-ordination of minor or smaller muscles of the body such as those of fingers or toes. Examples of fine motor functions include ability to pick up a peanut from the ground or scribble on a piece of paper [42]. Physical development that is acquisition of motor skills progresses from general to specific. The control of major muscles and gross motor skills such as walking or running, are acquired first followed by minor muscle and fine motor skills, such as picking up a pencil.

Language development is the ability to communicate. Infants initially communicate their needs through crying, followed later on with single words, followed by small sentences. By the age of three years, children are able to communicate through larger number of words. Cognitive development relates to the mental development of a child. Cognition refers to the process of gaining knowledge through thoughts, experiences and senses. A baby is not born with cognitive abilities at birth. During the course of development, cognitive skill-sets are acquired by children through surrounding experiences, own senses and own thoughts. Social-emotional development is another important area of human development. Social development refers to behaviour of a child with respect to accepted social culture. During infancy and early childhood, children do not form stable relationship bonds. But as age advances they can make friends, understand socially acceptable behaviours and etiquettes. Emotional development of an infant also evolves. For example, an infant can express basic emotions such as distress and delight. Gradually, other emotions such as joy, happiness, fear, anger, and sorrow develop, as well as ways of expressing emotions. All these areas of development are interrelated and interdependent on each other. A deviation in any one area can hamper overall development of a child.

Development occurs at variable rates and over a variable period of time, but is governed by some underlying principles. The rate of development is most rapid during early childhood [49]. Human physical development is cephalocaudal, that is development proceeds from the head downward. That means babies gain head control first, i.e. ability to lift up the head followed by coordination of upper extremities and finally the coordination of lower extremities. At the same time, development is proximodistal, that is the baby gains coordination of arms first and later of the hands and the fingers. Similarly, legs develop first then the feet and the toes. The spinal cord develops first before the peripheral structures. Development is largely dependent on biological maturation and learning new abilities. These are determined by the changes in the brain and the nervous system over a period of time. The rate of synaptogenesis determines the rate of maturation. The child must mature to a certain point before progressing to a new level. A stimulating environment, and positive experiences during early childhood enable the child to achieve optimal developmental potential.

Human development proceeds from simple to more complex. The ability to classify and find relationship between two objects is the most essential and primitive state in development. Children find simpler relationships between two objects initially and as age advances, they are able to proceed to find more complex relationships. Children utilize their cognitive skills for reasoning and problem solving. For example, younger children can identify the similarity in shape between an apple and an orange, and distinguish between them on the basis of colour. After attaining next maturation level, children achieve the ability to categorize items such as food and fruit.

In addition to being a continuous process, development is a sequential process, with the first level of maturation laying the foundation for other levels of maturation. Thus, children go through a continuous process of acquiring and adding sets of skills. Despite these general rules, every child is unique and the rate of growth and development is different for each child. However, the sequence and pattern remain the same. Hence, there is no fixed age at which a particular skill set is acquired, but there are age ranges for a particular developmental task [1, 20, 44].

Milestones of Development

"Developmental milestones are a set of behaviours, skills or abilities that are demonstrated at specified ages during infancy and early childhood in typical development" [5]. Table 13.1 summarizes the age at development of gross and fine motor, social/emotional, language and cognitive milestones [8, 20]. The achievement of milestones is influenced by a number of factors. For example, the achievement of milestones remains more or less the same among boys and girls in the first year of life, but thereafter, gender based differences in development emerge. Cultural, environmental and parental factors play an important role in achievement of milestones like independence, walking or climbing stairs [15]. A protective environment, where the child is not allowed to do things independently, delays the achievement of milestones. It is essential to understand these differences or else a child, without any delay may be at a risk of receiving early intervention services without need [50].

Figure 13.1 summarizes the early warning signs of developmental delay. These are, lack of response to sound, does not smile and cannot hold head by the second month of life; trouble moving eyes, inability to raise hand to mouth, does not make sounds by four months of age; appears floppy/stiff, does not roll and does not reach for objects by six months of age; does not sit, does not recognize familiar faces, does not babble, does not play by nine months of age; and does not crawl, stand, point to

Table 13.1 Milestones of development

Age	Gross motor	Fine motor	Social/emotional	Language/communication	Cognitive
2–3 months	Can hold up head, pushes up when lying in prone position	Movement of hands	Social smile, recognizes mother, anticipates feeds, can briefly calm herself	Alert to sound, coos, gurgling sounds	Begins to follow objects with eyes, pays attention to faces
4 months	Holds head steady, can roll over	Bidextrous reach	Smiles spontaneously, copies facial expressions	Begins to babble, laughs aloud	Hand eye coordination—sees and reaches for a toy with one hand, responds to affection
6 months	Sits with support, rolls in both directions	Unidextrous reach	Stranger anxiety	Monosyllables (ba, pa, da), responds to sound by making sounds	Shows curiosity to things around, tries to reach at them, transfers object from one hand to other
9 months	Sits without support, crawls, pulls to stand	Immature pincer grasp	Waves bye-bye	Bisyllables (baba, papa, dada), copies sounds	Transfers object from one hand to other
12 months	Stands without support, walks with support	Mature pincer grasp	Responds when called by name, cooperates in dressing, shows fear in some situations	One to two words with meaning	Explores objects by banging, shaking, throwing, follows simple directions
18 months	Walks alone, climbs stairs, explores , drinks from a cup	Scribbles, builds tower of 2–3 blocks	Copies caregivers in task, clings to caregivers in new surrounding or situations, may have temper tantrums, plays	8–10 single words, says "no" and shakes head "no"	Points to one body part, knows what common things are for; e.g. spoon, brush, phone

(continued)

Table 13.1 (continued)

Age	Gross motor	Fine motor	Social/emotional	Language/communication	Cognitive
2 years	Walks upstairs, jumps, runs, kicks and throws ball, stands on tiptoe	Builds tower of 6 blocks, makes and copies circular and vertical strokes	Copies others, excited in company of other children, shows independence, defiant behaviour, asks for food, expresses toilet need	2–3 word sentences, uses pronouns, repeats words, follows simple instructions, points to things	Sorts shapes and colours, finds things hidden under 2 or 3 covers, might use one hand more than other, names items in books
3 years	Runs, climbs well, walks upstairs with alternate feet, rides tricycle	Builds tower of 9 blocks, copies circle	Shares toys, knows name, gender, understands mine, his/her, shows concern and wide range of emotions	Can name familiar things, says first name, understands words like in, on, under, converses with 2 or 3 sentences	Can work with buttons, moving parts, does puzzle with 3–4 pieces, understands concept of two
4 years	Hops on one foot, walks downstairs (alternate feet)	Copies cross, bridges with blocks, uses scissors, plays card game	Enjoys company of other children, doing new things, plays cooperatively, attends toilet	Says songs or stories, correctly uses he/she	Understands concept of counting, understands time, same/different, remembers parts of a story
5 years	Hops, may be able to skip, summersault, swings, climbs	Copies triangle, and other shapes, able to draw a person with minimum 6 body parts	Wants to be like friends, wants to please them, likes to dance, act, sing, knows gender, shows more independence, helps in household tasks	Clear speech, tells simple story, uses full sentences, uses future tenses	Counts 10 or more things, knows about things used daily like food, money

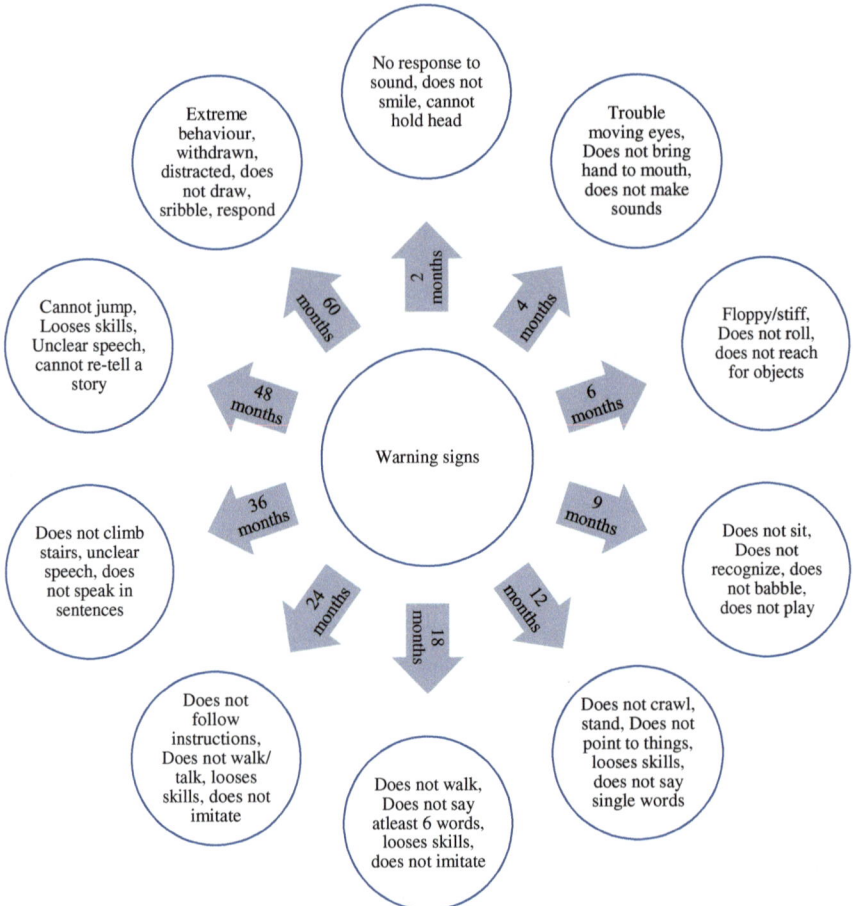

Fig. 13.1 Warning signs of delayed development

things and does not say single words by one year of age. The warning signs after this age are inability to walk, say at least six words, and unable to imitate at 18 months of life. Losing skills at any point is a major warning sign. Inability to follow instructions, lack of verbal abilities, communication, mobility issues, and extreme behaviour, or appearing withdrawn and distracted are all early signs of developmental delay and disabilities.

Factors Affecting Growth and Development of a Child

Early childhood development is determined by both intrinsic and environmental factors (Fig. 13.2) [47, 48, 50]. Brain development initiates from conception, with

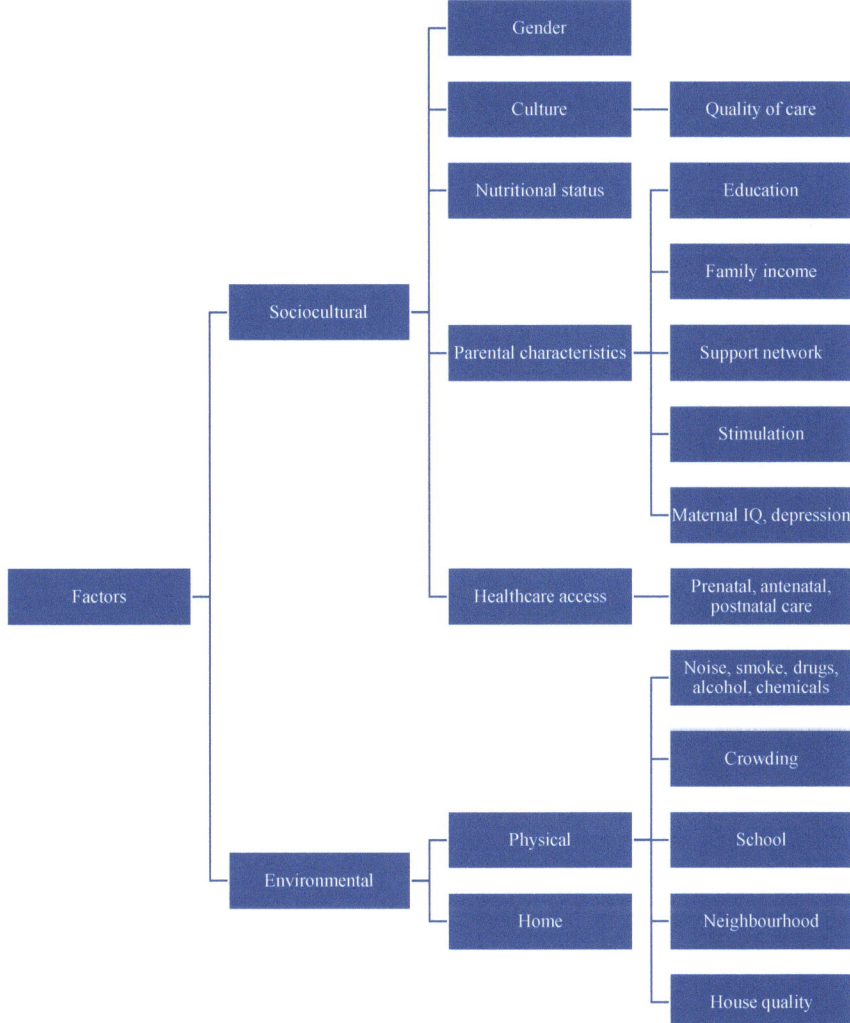

Fig. 13.2 Factors affecting growth and development

synaptogenesis (that is the establishment of synapses, the structure that facilitates communication between neurons) accelerating till the age of three years [22]. Synaptogenesis has a pivotal role in child development, determining language skills and cognitive abilities [7]. Positive and stimulating experiences and nurturing care received during this period lay the foundation for healthy child development. Socioeconomic factors are important influencers of child development, as education, occupation and family income determine the ability to provide quality education, stimulating environment and nutritious food which are basic needs for optimum child development. Engagement of children in income generating activities among the lower socio-economic strata adversely influences healthy child development. As primary caregivers, parental characteristics are equally important. During early years, children need love and care from parents. The amount and quality of time parents spend with children is determined by parents education, occupation and emotional status.

Environmental factors are influencers of child development. Residence in urban areas where there are more facilities in terms of hospitals, schools and recreational areas makes a difference in a child's development, as compared to rural areas where access to healthcare facilities and schools is difficult. Growing up in unhygienic environments influences child development with chances of repeated infections leading to bouts of morbidity and malnutrition. In the environment of urban slums, overcrowding, noise, lack of adequate space for children to play affects development in a negative way. Other factors also influence development. For example, gender is often a determinant, as in many countries including India, different attitudes prevail towards raising a girl child. There are evidences of girl children being neglected and deprived of opportunities for better education, food and clothing. Cultural and religious factors also determine child development. Different religions have different values and ways of living. Children usually follow the customs which their family follows and make own beliefs regarding many day-to-day living activities.

These factors are of utmost importance in a country with as much linguistic and cultural diversity as that of India. Intra-state differences in language and vocabulary may be very different, and language acquisition may be different at different ages. Similarly, an urban–rural difference is an important influencer. For example, urban children may have better cognitive abilities, whereas rural children may have better motor abilities. This implies that though belonging to a common country, there are different factors such as religion, cultural sub-groups, economic class, neighbourhood qualities which have direct or indirect impact on the growth and development of a child [14].

Early Childhood Intervention

Delayed developmental milestones are indicators of underlying health complications. Early intervention is a set of coordinated services and support systems that are available to infants and young children with developmental delays and disabilities, that aim at maximising the developmental and health outcomes [13]. Early intervention is required for newborns and children with developmental disabilities such as intellectual, visual, hearing, speech, locomotor, behavioural, social and learning difficulties. Very premature and low birth weight infants, whose survival has improved with advanced technologies, are at a higher risk of developmental difficulties. These children are another target group for early intervention. Early intervention basically helps these children and their families in learning basic life skills. Studies have shown that early intervention produces moderate to large effects on cognitive and social development [41]. Early behavioural intervention normalizes brain activity of autistic children, which improves social behaviour and has positive long term outcome [9].

Based on the need of the child, early intervention may be in the form of physical and occupational therapy, speech therapy, behavioural education and play therapy (Fig. 13.3). Early intervention services work closely with medical, nursing and psychological services, audiology and vision services and nutritional services. One of the key features of early intervention services is counselling and empowering families on caregiving. Parent-focussed programs are found to be effective not only for the child with developmental delay but also for other siblings as there are changes in the family environment due to new caregiving needs [4]. Children are linked up to appropriate assistive technology, and frequently, transportation issues are also addressed. The goal of early intervention can be restorative, i.e. restoring existing developmental issues, and preventive, i.e. preventing occurrence of developmental issues.

Primary prevention in early intervention activities aims at reduction in the risk factors that are responsible for developmental delay and disabilities. Examples include addressing malnutrition related to low birth weight and prematurity. Secondary prevention aims at reducing the impact of disability and shortening its duration. Stimulation, therapies and various forms of strategies and interventions are

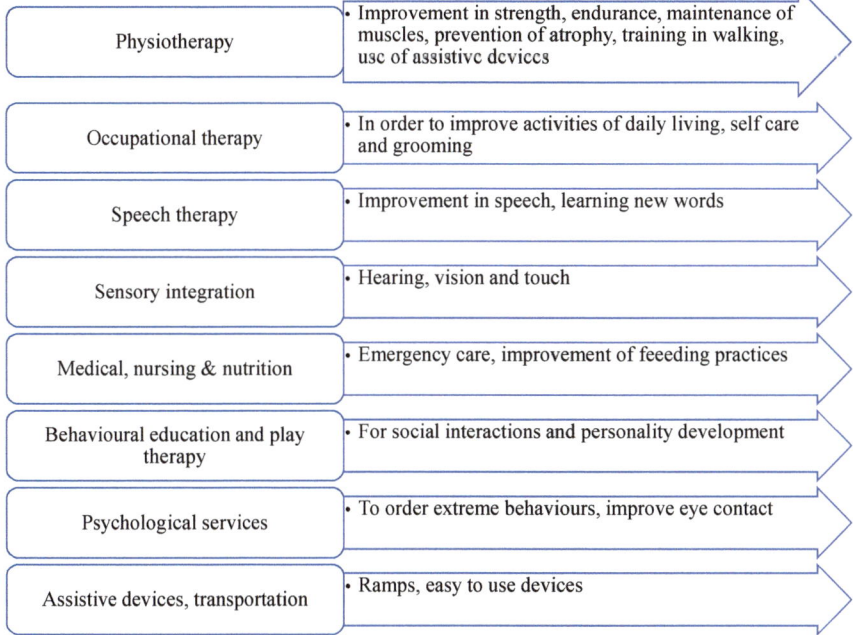

Fig. 13.3 Early intervention services

applicable at this stage of prevention. Tertiary prevention aims at preventing deterioration of the condition that may result in complications and hospitalization or immobilization.

Early intervention services initiated in the 1960s, focusing on children with developmental difficulties. Initially, care for children with disabilities was institutional. Emerging evidences, however, indicated that children staying at institutions or orphanages or in single parent families showed poor cognitive development, neurobiological and behavioural problems. They also experienced poor sensory and linguistic stimulation thus leading to poor attachment and socio-emotional development [16]. Studies indicated improvement in cognitive development if children living at institutions were moved earlier to foster care or parent support. With these emerging evidences, in the 1980s and 1990s, programs became family centred as the importance of family on child development was realized. Currently, early intervention has shifted in focus from school-based to home-based programs, with parents sharing a major responsibility in providing appropriate and recommended interventions to the child [21]. Additionally, early intervention also reduced the costs incurred by parents (Fig. 13.4) [17, 19].

Early intervention begins with an evaluation that helps in identifying the areas in which the child requires assistance and therapy. The child is further provided with tailored services relevant for his or her further development. Services are family oriented, so that the concerns of the family are addressed and the parents are counselled to help them understand the needs of the child, and the role of early intervention in the child's development. Early intervention is more beneficial and cost-effective

Fig. 13.4 Benefits of early intervention

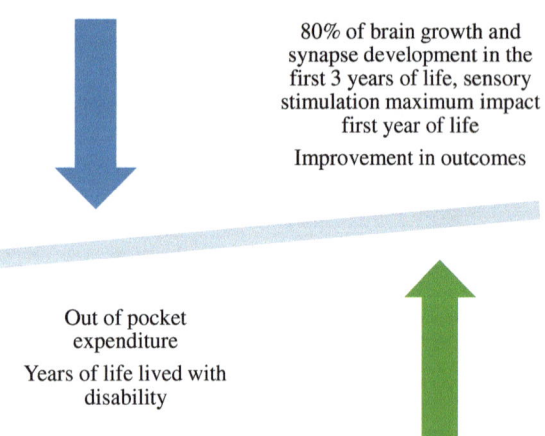

if provided early than at a later stage in life, with studies suggesting that that the impact of early intervention for a child with developmental disability is highest at 18 months of age [2, 6, 21].

Early Intervention Strategies

While designing early intervention strategies, it is essential to focus on components of intervention and mode of delivery for better compliance. Early intervention can be provided at home, day care, institutions, schools, play groups or centres specifically meant for providing early intervention. Other options are parent-child groups or family support groups. Group development interventions may also be given, where the child is provided intervention in a group of two or more children. Providing intervention in the natural environment and targeting everyday experiences are more effective. A review of early intervention in autism shows that interventions that are delivered for more than 20 hours per week and those that are individualized increase the development of children. Few points that need to be considered while designing any intervention program are as follows [3, 40].

- Developmental timing: interventions that begin at an earlier age and last longer are more effective than those that start later.
- Program intensity: intensive programs with more number of hours produce larger positive effects.
- Direct provision of learning experiences: direct contact and combination of centre-based and home-based treatment.
- Program breadth and flexibility: interventions that are comprehensive and multidisciplinary are more effective.
- Individual difference in program/Individualized care: are essential and beneficial.
- Environment: children's environment before and after intervention plays an important role in development.
- Child and parent needs have to be the focus of the programme.
- Evidence based practice has to be the centre of all intervention activities.
- Most importantly, the programme has to be embedded within local delivery systems in order to ensure sustainability.

The goals of early intervention are shown in Fig. 13.5. Cost of early intervention varies from setting to setting. It depends on the services offered and availed and also the competency of the staff. The higher qualified and trained the staff, the higher is the cost.

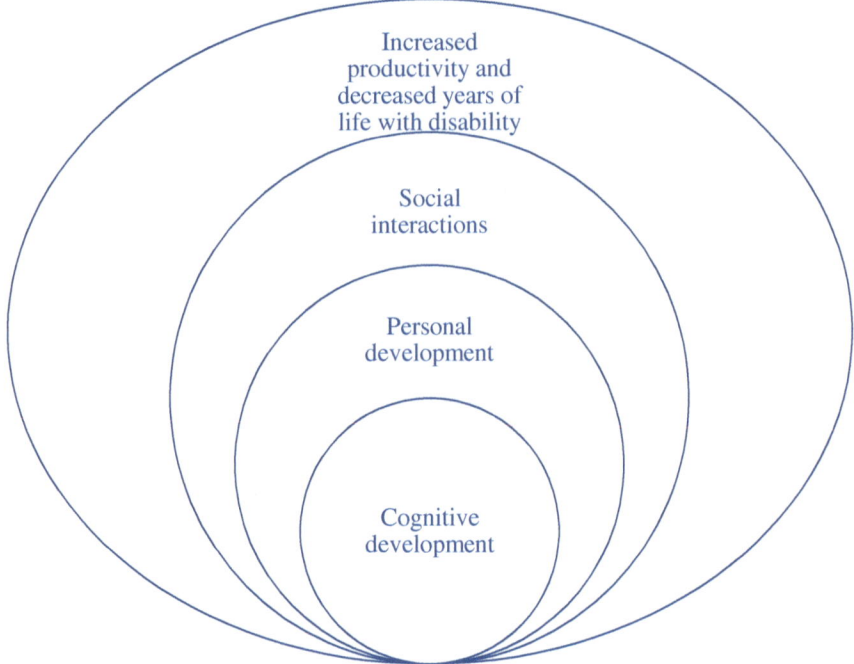

Fig. 13.5 Impact of early intervention

Early Intervention Services in India

In India, early intervention is mentioned under several policies. These include policies of the Ministry of Social Justice and Empowerment, which is the key ministry entrusted with the welfare of persons with disabilities [11], the Ministry of Health and Family Welfare, which has a screening and early intervention programme, and the Education Ministry, which calls for early intervention in its policies. Figure 13.6 enlists the various policies to promote early childhood development in India [26–32].

District Early Intervention Centres (DEICs) have been established by the Ministry of Health and Family Welfare for providing referral support to children with predetermined health conditions. The plan intends a team consisting of a paediatrician, physiotherapist, audiologist and speech therapist, psychologist, optometrist, lab technician, special educator, social worker and dentist for providing services. A programme manager has the responsibility of coordinating multiple therapies and guiding parents to required hospital departments. All newborns delivered at government hospitals, including those admitted in Special New-born Care Units, children in postnatal wards and paediatric wards, are to be screened for hearing, vision and congenital heart disease before discharge. Children born sick or preterm or with low birth weight or any birth defect are followed up. Records are maintained and are followed for all

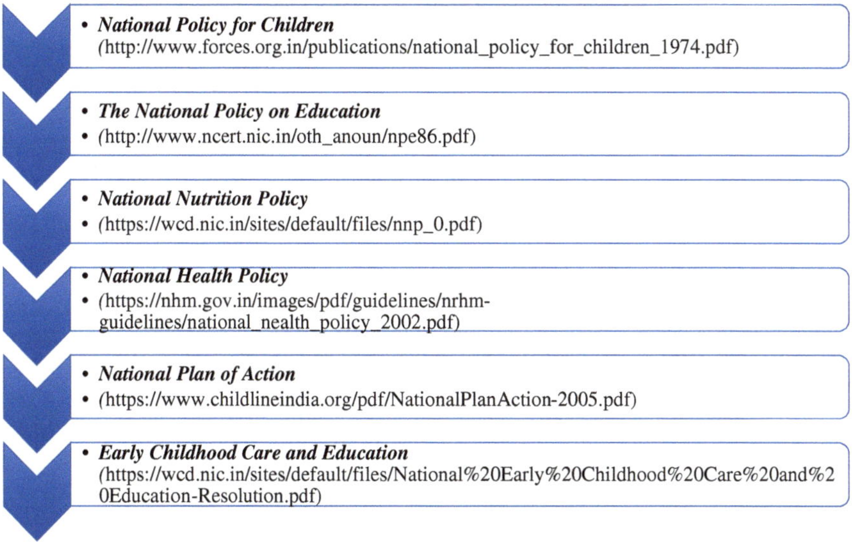

- **National Policy for Children**
  (http://www.forces.org.in/publications/national_policy_for_children_1974.pdf)

- **The National Policy on Education**
- (http://www.ncert.nic.in/oth_anoun/npe86.pdf)

- **National Nutrition Policy**
- (https://wcd.nic.in/sites/default/files/nnp_0.pdf)

- **National Health Policy**
- (https://nhm.gov.in/images/pdf/guidelines/nrhm-guidelines/national_nealth_policy_2002.pdf)

- **National Plan of Action**
- (https://www.childlineindia.org/pdf/NationalPlanAction-2005.pdf)

- **Early Childhood Care and Education**
  (https://wcd.nic.in/sites/default/files/National%20Early%20Childhood%20Care%20and%20Education-Resolution.pdf)

Fig. 13.6 Policies supporting early childhood development in India

children with developmental delay [46]. DEICs have been established in over 90 districts across the country (Chap. 12).

A rapid assessment of two DEICs in Chhattisgarh showed that beneficiary feedback was poor for one, due to lack of staff and required infrastructure. The other DEIC was well equipped, and recommendations for developing it as a model DEIC for Chhattisgarh were given [39]. A case study from Vishakhapatnam identified similar factors, that is the need to improve infrastructure and replace damaged equipments. Shortage of staff was another concern [35]. These findings were similar to the study from Odisha [37]. A study from Madhya Pradesh also reported dissatisfaction among beneficiaries, primarily as only few cases could be managed by the DEIC [38, 53]. A study from Pune identified that service uptake was poor because these were provided at district centres [18]. Non-compliance attributed to loss of wages, lack of time due to competing family issues, long travelling time, difficulty in transporting child with disability, failure to identify the health issue and belief that it will be resolved as the child grows old. Reasons attributed to health services were irregular time of visits by specialists, poor communication with healthcare professionals, inability to comprehend what is being told and where to go [18]. A long-term multi-sectored engagement is required for addressing the issue of disabilities in children. Caregivers, health personnel, government bodies are important stakeholders. Caregiver's role has been considered as pivotal. However, parental understanding of the nature of health issue, available treatment options, effectively following the treatment regimens, financial constraints, quality of counselling provided by health professionals are some of the important factors which determine the adherence of caregivers to medical advice/intervention/services [45]. Maternal education and proximity to intervention facility are also some determining factors [10]. Frequent monitoring and evaluation of DEIC, availability and working condition of the screening/evaluation equipments, screening tools, lab equipments, strengthening infrastructure and referral system, capacity building, awareness and training of staff, availability of specialists, and regular follow-up will improve DEIC functioning [37]. Use of social media as a tool to provide information on the DEIC will help in creating awareness and thus influence the use of services [43]. The use of telemedicine in delivering services should be considered for underserved districts [39].

In addition to the DEICs under the RBSK programme, early intervention centres have been established by the National Trust for Welfare of Persons with Autism, Cerebral palsy, Mental Retardation and Multiple Disabilities, a body established under the Ministry of Social Justice and Empowerment [51]. This Ministry is empowered with social welfare of persons with disabilities. The Rights of Persons with Disability Act 2016 aims at ensuring protection of rights, equal opportunities and full participation for persons with disabilities. Chapter III (Education) of the Act mentions the need for early intervention and early detection of disabilities [52].

One of the programmes offered by the National Trust is Disha, which is an early intervention and school readiness programme for children aged 0–10 years. The objectives of the programme are to provide therapies, training and support to families along with day care facilities to children up to 4 hours per day. The programme is for children in the age group 0–10 years with four disabilities (mental retardation, autism, cerebral palsy and multiple disabilities). The programme is implemented through registered voluntary organizations, and mandates qualified personnel, that is an Early Intervention Therapist or a special educator, physiotherapist or occupational therapist, counsellor and speech therapist. Vikaas is a day care programme for enhancing interpersonal and vocational skills for persons with disabilities. Vikaas centres also provide respite care, that is it helps family members of children with disabilities to have some time off to carry out other responsibilities.

District-Based Model for Early Intervention for Children with disabilities

Although available, early intervention services in India are limited, under-developed with little coordination among services and the ministries offering these services. At the community level, access to early intervention in India is plagued by both parent- and provider-related factors. Firstly, the lack of awareness about early intervention for childhood disability is apparent from a paucity of literature in the area. Studies identify parent-related factors to include beliefs about disability, lack of awareness about care for children with disabilities, lower education, family customs and traditions, transportation, financial constraints, early intervention centres being concentrated in urban areas thus missing target population, and parents perception that these centres lack trained professionals [33, 36].

There are several provider-related factors. A study conducted in Gujarat, India, to understand the perspective of paediatrician on early intervention showed that paediatricians referred the child mainly for three reasons, viz. sensory impairment (hearing or vision), abnormal muscle tone or loss of developmental milestones. The barriers to referral encountered by providers included lack of knowledge about referral services, and limited knowledge about options for treatment at affordable costs. Among providers, 28% reported that they were not confident about screening for a disability. Also referred parents do not report at the centre early, and only visit when the condition has become severe, showing lack of awareness among the parents [12]. These factors, that are observed globally, identify the need to create awareness among both providers, parents and other family members regarding the danger signs of a developmental disability, and the benefits of early interventions.

The feasibility of bringing health and early intervention services closer to communities was demonstrated in a district-based model by Nair and colleagues [24]. In this study, community-based health workers, Accredited Social Health Activists (ASHA), were trained in screening children, using two validated tools developed by this research group. These two tools were the Trivandrum Developmental Screening Chart [23] and the Language Evaluation Scale Trivandrum [25]. Both these tools were validated against the Denver Developmental Screening Test and Receptive Expressive Emergent Language Scale. Following training of the ASHAs, 101,438 children less than 6 years of age were screened, of which 1329 children were further evaluated by specialists at Developmental Evaluation Camps organized at Community Health Centres. Among screened children, 57% were identified with either developmental delay (49.89%), speech and language delay (24.98%), multiple disabilities (22.95%), intellectual disability (16.85%), cerebral palsy (8.43%), hearing impairment (5.12%), seizure disorders (3.99%), visual impairment (3.31%), neuromuscular disorders (1.35%) and autism (1.28%). Individualized treatment plans were offered to children by multidisciplinary teams. The project developed and offered education materials not only for mothers, but also for healthcare providers in the district.

Conclusion

Children with disabilities can be identified through delayed developmental milestones. Early intervention services provide the opportunity for maximizing the health and developmental potential of children with developmental delays or disabilities. In India, policies for early intervention are not coherent. Services are also not coordinated. Considering the magnitude of childhood disability in the country (Chap. 8), there is a need to focus on developing integrated services for children and their families. The available literature shows that there is a paucity of evidence, and there are still many unanswered questions on developing the most cost-effective system for providing early intervention. Research to understand how to identify those that require early intervention, the duration of early intervention, approaches and their benefits are needed to inform the development of a model for early intervention in resource-limited settings.

References

1. Adolph KE, Franchak JM (2017) The development of motor behavior. Wiley Interdiscip Rev Cogn Sci 8:1–30
2. Allen G (2011) Early intervention: the next steps an independent report to her majesty's government by Graham Allen MP. The Stationery Office. Available at https://assets.publis hing.service.gov.uk/government/uploads/system/uploads/attachment_data/file/284086/early-intervention-next-steps2.pdf. Accessed Feb 2019
3. Baker BL, Feinfield KA (2003) Early intervention. Curr Opin Psychiatry 16:503–509
4. Barlow J, Bergman H, Kornør H et al (2016) Group-based parent training programmes for improving emotional and behavioural adjustment in young children (review). Cochrane Database Syst Rev 1–128
5. Beighley JSMJL (2013) Developmental milestones. In: Volkmar FR (ed) Encyclopedia of autism spectrum disorders. Springer, New York
6. Blauw-Hospers CH (2005) A systematic review of the effects of early intervention on motor development. Dev Med Child Neurol 47:421–432
7. Bruer J (1998) The brain and child development. Public Health Rep 113:368–397
8. Child developmental screening checklist. Centre for Disease Control and Prevention (CDC). https://www.cdc.gov/ncbddd/actearly/pdf/parents_pdfs/MilestonesChecklists.pdf. Accessed 9 July 2020
9. Dawson G, Jones EJH, Merkle K et al (2012) Early behavioral intervention is associated with normalized brain activity in young children with autism. J Am Acad Child Adolesc Psychiatry 51:1150–1159
10. De Souza N, Sardessai V, Joshi K, Joshi V, Hughes M (2006) The determinants of compliance with an early intervention programme for high-risk babies in India. Child Care Health Dev 32:63–72
11. Department of Empowerment of Persons with Disabilities (Divyanjan). Ministry of Social Justice and Empowerment, Government of India. https://disabilityaffairs.gov.in/content/page/statutory-bodies.php. Accessed 9 July 2020
12. Desai PP, Mohite P (2011) An exploratory study of early intervention in Gujarat State, India: pediatricians' perspectives. J Dev Behav Pediatr 32:69–74
13. Early intervention. Centre for Disease Control and Prevention (CDC). https://www.cdc.gov/ncbddd/actearly/parents/states.html. Accessed 9 July 2020
14. Georgiadis A, Benny L, Thuc L et al (2017) Growth recovery and faltering through early adolescence in low- and middle-income countries: determinants and implications for cognitive development. Soc Sci Med 179:81–90
15. Gupta A, Kalaivani M, Gupta SK et al (2016) The study on achievement of motor milestones and associated factors among children in rural North India. J Family Med Prim Care 5:378–382
16. Humphrey K, Gleason M, Drury S et al (2015) Effects of institutional rearing and foster care on psychopathology at age 12 years in Romania: follow-up of an open, randomised controlled trial. Lancet Psychiatry 2:625–634
17. Jacobson JW, Mulick JA, Green G (1998) Cost-benefit estimates for early intensive behavioral intervention for young children with autism—general model and single state case. Behav Intervent 13:201–226
18. Kar A, Radhakrishnan B, Girase T, Ujagare D, Patil A (2020) Community-based screening and early intervention for birth defects and developmental disabilities: lessons from the RBSK programme in India. Disabil Inclus Dev 31:30–46

19. Kaur P, Chavan BS, Lata S et al (2006) Early intervention in developmental delay. Indian J Pediatr 73:405–408
20. Kliegman ST, Blum G et al (2016) Nelson textbook of pediatrics, 20th edn. Elsevier Inc
21. Majnemer A (1998) Benefits of early intervention for children with developmental disabilities. Semin Pediatr Neurol 5:62–69
22. Munno DW, Syed NI (2003) Synaptogenesis in the CNS: an odyssey from wiring together to firing together. J Physiol 552(1):1–11
23. Nair MKC, George B, Padmamohan J et al (2009) Developmental delay and disability among under-5 children in a rural ICDS block. Indian Pediatr 46:S75–S78
24. Nair MKC, Nair GSH, Beena M et al (2014) CDC Kerala 16: early detection of developmental delay/disability among children below 6 y—a district model. Indian J Pediatr 81:S151–S155
25. Nair M, Nair H, Mini A et al (2013) Development and validation of Language Evaluation Scale Trivandrum for children aged 0–3 years—LEST (0–3). Indian Pediatr 50:463–467
26. National Early Childhood Care and Education (ECCE) Policy (2013) Ministry of Women and Child Development, Government of India. https://wcd.nic.in/sites/default/files/National_Early_Childhood_Care_and_Education-Resolution.pdf. Accessed 9 July 2020
27. National Health Policy, Government of India (2002) https://nhm.gov.in/images/pdf/guidelines/nrhm-guidelines/national_nealth_policy_2002.pdf. Accessed 9 July 2020
28. National Health Policy, Government of India (2017) https://www.nhp.gov.in/nhpfiles/national_health_policy_2017.pdf. Accessed 3 Feb 2021
29. National Nutrition Policy (1993) Department of Women & Child Development, Government of India. https://wcd.nic.in/sites/default/files/nnp_0.pdf. Accessed 9 July 2020
30. National Plan of Action for Children (2005) Department of Women and Child Development, Government of India. https://www.childlineindia.org/pdf/NationalPlanAction-2005.pdf. Accessed 9 July 2020
31. National Policy for Children (1974) Government of India. https://www.forces.org.in/publications/national_policy_for_children_1974.pdf. Accessed 9 July 2020
32. National Policy on Education (1986) Government of India. https://www.nccrt.nic.in/oth_anoun/npe86.pdf. Accessed 9 July 2020
33. Odom SL, Wolery M (2003) A unified theory of practice in early intervention/early childhood special education: evidence-based practices. J Spec Educ 37:164–173
34. de Onis M, Garza C, Victoria C et al (2004) The WHO multicentre growth reference study: planning, study design, and methodology. Food Nutr Bull 25:S15–S26
35. Pagolu KR, Rao TR (2020) Assessment of institutional and management capacities against health conditions in District Early Intervention Centre (DEIC). Visakhapatnam
36. Pang Y, Richey D (2005) A comparative study of early intervention in Zimbabwe, Poland, China, India, and the United States of America. Int J Spec Educ 20:122–131
37. Panigrahy B, Swain A (2019) A cross-sectional study to evaluate the functioning and infrastructure of mobile health teams and DEIC (District Early Intervention Centre) at Koraput District of Odisha under Rastriya BAL Swasthya Karyakram (RBSK). World J Pharm Med Res 5:165–172
38. Parmar S, Bansal SB, Raghunath D, Patidar A (2016) Study of knowledge, attitude and practice of AYUSH doctors, evaluation of MHTs working in RBSK and client satisfaction. Int J Community Med Public Health 3:2186–2190
39. Prabhu SA, Shukla NK, Roshni MS (2021) Rapid assessment of Rashtriya Bal Swasthya Karyakram program implementation and beneficiary feedback at two district early intervention centers in Chhattisgarh State in India. Curr Med Issues 19:3–7
40. Ramey CT, Ramey SL (1998) Early intervention and early experience. Am Psychol 53:109–120
41. Rao N, Sun J, Wong JMS et al (2014) Early childhood development and cognitive development in developing countries: a rigorous literature review. Department of International Development
42. Ruffin NJ (2019) Human growth and development—a matter of principles
43. Sahoo S (2017) Requirement of engaging social media for health communication: a case study on Rashtriya Bal Swasthya Karyakram (RBSK) in District Early Intervention Centre (DEIC), Jaipur district in the state of Odisha (India). Int J Manag Humanit 4:7–12

44. Samuelson LL, Mcmurray B (2017) What does it take to learn a word? Wiley Interdiscip Rev Cogn Sci 8:1–16
45. Santer M, Ring N, Yardley L, Geraghty AW, Wyke S (2014) Treatment non-adherence in pediatric long-term medical conditions: systematic review and synthesis of qualitative studies of caregivers' views. BMC Pediatr 14:1–10
46. Setting up district early intervention centres. Operational guidelines (2014) Rashtriya Bal Swasthya Karyakram. Ministry of Heath & Family Welfare, Governement of India. https://nhm.gov.in/images/pdf/programmes/RBSK/Operational_Guidelines/Operational%20Guidelines_RBSK.pdf. Accessed 9 July 2020
47. Shonkoff J, Meisels S (2000) Handbook of early childhood intervention, 2nd edn. Cambridge University Press
48. Shonkoff JP, Phillips DA (2000) From neurons to neighborhoods: the science of early childhood development. National Academy Press, Washington
49. Singh R, Bisht N, Parveen H (2019) Principles, milestones and interventions for early years of human growth and development: an insight. Int J Curr Microbiol Appl Sci 8:181–190
50. Srinivasan R, Johnson B, Gan G et al (2018) Similarities and differences in child development from birth to age 3 years by sex and across four countries: a cross-sectional, observational study. Lancet Glob Health 6:e279–e291
51. The National Trust Act (1999) Rules and regulations. Ministry of Social Justice and Empowerment, Government of India. https://thenationaltrust.gov.in/upload/uploadfiles/files/act-englsih.pdf. Accessed 9 July 2020
52. The Rights of Persons with Disabilities Act (2016) An act to give effect to the United Nations Convention on the Rights of Persons with Disabilities and for matters connected therewith or incidental. https://www.indiacode.nic.in/bitstream/123456789/2155/3/A2016-49.pdf. Accessed 9 July 2020
53. Tiwari J, Jain A, Singh Y, Soni AK (2015) Estimation of magnitude of various health conditions under 4Ds approach, under RBSK programme in Devendra Nagar block of Panna district, Madhya Pradesh, India. Int J Community Med Public Health 2:228–233
54. Touwslager RNH, Gielen M, Derom C et al (2011) Determinants of infant growth in four age windows: a twin study. J Pediatr 158:566-572.e2
55. WHO Child Growth Standards (2006) Methods and development. World Health Organization. https://www.who.int/childgrowth/standards/Technical_report.pdf. Accessed 9 July 2020

Part V
Quality of Life

The three articles in Part V discuss the impact of birth defects and disabilities on childrens, caregivers and families. The first article (14. "Birth Defects Stigma") discusses the stigmatizing consequences of birth defects, associated with disability, prenatal and genetic testing, and the dilemma of terminating an affected pregnancy. The next article (15. "Parenting a Child with a Disability : A Review of Caregivers' Needs in India and Service Implications") points to a paucity of studies documenting the needs of caregivers of children with birth defects and developmental disabilities in India and other low- and middle-income countries. The final article discusses quality of life and psychosocial issues of caregivers, using a specific case study of mother's of children with haemophilia (16. "Quality of Life and Psycho-social Needs of Caregivers of Children with Birth Defects : A Case Study of Haemophilia").

Chapter 14
Birth Defects Stigma

Anita Kar

Abstract Stigma is the shame and disgrace associated with a characteristic that is socially unacceptable. Three biomedical characteristics are drivers of birth defects stigma. Firstly, children with birth defects are born with disabilities, or with chronic illness, which are highly stigmatized traits. Secondly, birth defects are congenital/hereditary in nature, implying a parental or familial role in transmission of the stigmatized characteristic. Thirdly, prenatal diagnostic technology can detect a carrier of a genetic disorder, labelling the individual's risk of transmitting a disabling condition to the next generation. Prenatal diagnostic technology can detect a malformed foetus or detect a foetus at risk of a genetic condition, confronting parents with the highly stigmatized healthcare choice of abortion (where legally permitted). This chapter expands these biomedical characteristics of birth defects and the three intersecting stigmas that constitute birth defects stigma. Using selected examples from India, the article illustrates how well-intentioned public health activities have the potential to stigmatize individuals and communities.

Keywords Stigma · Birth defects · Congenital · Genetic technology · Prenatal · Carrier · India

Stigma

'Stigma is an attribute that is deeply discrediting … that reduces someone from a whole and usual person to a tainted, discounted one'. This description of stigma was proposed by Goffman (1963) in *Stigma: Notes on the Management of Spoiled Identity*; the seminal work on conceptualizing stigma, its consequences and its drivers [1]. Stigma is the consequence of the collective response to characteristics that are considered undesirable. The drivers of stigma are traits such as disease (leprosy), disability (intellectual impairment), behaviour (autism) or social attributes like race, ethnicity, religion, gender and economic status. Goffman described stigma as a 'situation of the individual who is disqualified from full social acceptance'. He emphasizes

A. Kar (✉)
Birth Defects and Childhood Disability Research Centre, Pune 411020, India

© Springer Nature Singapore Pte Ltd. 2021
A. Kar (ed.), *Birth Defects in India*,
https://doi.org/10.1007/978-981-16-1554-2_14

on dehumanization as a part of stigma [1], arguing that the manifestations of stigma arise as a consequence of labelling and stereotyping due to the perceived negative evaluation of the characteristic.

Stigmatization as a social process reflects the disapproval of a person or a group of individuals based on characteristics that are understood to be different from the existing cultural and social customs. The process of stigmatization is brought about by attitudes, which are a set of beliefs and feelings that predispose individuals to behave in a specific manner. They foster discrimination that may be defined as the act or a behaviour where individuals or groups are treated with prejudice, thereby endorsing the negative stereotype [2]. Stigma has an affectational quality in that it can be 'felt'. The targeted individual may feel ostracized or victimized, but be unable to substantiate the instances of discrimination. Enacted stigma is when discriminatory acts are committed against the individual [3].

Certain groups of individuals are at a higher risk of being stigmatized as they contrast in characteristics or practices from those of the majority. Goffman describes three types of stigmatizing conditions, 'tribal identities' (such as race, gender, nationality), 'blemishes of individual characteristics' (e.g. addiction) and 'abominations of the body' (such as disabilities) (cited in [4]). Birth defects fall in the latter category, as children are born disabled or with medical conditions which require frequent hospitalizations. The disability and disablement caused by some common birth defects like Down syndrome, congenital heart defects, congenital deafness, autism, muscular dystrophy and others are described in other chapters of the book.

Stigma affects the same arenas of life globally. Despite the diversity of settings, stigma can impact education, employment, marriage, interpersonal relationships, mobility, housing, leisure activities and participation in religious and social gatherings [5]. Stigma affects psychological well-being, causing feelings of shame, blame, stress, anxiety and hopelessness. Reduced acceptance, discrimination and rejection lead to social isolation and the marginalization of the stigmatized. Stigmatizing social experiences can lead to suicidal ideation and attempts.

Stigma acts as a barrier in accessing health and welfare services. It is, therefore, a major cause of health inequalities [6]. Its consequences have been well documented for a range of conditions such as HIV, leprosy, epilepsy, intellectual disability, physical disability, deafness and others. Stigma has been attributed to several congenital conditions including congenital physical disability [7], limb reduction deficiencies [8], facial differences [9], intellectual disability [10], epilepsy [11], congenital deafness [12], congenital heart defects [13], cystic fibrosis [14], cleft lip and palate [15], cerebral palsy [16], albinism [17], disorders of sexual development [18], harlequin icthyosis [19], Zika virus-associated microcephaly [20], Down syndrome [21], autism spectrum disorders [22], spina bifida [23], developmental dysplasia of the hip [24], sickle cell anaemia [25], thalassemia [26], bleeding disorders [27], Marfan syndrome [28] and achondroplasia [29].

Across disease conditions, whether congenital or acquired, stigma is responsible for the denial of care, or care of sub-optimal quality, or provision of care through less competent people, physical and verbal abuse, or longer waiting times [30, 31]. There are several public health arguments for the lack of services for children with disabilities or medical complexities, but stigma appears to be an underlying factor. As Sartorius points out, '… stigma reduces the value of the person … Medications … (and services) … are not considered expensive because of their cost but because they are meant to be used in the treatment of people who are not considered to be of much value to the society …' [32]. Such stigmatizing attitudes and discrimination in health-care settings are barriers to basic and routine healthcare services, including diagnosis and treatment. Early diagnosis and appropriate treatment can prevent complications, and denial of services influence prognosis. For congenital conditions, the lack of access to care results in the progression of disabilities and further complications from existing co-morbidities. Lack of care and the resulting complications enhance prejudice and negative behavioural reactions, fostering greater social distance and social withdrawal [33].

Birth Defects Stigma

The drivers of birth defects stigma arise from two attributes of these conditions [34]. The first attribute is *disability and disablement*, as children with birth defects are born with or detected with disabilities or chronic health conditions, early in childhood. Disability is one of the most stigmatized traits [35], and children with disabilities are perceived as different from 'normal', healthy children. Prejudice, stereotyping and low expectations from people with disabilities result in their exclusion. Self-stigma, where people with disabilities endorse cultural stereotyping by believing that they will be undervalued in society, is widely prevalent. Courtesy stigma, that is, the stigma that affects parents, siblings and other family members, is also common among families with a person with a disability [36].

The second equally stigmatizing attribute of birth defects is the *congenital* nature of these conditions. As children are born with disabilities, a parental role for the child's disfigurement and ill-health is implied. Even though the factors causing most congenital anomalies are unknown, a parental, most often, maternal role in causing the malformation is assumed. Genetic disorders and some chromosomal disorders may be hereditary. This trans-generational nature of birth defects is responsible for stigma, driven by the erroneous belief that all other family members are likely to transmit these stigmatizing traits to the next generation.

Congenital/hereditary conditions as socially unacceptable traits emerge from the eugenics movement that arose in the early twentieth century in the USA. It was based on the belief that individuals with congenital/inherited conditions were 'weak and feeble' and a threat to the creation of a healthy society. The abuse of the rights of individuals with genetic conditions and coercive policies such as the compulsory sterilization of their parents were supported by legislation in the USA in the early twentieth century [37]. These thoughts formed the core of the eugenics philosophy of Nazi Germany in the late 1930s [37, 38]. All children with 'Serious Hereditary and Congenital Illnesses' were registered. Doctors and midwives were to report children under three years of age with serious hereditary diseases such as 'idiocy', Down syndrome, microcephaly, hydrocephaly and malformations, as well as children who were blind, deaf, paralysed or spastic [38]. Under the belief that children with birth defects were a burden to society, an 'alternative solution' of extermination of such children was practiced.

A third driver that may foster the process of stigmatization of birth defects is *technological developments*, specifically prenatal diagnostic and carrier detection technologies. Ultrasound is a commonly used prenatal test, that can detect a malformed foetus, causing guilt and distress to parents. In situations where the foetus may be at risk, genetic testing can detect a foetus with an inherited disorder. Where legally permitted, detection of these conditions confront prospective parents with the choice of abortion, another stigmatizing healthcare choice. Another application of genetic testing is in carrier detection. Testing leads to the identification of a carrier of an inherited condition [40, 42]. These issues are further described below.

Birth Defects and the Dimensions of Stigma

Nine different stigma dimensions have been described, which attempt to explain the way stigmas differ from each other [39–42]. Summarized in Table 14.1, they suggest multiple ways by which stigma can develop. All these dimensions of stigma are evident in the lived experience of birth defects. *Peril* describes the perceived dangerousness of the characteristic, which may be viewed as frightening, unpredictable and strange. An adult with intellectual disability may appear threatening to other adults. A more extreme example is albinism in sub-Saharan Africa. Albinism is a congenital condition where there is complete or partial absence of pigmentation. This characteristic sets apart individuals with albinism from others. Stigma and discrimination can lead to extreme beliefs, such as magical powers of albino body parts. Such extreme

Table 14.1 Dimensions of birth defects stigma

Dimension	Explanation	Birth defect stigma
Peril	Perceived dangerousness of the characteristic (frightening, unpredictable, strange)	An adult with an intellectual disability
Aesthetics	Displeasing	Skeletal deformities, orofacial cleft
Concealability	Visibility of the characteristic	Carrier of a genetic disorder
Origin	Origin of the characteristic	'Born that way' [44]
Controllability	Individual capacity to control the characteristic	ADHD
Pity	Sympathy	Duchenne muscular dystrophy
Course, stability	Likelihood of the person to recover	Bleeding episodes of haemophilia
Disruptiveness	Likely to be indisciplined	ADHD, intellectual disability, autism

consequences of stereotyping and stigma have led to murder of individuals with albinism in some African countries [43].

Aesthetics is another dimension of stigma, highlighting the displeasing nature of the attribute. Children with unrepaired orofacial clefts, or with limb or skeletal deformities may evoke a sense of revulsion, leading to limited interaction and isolation. Yet another dimension of stigma is its *origin*, that is whether the characteristic is present since birth ('born that way') or if it is acquired ('became that way') [44]. Often, acquired disability, such as those caused by road traffic injuries for example, is less stigmatized than congenital disability [44].

Associated with these dimensions is that of *pity*, which evokes sympathy and solicitousness. For example, a family with a child with Duchenne muscular dystrophy, a progressively degenerative condition resulting in premature death, may be marked as unfortunate or pitiable, thereby fostering stigma. *Course* identifies whether the person is likely to recover from the condition. The belief that a disabled child or adult will never be 'normal' and therefore is not a worthwhile investment is one of the major reasons for the marginalization of people with disabilities from social participation. *Stability* is the understanding of whether the person can recover from the condition while *controllability* is the understanding of whether the stigmatizing behaviour can be controlled by the person. Undiagnosed attention deficit hyperactivity disorder (ADHD) or learning disorders are examples where failure to identify these behaviours as constitutional, rather than intentional leads to contempt and exclusion. *Disruptiveness* describes the dimension of whether the characteristic can disrupt relationships and social integration. Children with intellectual disability, ADHD or autism do not fit into the social behavioural norms, leading to avoidance and isolation.

Concealability, that is the visibility of the characteristic, is another important dimension of stigma. As compared to visible birth defects, a positive genetic test can be concealed by the individual. The result of the test may involve feelings of shame and a negative self-image. The fear of discovery may place a strain on interpersonal relationships. Thus, multiple pathways, functioning independently or simultaneously, can lead to birth defects stigma.

Public Health Activities and Birth Defects Stigma

Public health activities can unintentionally precipitate stigma [45]. Figure 14.1 illustrates some public health interventions for birth defects, that operate across the life course of individuals. Genetic testing/screening are key public health activities, with the primary purpose of preventing and controlling genetic conditions [46]. Medical genetic services (that are offered to individuals at risk of disorders) and public health genetic screening programmes (that screen communities for genetic disorders) identify individuals who have the potential to transmit a genetic disorder to the next generation. These tests are offered during the preconception period, so that individuals are aware of their carrier status. The public health rationale is to empower individuals to make informed reproductive choices, but carry the risk of causing distress to individuals [47].

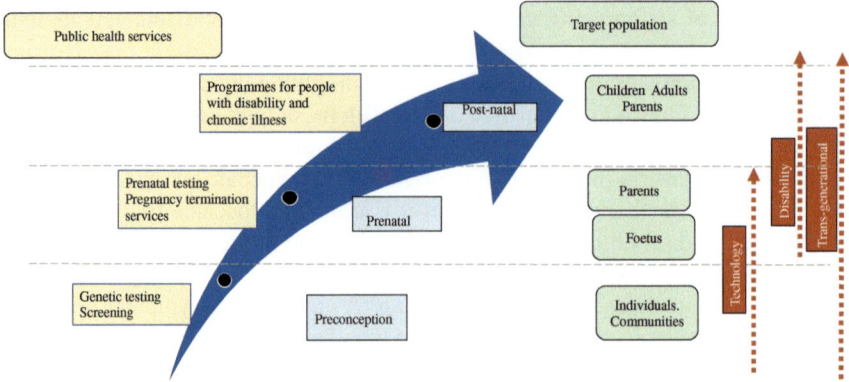

Fig. 14.1 Birth defects stigma across the life course. Public health services for birth defects are offered at various stages of the lifecycle. In the preconception period, carrier detection services are offered as genetic testing services for individuals at risk and screening programmes for at-risk communities. A positive carrier status enables the individual to make appropriate reproductive decisions, but carries a high risk of stigma. During pregnancy, prenatal testing, with elective termination of pregnancy for either foetal anomaly or a positive genetic test, is another public health service. Termination of pregnancy (abortion) is a reproductive decision that carries a high risk of perceived, internalized or felt stigma. Disability services offered to children, adults and their families carry with it a risk of stigmatization of beneficiaries of these services. The dotted arrows depict the drivers of birth defects stigma, which are technology in the preconception and prenatal periods, disability stigma that spans the entire life course initiating from the prenatal period, and the trans-generational nature of birth defects, manifesting as the stigma associated with knowledge of being a carrier, or with the status of being a parent of a child with a disability

The next driver of birth defects stigma occurs in the prenatal period with the detection of a malformed foetus through ultrasonography or diagnosis of a foetus with a genetic disorder through genetic testing. It places prospective parents with either the possibility of a child with a birth defect, or where legally permitted, in confronting abortion, an equally stigmatizing condition. In the post-natal period, programmes for persons with disabilities need to be sensitive in order to prevent stigmatization.

Genetic Testing, Genetic Screening and Stigma

Genetic testing is used for confirmatory diagnosis of individuals with clinical signs and symptoms of a genetic disorder (diagnostic genetic testing), or for carrier testing, that is identifying individuals who are at an increased risk of being carriers of a recessive genetic disorder [46]. As a public health measure, genetic testing intends to apprise individuals on carrier status, thereby enhancing reproductive autonomy and facilitating informed reproductive choices.

A carrier is a healthy individual without signs or symptoms of the condition, but at risk of transmitting the condition to the next generation. Carrier testing is offered to at risk individuals, such as those with a family history of a genetic disorder. Carrier testing is most often done in retrospect, after the birth of a baby with a genetic condition. In low- and middle-income countries, for example, severe haemophilia may not be diagnosed due to the low awareness about the condition and early mortality due to lack of availability and access to appropriate medical care. As such, there may not be a diagnosis of the disorder in antecedent generations of a family. Confirmatory diagnosis of haemophilia in a family member frequently leads to understanding of the disorder in other family members. (Several single gene disorders are sporadic, caused by de novo mutations, that is the mutation is not present in any other family member.)

Figure 14.2 explains carrier status, laying the background to understanding how testing is a driver of birth defects stigma [47–49]. An X-linked recessive disorder like haemophilia is transmitted from an individual with the disorder, through the daughter to the grandson. The daughter, as the carrier of the disorder, does not have any signs and symptoms of haemophilia, but each male conceptus has a 50% risk of the disorder (Fig. 14.2a). X-linked recessive disorders have significant gender implications, as the medical condition is transmitted by the mother to the son. Women carriers of X-linked recessive conditions experience guilt and sorrow, while female siblings of the patient are stigmatized as they may be carriers of the condition [50]. Unlike X-linked recessive conditions, the union of two individuals who are carriers of an autosomal recessive disorder, such as sickle cell disease, carries a 25% risk that each pregnancy may result in an affected birth. Again, the carrier may have mild signs or may not have any signs or symptoms of sickle cell disease [51].

Carrier testing is routinely offered for many common single-gene disorders. Expanded carrier screening panels for several autosomal and X-linked recessive disorders have also been developed, which screen individuals who are not at a high risk of a specific disorder [52], raising several ethical issues [53]. After a genetic condition has been diagnosed in a family, cascade screening/testing may be offered to family members likely to be carriers of the condition. Cascade screening has the potential of stigmatization, as the identity of the child with the disorder needs to be revealed to other members of the family.

In contrast to genetic testing, genetic screening is when a test is offered to individuals in a community who are a priori not at an individual higher risk of a genetic disorder and have not sought medical attention. However, the community is considered to be at risk, due to higher prevalence of a specific disorder. Examples of genetic screening programmes include screening for sickle cell disease among African-Americans and thalassemia screening in Mediterranean and South Asian countries. Individuals screened through this process are offered further tests for confirmatory diagnosis.

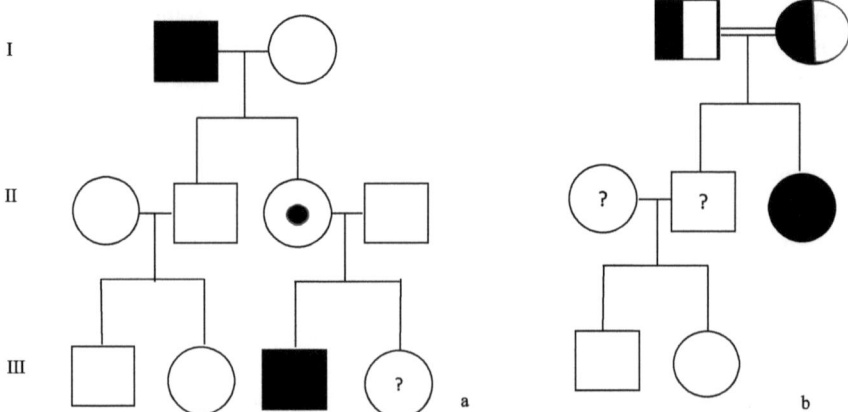

Fig. 14.2 Carrier of a genetic disorder. a Pedigree showing the transmission of an X-linked recessive disorder, such as haemophilia. In this example, the father (dark box in generation I) transmits the mutation to the daughter (generation II, circle with smaller black circle), but not the son (open box, generation II). The daughter is a carrier and has no symptoms of haemophilia. For every pregnancy, there is a 50% risk of haemophilia in male children (dark box in generation III). Genetic testing of the daughter in generation III (circle with question mark) will determine whether the individual is a carrier or not. **b** Transmission of an autosomal recessive condition like sickle cell disease. The parents in generation I are related and are carriers of the mutation causing sickle cell disease (half-shading). The double line in generation I indicates a consanguineous marriage, that is, marriage among blood relatives. As carriers have mild or no symptoms, they may be unaware of their carrier status till a screening test is conducted. In case of marriage between carriers, each pregnancy carries a 25% risk of an affected birth (generation II daughter, dark circle), 25% risk of a normal birth, that is the child will not be a carrier of sickle cell disease, while there is a 50% risk that the pregnancy will result in a carrier of the mutation. In generation II, testing determines whether the son (square box with question mark) is a carrier or not.

As public health measures, genetic testing and screening and carrier testing and carrier screening carry with them a high risk of stigmatization [46]. A positive carrier test result evokes a series of psychological reactions that range from shock and denial, anger, guilt and depression, till ultimately, psychological homeostasis sets in. Knowledge of carrier status is related to a poorer perception of health. Even though carrier status is not apparent to others, the concealability of carrier status is responsible for the stigmatization process. Shame and fear of discovery lead to concealment, compromising relationships and resulting in a life-long fear of discovery. In situations where privacy and confidentiality protocols are not in place, public knowledge of carrier status may lead to discrimination. Often, the discrimination is reinforced by the inability to distinguish between the carrier state and disease. In societies which lack gender-equity, the burden of carrier detection may be more pronounced on women than on men. Courtesy stigma, that is the stigmatization of all members of a family who may not be carriers, may deter marital alliances for other members [46]. Overall, the risk of discrimination is higher for female carriers of a genetic mutation, as well as other members of a family during the process of formalization of marriage.

Genetic screening is implemented as a premarital intervention, with the public health objective of prevention of genetic conditions. However, screening is plagued with several social and ethical ramifications. Carrier screening has been offered for life-limiting disorders such as cystic fibrosis (USA, Australia, Italy), beta-thalassemia (in Middle-eastern countries), and Tay-Sachs disease (among Ashkenazi Jews). One of the most successful programmes for controlling a genetic condition has been community screening strategies for haemoglobinopathies (thalassemias and sickle cell anaemia) [54]. Although primarily restricted to the Mediterranean, Asian and sub-Saharan African regions, thalassemia has now spread to Europe and North America through migrants. In 2006, the World Health Organization encouraged nations to develop an integrated programme for the prevention and management of haemoglobinopathies [55]. Since then, several screening programmes have been established.

Carrier screening for haemoglobinopathies was first introduced in some Mediterranean countries where the disorder was highly prevalent (Sardinia, Cyprus and Greece). The testing was offered as hospital and community-based voluntary screening services. Within two decades, these countries reported significant reductions in community prevalence of these disorders. In Greece, the numbers of children born with thalassemia reduced from 120–130 per year to 15 per year [56]. Community genetic screening has also been implemented in the Middle East, where high rates of consanguinity contribute to a high prevalence of haemoglobinopathies. Premarital screening is mandated in countries like Iran and Saudi Arabia. Several South Asian countries have national policies on controlling haemoglobinopathies through voluntary genetic testing programmes.

Another community-based carrier screening programme is the Dor Yeshorim carrier matching programme, implemented by the ultra-Orthodox Ashkenazi Jewish community in the USA, Europe and Israel [57]. The community has a high prevalence of Tay-Sachs disease. The religion does not permit abortion unless the mother's life is at risk. The Dor Yeshorim strategy is based on screening of school children, followed by appropriate match-making to avoid affected births. Children screened at schools are not provided test results, but a number which can be used to access information on the test results of both prospective partners prior to finalization of marriage. The test results are revealed as whether the prospective marriage is compatible or not. With endorsement by religious leaders, the Dor Yeshorim programme has become a standard prerequisite, despite ethical concerns of violation of autonomy (an individual's right to testing, versus imposition of testing) and non-directiveness (marriage between carriers is indirectly prohibited).

Carrier screening has the potential of stigmatization of carriers within communities. In the early 1970s, a screening programme for sickle cell disease among African-Americans led to widespread discrimination against this community in the USA, with insurance agencies refusing health and life insurance. The discrimination was reinforced by the lack of understanding on the difference between sickle cell trait (that is a carrier of sickle cell disease) and sickle cell disease.

That screening programmes need to be thought out carefully is illustrated by a study from India, in which premarital screening for beta-thalassemia was spearheaded by several government research institutions across the country. The studies screened large populations of school and college children aged between 11 and 18 years, with the purpose of informing families about the child's risk of being a carrier of thalassemia. In a follow up study, Colah et al. traced children, twenty years after the carrier testing was conducted [58]. Seventy-five percent of respondents could recall that some medical tests were done and 27% specifically recalled that they were carriers of beta-thalassemia. Of note was that none had disclosed their carrier status at the time of marriage, but 11 individuals reported that their spouse had also been tested. One child with thalassemia major was born. Thus, premarital testing can stigmatize individuals, leading to concealment of carrier status, defeating the purpose of the public health programme to prevent and control the disorder [59].

Another example of a well-intentioned public health programme that has a high potential for stigmatizing families is the sickle cell testing programme being carried out by some state governments in India. Sickle cell disease is highly prevalent in endogamous tribal populations. In order to prevent and control the condition, population screening followed by marriage counselling is conducted. Carriers are provided with a yellow-white card, a person testing negative is given a white card and a patient is given a yellow card [60] (Chap. 11, Fig. 11.2). Coloured cards are simple measures to explain carrier status to poorly literate individuals, and enhance awareness about the risk of marriage between carrier individuals. Despite the simplicity, providing coloured cards violates ethical principles, as it risks exposing individuals to stigma and discrimination. Selective screening also has the potential of stigmatizing and labelling communities [61]. Often, in public health programmes, privacy and confidentiality measures to safeguard patients are forgotten in instances such as public documentation. In order to promote the uptake of services, and promote confidence among users of these programmes, identities of carriers may be revealed by public health services.

Prenatal Testing and Abortion Stigma

Prenatal testing determines whether a foetus is at an increased risk of a congenital malformation or has inherited a genetic condition from a parent/parents. The detection of a malformed foetus or a foetus with a genetic condition destroys the social expectation that a pregnancy will result in a healthy outcome [34]. The prenatal detection of a foetal malformation stigmatizes the woman, distinguishing her from the majority of women who have healthy babies. The woman experiences guilt, shame and distress.

In countries where abortion is permissible, the next stigmatizing consequence is related to the decision to terminate the pregnancy. Abortion stigma is defined as 'a shared understanding that abortion is morally wrong and/or socially unacceptable' [62]. Studies describe the experience of abortion stigma as perceived stigma, that is, fears or expectations of being stigmatized, internalized stigma which is self-judgement or negative feelings about ones abortion and felt stigma which is experiencing negative treatment for having an abortion [63]. A systematic review of abortion stigma reported that women identified abortion as socially unacceptable and personally discrediting, perceiving stigma from several sources, such as family members, society, medical institutions and religious institutions [63]. Women expected to be discriminated and experienced fear of social judgement. Women who felt stigmatized by abortion were more likely to conceal the event. The need for secrecy was associated with increased psychosocial distress and social isolation. In a few instances, women perceived stigma from medical providers.

In India, the Medical Termination of Pregnancy Act 1971, legalizes abortion till 24 weeks of pregnancy. Beyond the legal period, abortion is permitted with judicial approval, only in situations where it is necessary to save the life of a woman, or upon detection of a severe congenital anomaly [64] . Detection of a malformation late is pregnancy is a distressing situation, as confidentiality is jeopardized when parents need to seek legal redress to terminate the pregnancy. Abortion stigma is widespread in India, with a study conducted across six states in the country reporting that half of all women choosing an abortion, reported stigma as one of the major barriers to accessing abortion services [64].

One of the stigmatizing attributes of termination of a pregnancy for a foetal anomaly is the associated ethical and moral debate on the rights of persons with disabilities. In India, where subsidized services for children with disabilities is limited, societal attitudes were evident in a very public debate on abortion of a foetus with a severe congenital anomaly (a complex congenital heart defect) [65]. The woman had appealed to the courts for permission to terminate the pregnancy in the 25th week of gestation. Public opinion was divided over whether it was ethical to end the life of the foetus because of the disability, versus the rights of the woman/parents to choose to terminate the pregnancy in order to prevent the suffering of the child. In the very public debate, not only was abortion identified as deviant behaviour of the woman, but the parents decision to not continue the pregnancy on the basis of the child's disability appeared morally unacceptable to the public at large.

Disability

A birth defect may be detected by ultrasonography or by genetic testing during the prenatal period. Majority of congenital anomalies are detected at birth due to a defect or deformation, through newborn screening programmes [66], or in early childhood through manifestation of failure to thrive or clinical symptoms of the disorder. Disability is common to all birth defects and is one of the most stigmatizing of all conditions [67]. Congenital disfigurement has a sustained and profound stigmatizing impact on the lives of individuals, parents and siblings [34]. The social and psychological impact of congenital anomalies enhances the intensity of feelings and experiences of stigma and isolation. The major underlying factors are the cosmetic consequences of the physical appearance, and the deviation from routine experiences and an expected normal life course.

Perceptions and attitudes towards disability have improved over the years, but negative attitudes towards persons with disabilities still persist, priarily due to negative attitudes and inaccurate beliefs about disability [68]. These attitudes result in stigma, which leads to discrimination and difficulties in social participation. The consequences of disability are serious. Instances such as infanticide, paternal abandonment, lack of inclusion of disabled individuals in civil registration systems, violence, violent cures, restricted participation and ostracism are well documented [2].

People with disabilities face many barriers in low- and middle-income countries, with stigma contributing to this exclusion [67]. Factors that trigger disability stigma are a lack of understanding and misconceptions regarding the causes of disabilities, frequently influenced by cultural or religious beliefs. Lack of awareness about the capabilities of people with disabilities and their marginalization from partaking in all spheres of society further reinforce negative stereotypes. Disability stigma varies with the type of disability. Intellectual and sensory impairments are more stigmatized than locomotor impairments. It varies by gender, socio-economic status and is more pronounced in rural areas. Thus, pre-existing modes of discrimination and exclusion intensify in their performance towards the disabled.

Parents and Caregivers

Parental reactions of guilt, denial and chronic sorrow lead to feelings of loss of image, loss of social standing, a sense of failure, shame, embarrassment and guilt. The 'undesired differentness from what had been expected' [1] leads to stigma. Parental responses may vary between denying, diminishing or concealing the impact of the disability, for example by avoiding social contact or seeking information to legitimize the existence of the defect [34]. Parents encounter difficulties in disclosing the disability. As Sensky writes, '… they must decide when, how, and what to tell family and friends about their child. Should they attempt to disguise the abnormality, or to distract attention from it by emphasising the child's healthy attributes? They might even consider trying to ignore the stigma and attempting to behave as though nothing were wrong …' [7].

The social reaction to disability, that is anxiety or discomfort in the company of a child with a disability, is a driver of stigma. Not knowing how to appropriately respond to a child with a disability leads to social distancing from the family, reinforcing parental isolation and stigma. Families may distance themselves from the cause of their stigma by abandoning the child, or having the child adopted or institutionalized [68].

In settings where there are no birth defects services, parents remain unaware about the cause of the disability and options for care and rehabilitation. The child is perceived as an economic burden. Without disability sensitization programmes, children are excluded from education and skill development programmes. The brunt of care for children with disabilities also falls unfairly on women [67]. Paternal desertion is frequent. Disability may be associated with severe expressions of gender discrimination.

Stigma Prevention in a Birth Defects Service

A public health programme on birth defects consists of medical care that can correct deformity, improve life span and improve the quality of life. Prevention services target severe congenital/hereditary disorders and empower individuals with prenatal test results with the hope that this information can lead to informed reproductive choices. Non-directive genetic counselling can ensure that parents are made aware of the emotional, physical and financial implications of caring for a child with a disability, available options and needed preparations.

Stigmatization of carriers is of special concern. Policymakers and programme managers expect compliance from carriers of a genetic disorder. They '… have an expectation that pregnant women will (and, by implication, 'should') participate in screening designed to identify pregnancies in which the foetus has Down syndrome and/or a serious malformation. The pressure exerted in such screening programmes, arising from the very existence of the screening programme, can amount almost to coercion, whatever the wishes and intentions of the individual professionals involved are …' [50]. However, as the example of the sickle cell marriage counselling intervention in India shows, the well-intentioned programme may end up stigmatizing the very beneficiaries it is intended for.

There are several important stigma prevention strategies. Public education on birth defects, their prevention, options for medical care and awareness about rehabilitation therapies should be core components of a birth defects programme. Stigma prevention activities need to include disability sensitization sessions, in order to ensure that children with disabilities are not viewed as objects of pity, charity and as a social burden. As an illustration, a tribal woman living in a remote hamlet in the Indian state of Madhya Pradesh cut off the extra digits of her female infant who was born with polydactyly. The mother perceived that such a disfigurement would hamper the child's chances of marriage. She tried to stem the bleeding with cow dung, but the infant died [69]. This incident explains the importance of public education on common birth defects, which is likely to enhance the inclusion of children with disabilities in society.

Another important public health activity is to increase public awareness about the benefits of carrier testing, that can decrease stigma and discrimination and increase the uptake of these services. In gender unequal societies, where the burden of stigma is associated with women, it is imperative to explain the causation and transmission of birth defects and hereditary disorders, so that women are not blamed for the birth of a child with a disability. Public education programmes on birth defects should be commensurate with the literacy level of communities and should ensure that the communication does not provoke stigma, guilt and sadness. Community genetic programmes should not only ensure that the advice is non-directive, but also ensure privacy and confidentiality to protect individuals against stigma.

While public education is important, reducing stigma also rests with healthcare staff in the birth defects programme. Especially in low- and middle-income countries where health service staff may not be exposed to genetics, education on family history and disease, and genetic counselling is imperative. Staff working in birth defects services needs to be aware of their personal attitudes and biases and be sensitive to the unintentional risk of expressing a stigmatizing response while dealing with parents or prospective parents. Training in active listening and reflection programmes is needed to keep staff sensitized. Birth defects stigma prevention should be supported by guidelines and protocols on sensitive communication and non-discriminatory practices.

References

1. Goffman E (2009 [1968]) Stigma: notes on the management of spoiled identity. Penguin, Harmondsworth
2. National Disability Authority. Literature review on attitudes towards disability. Disability series 9. Available at https://www.ucd.ie/issda/t4media/0048-01%20NDA_public_attitudes_disability_2006_literature_review.pdf. Accessed Feb 2019
3. Jacoby A (1994) Felt versus enacted stigma: a concept revisited: evidence from a study of people with epilepsy in remission. Soc Sci Med 38(2):269–274
4. LeBel TP (2008) Perceptions of and responses to stigma. Sociol Compass 2(2):409–432
5. Van Brakel WH (2006) Measuring health-related stigma—a literature review. Psychol Health Med 11(3):307–334
6. Hatzenbuehler ML, Phelan JC, Link BG (2013) Stigma as a fundamental cause of population health inequalities. Am J Public Health 103(5):813–821
7. Sensky T (1982) Family stigma in congenital physical handicap. Br Med J (Clin Res Ed) 285(6347):1033–1035
8. Krantz O, Bolin K, Persson D (2008) Stigma-handling strategies in everyday life among women aged 20 to 30 with transversal upper limb reduction deficiency. Scand J Disabil Res 10(4):209–226
9. Masnari O, Landolt MA, Roessler J (2012) Self- and parent-perceived stigmatisation in children and adolescents with congenital or acquired facial differences. J Plast Reconstr Aesthet Surg 65(12):1664–1670
10. Golberstein E, Eisenberg D, Gollust SE (2008) Perceived stigma and mental health care seeking. Psychiatr Serv 59(4):392–399
11. Jacoby A (2002) Stigma, epilepsy, and quality of life. Epilepsy Behav 3(6):10–20
12. Becker G (1981) Coping with stigma: lifelong adaptation of deaf people. Soc Sci Med B Med Anthropol 15B(1):21–24
13. McMurray R, Kendall L, Parsons JM (2001) A life less ordinary: growing up and coping with congenital heart disease. Coronary Health Care 5(1):51–57
14. Pizzignacco TMP, Mello DFD, Lima RAGD (2010) Stigma and cystic fibrosis. Rev Lat Am Enfermagem 18(1):139–142
15. Adeyemo WL, James O, Butali A (2016) Cleft lip and palate: parental experiences of stigma, discrimination, and social/structural inequalities. Ann Maxillofac Surg 6(2):195–203
16. Read SA, Morton TA, Ryan MK (2015) Negotiating identity: a qualitative analysis of stigma and support seeking for individuals with cerebral palsy. Disabil Rehabil 37(13):1162–1169

17. Brocco G (2016) Albinism, stigma, subjectivity and global-local discourses in Tanzania. Anthropol Med 23(3):229–243

18. Meyer-Bahlburg HF, Khuri J, Reyes-Portillo J, New MI (2017) Stigma in medical settings as reported retrospectively by women with congenital adrenal hyperplasia (CAH) for their childhood and adolescence. J Pediatr Psychol 42(5):496–503

19. Sharma ML (2017) Harlequin ichthyosis: case report of a rare disorder and stigma attached to it. J Med Sci Clin Res 5(6):22910–22914

20. Howells ME, Pieters MM (2018) "The mosquito brings the sickness": local knowledge, stigma, and barriers to Zika prevention in rural Guatemala. In: Maternal death and pregnancy-related morbidity among indigenous women of Mexico and Central America. Springer, Cham, pp 567–581

21. Jain R, Thomasma DC, Ragas R (2002) Down syndrome: still a social stigma. Am J Perinatol 19(02):99–108

22. Farrugia D (2009) Exploring stigma: medical knowledge and the stigmatisation of parents of children diagnosed with autism spectrum disorder. Sociol Health Illn 31(7):1011–1027

23. Appleton PL, Minchom PE, Ellis NC, Elliott CE, Böll V, Jones P (1994) The self-concept of young people with spina bifida: a population-based study. Dev Med Child Neurol 36(3):198–215

24. Cox SL, Kernohan WG (1998) They cannot sit properly or move around: seating and mobility during treatment for developmental dysplasia of the hip in children. Pediatr Rehabil 2(3):129–134

25. Bediako SM, Lanzkron S, Diener-West M, Onojobi G, Beach MC, Haywood C Jr (2016) The measure of sickle cell stigma: initial findings from the improving patient outcomes through respect and trust study. J Health Psychol 21(5):808–820

26. Tsiantis J, Dragonas TH, Richardson C, Anastasopoulos D, Masera G, Spinetta J (1996) Psychosocial problems and adjustment of children with β-thalassemia and their families. Eur Child Adolesc Psychiatry 5(4):193–203

27. Barlow JH, Stapley J, Ellard DR (2007) Living with haemophilia and von Willebrand's: a descriptive qualitative study. Patient Educ Couns 68(3):235–242

28. Peters KF, Apse KA, Blackford A, McHugh B, Michalic D, Biesecker BB (2005) Living with Marfan syndrome: coping with stigma. Clin Genet 68(1):6–14

29. Ablon J (1981) Dwarfism and social identity: self-help group participation. Soc Sci Med B Med Anthropol 15(1):25–30

30. Carlisle C, Mason T, Watkins C, Whitehead E (eds) (2001) Stigma and social exclusion in healthcare. Routledge, London

31. Nyblade L, Stockton MA, Giger K, Bond V, Ekstrand ML, McLean R (2019) Stigma in health facilities: why it matters and how we can change it. BMC Med 17(1):25

32. Sartorius N (2007) Stigmatized illnesses and health care. Croat Med J 48(3):396

33. Werner S (2015) Public stigma and the perception of rights: differences between intellectual and physical disabilities. Res Dev Disabil 38:262–271

34. Farrell M, Corrin K (2001) The stigma of congenital abnormalities. In: Carlisle C, Mason T, Watkins C, Whitehead E (eds) Stigma and social exclusion in healthcare. Routledge, London, pp 51–62

35. Fine M, Asch A (1988) Disability beyond stigma: social interaction, discrimination, and activism. J Soc Issues 44(1):3–21

36. Ali A, Hassiotis A, Strydom A, King M (2012) Self-stigma in people with intellectual disabilities and courtesy stigma in family carers: a systematic review. Res Dev Disabil 33(6):2122–2140

37. Christianson AL (2013) Attaining human dignity for people with birth defects: a historical perspective. S Afr Med J 103(12):1014–1019

38. Allen GE (1989) Eugenics and American social history, 1880–1950. Genome 31(2):885–889

39. Lifton RJ (1986) The Nazi doctors medical killing and the psychology of genocide. https://web.archive.org/web/20061004112317/, https://www.holocaust-history.org/lifton/contents.shtml. Accessed Feb 2019

40. Rose NC, Wick M (2018) Carrier screening for single gene disorders. Semin Fetal Neonatal Med 23(2):78–84
41. Pachankis JE, Hatzenbuehler ML, Wang K, Burton CL, Crawford FW, Phelan JC, Link BG (2018) The burden of stigma on health and well-being: a taxonomy of concealment, course, disruptiveness, aesthetics, origin, and peril across 93 stigmas. Pers Soc Psychol Bull 44(4):451–474
42. Ahmedani BK (2011) Mental health stigma: society, individuals, and the profession. J Soc Work Values Ethics 8(2):1–16
43. Cruz-Inigo AE, Ladizinski B, Sethi A (2011) Albinism in Africa: stigma, slaughter and awareness campaigns. Dermatol Clin 29(1):79–87
44. Bogart KR, Rosa NM, Slepian ML (2019) Born that way or became that way: stigma toward congenital versus acquired disability. Group Process Intergroup Relat 22(4):594–612
45. Bayer R (2008) Stigma and the ethics of public health: not can we but should we. Soc Sci Med 67(3):463–472
46. World Health Organization (2006) Medical genetic services in developing countries: the ethical, legal and social implications of genetic testing and screening. https://www.who.int/genomics/publications/GTS-MedicalGeneticServices-oct06.pdf. Accessed Feb 2019
47. Billings PR, Kohn MA, De Cuevas M, Beckwith J, Alper JS, Natowicz MR (1992) Discrimination as a consequence of genetic testing. Am J Hum Genet 50(3):476
48. Markel H (1992) The stigma of disease: implications of genetic screening. Am J Med 93(2):209–215
49. Kenen RH, Schmidt RM (1978) Stigmatization of carrier status: social implications of heterozygote genetic screening programs. Am J Public Health 68(11):1116–1120
50. Clarke A (2016) Anticipated stigma and blameless guilt: mothers' evaluation of life with the sex-linked disorder, hypohidrotic ectodermal dysplasia (XHED). Soc Sci Med 158:141–148
51. James CA, Hadley DW, Holtzman NA, Winkelstein JA (2006) How does the mode of inheritance of a genetic condition influence families? A study of guilt, blame, stigma, and understanding of inheritance and reproductive risks in families with X-linked and autosomal recessive diseases. Genet Med 8(4):234–242
52. Henneman L, Borry P, Chokoshvili D, Cornel MC, van El CG, Forzano F et al (2016) Responsible implementation of expanded carrier screening. Eur J Hum Genet 24(6):e1–e12
53. Kihlbom U (2016) Ethical issues in preconception genetic carrier screening. Upsala J Med Sci 121(4):295–298
54. Goonasekera HW, Paththinige CS, Dissanayake VHW (2018) Population screening for hemoglobinopathies. Annu Rev Genomics Hum Genet 19:355–380
55. World Health Organization (WHO) (2006) Thalassaemia and other haemoglobinopathies. Rep. EB118/5. WHO, Geneva. https://apps.who.int/gb/archive/pdf_files/EB118/B118_5-en.pdf. Accessed Feb 2019
56. Loukopoulos D (2011) Haemoglobinopathies in Greece: prevention programme over the past 35 years. Indian J Med Res 134:572–576
57. Raz AE, Vizner Y (2008) Carrier matching and collective socialization in community genetics: Dor Yeshorim and the reinforcement of stigma. Soc Sci Med 67(9):1361–1369
58. Colah R, Thomas M, Mayekar P (2007) Assessing the impact of screening and counselling high school children for β-thalassaemia in India. J Med Screen 14(3):158–158
59. De Wert GM, Dondorp WJ, Knoppers BM (2012) Preconception care and genetic risk: ethical issues. J Community Genet 3(3):221–228
60. Epidemic Branch Commissionerate of Health Family Welfare and Medical Services Gandhinagar, Gujarat. Sickle cell anemia program manual. Available at https://nhm.gujarat.gov.in/Images/pdf/SickleCellAnemiaManual.pdf. Accessed Feb 2019
61. Chattoo S (2018) Inherited blood disorders, genetic risk and global public health: framing 'birth defects' as preventable in India. Anthropol Med 25(1):30–49
62. Hanschmidt F, Linde K, Hilbert A, Riedel-Heller SG, Kersting A (2016) Abortion stigma: a systematic review. Perspect Sex Reprod Health 48(4):169–177

63. Makleff S, Wilkins R, Wachsmann H, Gupta D, Wachira M, Bunde W (2019) Exploring stigma and social norms in women's abortion experiences and their expectations of care. Sex Reprod Health Matters 27(3):1661753
64. Shekhar C, Sundaram A, Alagarajan M, Pradhan MR, Sahoo H (2020) Providing quality abortion care: findings from a study of six states in India. Sex Reprod Healthcare 24:100497. https://doi.org/10.1016/j.srhc.2020.100497
65. Madhiwalla N (2008) The Niketa Mehta case: does the right to abortion threaten disability rights. Indian J Med Ethics 5(4):152–153
66. Pitt JJ (2010) Newborn screening. Clin Biochem Rev 31(2):57
67. Munyi CW (2012) Past and present perceptions towards disability: a historical perspective. Disabil Stud Q 32(2)
68. Susman J (1994) Disability, stigma and deviance. Soc Sci Med 38(1):15–22
69. India Today App Magazine (2018) Madhya Pradesh: newborn dies after mother chops off extra fingers, toes. 29 Dec 2018. Available at https://www.indiatoday.in/india/story/madhya-pradesh-newborn-dies-after-mother-chops-off-extra-fingers-toes-1419727-2018-12-29. Accessed Feb 2019

Chapter 15
Parenting a Child with a Disability: A Review of Caregivers' Needs in India and Service Implications

Charuta Gokhale

Abstract Birth defects and developmental disabilities are highly disabling conditions of childhood. Parents are primary caregivers for children with disabilities. Caregivers' needs are primarily documented from industrialized countries. This article documents the studies on medical and social sector service needs of caregivers of children with disabilities in India. As compared to high-income countries, literature on caregivers' needs is limited. The review identified 22 studies on needs of caregivers of children with locomotor and intellectual disability out of which 7 were conducted in India. Majority of the data were from qualitative inquiry of challenges of caregiving. The review identified that caregivers had inadequate information of the condition, its prognosis and management. Lack of teaching caregivers for managing activities of daily living of the child and guidance on schooling were highlighted as severe challenges in the available literature. Caregiver needs arose from the time of diagnosis, with caregivers confronting difficulties in accessing medical care and rehabilitation therapies, and in negotiating with school and vocational services. These challenges were compounded with economic difficulties. These needs arose from limited coordination between health care and social support service providers. The findings of this and other studies provided valuable inputs on the need for integrated, comprehensive care for children with disabilities in India.

Keywords Caregivers · Children with disabilities · India · Low-income countries · Needs · Health care · Social support

Introduction

Children born with disabilities have special care needs throughout life. These needs are related to accessing medical care and rehabilitation therapies, specialized therapies, schools, assistive aids, nursing care, respite care and supplemental income to support these additional expenses. The needs are dynamic and change over time,

C. Gokhale (✉)
State Program Manager, Global Fund supported project on Psychosocial Counseling of People with Drug Resistant Tuberculosis, Tata Institute of Social Sciences, Mumbai, India

© Springer Nature Singapore Pte Ltd. 2021
A. Kar (ed.), *Birth Defects in India*,
https://doi.org/10.1007/978-981-16-1554-2_15

going beyond medical care to needs arising from routine living experiences. Caregivers need to be well informed about the condition of the child, prognosis, availability of medical care and social welfare services. Several studies from developed and developing countries have identified the unmet need of information pertaining to illness, treatment modalities, routine care of the child, assistive aids and medical equipments, psychosocial aspects of caregiving, concerns about the future of the child, family relationships and peer communication [1, 2].

The service model of the UK National Health Service (NHS) can illustrate the complexity and diversity of medical and social service needs of children with disabilities (Fig. 15.1) [3]. A child with a disability or a lifelong medical condition may be suspected through routine screening, by home visitors or by parents themselves. A confirmatory diagnosis requires the development of a medical management plan, followed by a comprehensive and individualized care plan. This plan is developed by a team comprising of clinicians, specialists, physiotherapists, occupational therapists, speech therapists, psychologists, dieticians and other professionals as required. Integral to this plan are caregivers, who are provided information about the condition and its prognosis, and imparted skills for carrying out activities of daily living (ADL) of the child, including familiarization with the use of assistive aids and technology. Psychosocial counselling and support for dealing with the disturbed routine and stress associated with parenting are integral to these services. Using a continuum of care approach, health and therapy plans are integrated with social care plans to address educational needs. Financial support is provided as required for personal health budget for therapies, personal care, equipment, residential and respite care facilities and carers allowance. Thus, support for children with disabilities and disabling conditions is lifelong.

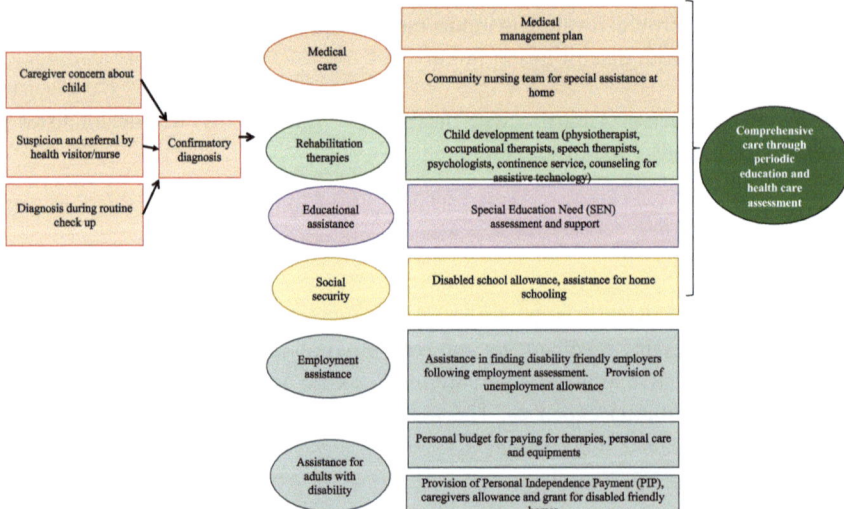

Fig. 15.1 Schematic diagram of medical and social welfare services for children with disabilities

The process of parenting a disabled child has profound effects on the quality of life (QoL) of caregivers. One of the major causes is the challenges faced by caregivers while availing the multitude of medical, educational and rehabilitation services for the child. Arranging finances where free of cost services are not assured, coping with the physical and emotional stress of caregiving, managing routine activities and taking care of siblings are additional burden. The effect of physical, mental and emotional health of caregivers, and the home environment of children with disabilities is well researched. A systematic review of studies of caregivers of children with autism from USA, Sweden, Australia, Jordan and Qatar showed that caregivers (both mothers and fathers) had significantly poor QoL than that of caregivers whose children did not have autism. Among parents, mothers were more affected than fathers with physical well-being being most affected [4]. A systematic review conducted by Pousada et al. identified 46 articles examining QoL of parents of children with cerebral palsy. These studies used both quantitative and qualitative methods along with mixed method approach to document higher levels of stress and depression among caregivers of children with cerebral palsy. Findings of these studies also showed the physical effect of caregiving, as caregivers expressed more physical complaints including musculoskeletal and lower back pain [5].

Evolution of Family Centred Care (FCC)

FCC is a health service approach for caring for children with disabilities, which ensures that not just the individual but the whole family as a unit is recognized as a care recipient [6]. Research on caregivers needs emerged in early 1980s in Europe as rehabilitation care shifted from an institution-based care model to FCC. The focus of rehabilitative intervention shifted from care for the child in an institutional setting to caregiving in the family environment, with parents as the primary caregivers.

In the middle of twentieth century, sick children used to be treated in hospitals without parents, possibly in lines with medical management strategies of the pre-antibiotic era [7]. This isolation was responsible for severe psychological stress among children as well as parents [8, 9]. Between 1920 through 1970, there were few sporadic attempts to develop family centred care in the USA and UK. Some studies were conducted which showed that prolonged hospitalization without families' involvement had adverse effects on children's health [9, 10]. World War II played a catalyst role in the emergence of FCC. Society witnessed separation of millions of children from their families and the resulting distress. Although medical care was reluctant to respond to it, it created an understanding about the family as a conducive environment for children.

The existing FCC model has evolved through various transient models of care. The care by parent model emerged around 1960 in the USA. It retained the responsibility of a child with parents. At the same time health professionals were seen as the primary medical caregivers [11]. 'Partnership in care' emerged in 1980, where care was provided by parents, taking educational support from the nursing staff [12]. The current FCC model relies on caring for children and their families within a 'medical home', where care for the child is a partnership between families and health and other services, with the primary care physician coordinating health services and therapies based on the needs of the family [13].

Several studies have shown that the family plays a critical role in addressing the needs of children with disabilities and family-based interventions have significant impact on improving the child's health outcomes. Reichow et al. [14] conducted a systematic review of non-specialist psychosocial interventions for children and adolescents with intellectual disability or low functioning autism spectrum disorder. Parents were trained in providing therapies, behavioural analytic intervention and improving their own well-being. The effect of training was studied on developmental milestones of the child, daily life, school performance, family life and favourable behavioural outcomes. Out of 9 studies identified, four were randomized controlled trials set in low- and middle-income countries (LMICs) (India and Vietnam). Findings showed that in case of behavioural interventions, best outcomes were observed for development and daily skills, particularly for children with severe cognitive disability. Effect was also observed in terms of better handling of behavioural problems, and better family outcomes. These studies however failed to show alleviation of maternal distress [14].

Kurani et al. conducted a study in Mumbai, India, where parents were actively involved in learning readiness programme and its impact was measured on social, language, behavioural and cognitive development of children with severe and profound intellectual disability. The aim of this programme was to train children in basic personal, social, imitation, communication and self-help skills. Using 'Parental Involvement/Engagement Scale' it was observed that parents having active engagement with the child had significantly better outcomes in the above-mentioned areas of development [15].

A study conducted in Ghana involved participatory training of 75 caregivers of children with cerebral palsy in the age group of 18 months to 12 years. Effects of training were studied on knowledge of the condition, QoL of caregivers, and child health indicators such as feeding practices. It was observed that following 12 months of training, there was an increase in knowledge about child feeding practices and improvement in emotional and physical health of the child. All these findings highlight the fact that child and family outcomes can be improved by involvement of caregivers [16].

In order to design family centred interventions, it is critical to identify caregivers needs while parenting a child with disability. This article reviews available evidence on medical and social sector service needs of caregivers of children with disabilities in India. It also documents data from other LMICs and compares them with studies conducted in industrialized nations.

Medical and Rehabilitation Needs of Caregivers of Children with Disability in Industrialized Countries

Several studies have documented the needs of caregivers of children with disability in industrialized countries. These studies identify caregivers needs across the lifecycle, from diagnosis till occupational rehabilitation of the child.

Needs During Diagnosis

Studies identified that during the process of diagnosis and development of the child, parents were challenged with the lack of information about the condition, lack of knowledge regarding developmental milestones and need for knowledge on the prognosis and child's future abilities [1, 17]. Need for counselling was expressed especially by caregivers of children with intellectual disability for coping with the feeling of shock and guilt.

Need for Information on Disability Services

As the parents transitioned from the initial trauma of diagnosis they werc challenged with inadequate information of existing condition-specific medical and rehabilitative services [17, 18]. Ineffective communication with healthcare providers was reported as a cause of misinformation about the condition [19]. Caregivers were unable to carry out ADL of the child. There was an unmet caregiving need for communication with the child, and feeding and bathing the child [20–22]. Studies reported needs of parental guidance of managing emotional and behavioural issues of not only children, but also their siblings [23]. Children with intellectual disability have poor oral hygiene mainly due to behaviour and sensory impairments. There was unmet needs for maintaining the child's oral health [22, 24].

A multicentric study was conducted in Canada to document the experiences and needs of families of children diagnosed with hearing loss, who were beneficiaries of the Universal Infant Hearing and Communication Development programme. Parents were reported to be appreciative about the early screening for hearing loss. Some parents faced difficulties during the process of diagnosis, especially in terms of number of visits. Families were satisfied with the technical experts at the clinic and family support worker responsible for counselling the families at the time of diagnosis. One of the common perceptions about the professional support was less priority given to a child with mild hearing loss. Parents also stated the need for emotional support along with professional support. Lack of guidance on importance of amplification and consensus about fitting the hearing aid caused lot of confusion among parents resulting in delay in intervention. Assistance for achieving hearing aid fitting was also found to be limited [25].

Needs of Information on Schooling, Recreation, Assistive Devices

Inadequate coordination of services from early intervention centre to school was a caregiving challenge documented in some studies [26, 27]. The ability of children with disabilities to participate in various social activities determines their well-being, development, and quality of life. Houtrow et al. documented factors associated with participation restriction in schools and social activities among children with disabilities of age 6–17 years in the USA. The findings suggested that the percentage of children with disabilities missing schools and not participating in the organized social activities was higher than that of healthy children. Low socio-economic status, functional limitations and poor health status of children were associated with participation restriction [28]. Piskur et al. in his scoping review documented parents' actions, challenges and needs while enabling participation of children with physical disability in various activities including schooling, recreational activities, community gatherings and household activities. Parents reported a range of challenges with the schooling system. Teachers were reported to be unaware about the condition and were ill equipped for managing behavioural issues of children [26, 29, 30].

Children with intellectual disability have major functional deficiencies in the area of communication and mobility. Access to age appropriate alternative and augmentative communication systems was a documented need [31]. Parents were challenged by lack of material support in the form of assistive aids, modalities for home modification and technical assistance for visual and auditory disabilities [20, 31]. Children with developmental disabilities in the USA receive Individualised Education Plan (IEP). In a secondary analysis of national data on children with disabilities, Lindsay et al. documented lack of IEP and lack of parental counselling regarding education and school-based rehabilitation therapy [19].

There are inadequate studies available on needs of caregivers of children with visual impairment. Limited evidence from developed countries report needs of professional support for holistic rehabilitation of disabled children. One such study conducted in Hong Kong collected data on information needs, service needs, psychosocial and relationship needs. Needs related to educational support were expressed on priority. Need of creating well-equipped special schools and social opportunity for children around the school was also expressed in a qualitative study [32].

Needs During Transition Beyond Schooling

Transition beyond schooling was also a major concern for some families [33]. There was an unmet need on guidance on appropriate leisure activity for the child [20, 21]. Challenges of obtaining information about meaningful activity for the child and career options have also been discussed in the literature [33]. Financial assistance for paying for medical care, transportation and therapies came out as an overarching need [34–37].

Financial Support

Caregivers reported a dearth of information about existing financial support programmes [29]. Buran et al. identified that the income of certain families was too high to become eligible for government financial assistance, but not high enough to be able to spend for the child's healthcare needs [36]. Sometimes families reported having paid out of pocket for medical care due to delay in reimbursement [26].

Parent and Family-Centred Needs

There was a need for emotional support, identified by caregivers reports of negative impact of the child's condition on their well-being. Specific needs expressed included interaction with other parents, help from family and friends and psychosocial support by professionals [17, 21]. Caregivers were worried about the acceptance and teasing of the child by peers, bullying and physical harm to the child which in turn could deter social participation [38, 39]. Addressing marital relationships was another identified need [40]. In some studies, caregivers expected involvement of family members in caring for the child and the need to be able to discuss problems with family members [21, 37, 41, 42]. Studies from the USA and Canada highlighted some additional parent-centred needs. Some parents expressed the need for acquiring skills for themselves while raising a child with a disability. These skills were that of time management and handling the ever-changing behaviour of the child. Information on legal rights, individual and family support options, making services available for geographically isolated regions on a regular basis, were some of the expectations stated in other studies [26, 33, 43–45].

Needs Related to Respite Care

Several administrative and programmatic barriers were perceived by caregivers in seeking family support services. In one of the studies conducted by Freedmon et al. families reported unavailability of respite care workers for geographically isolated communities. Respite care was also reported to be unavailable to families of children with severe disabilities. Caregivers did not receive support when the clinical diagnosis did not fit the eligibility criteria. Families with ethnic minorities who were unfamiliar with the language were challenged with lack of access to support services. Rehabilitation therapies ceased during summer breaks, which caregivers believed jeopardized the development of the child [26].

Medical and Rehabilitation Needs of Caregivers of Children with Disability in LMICs

In contrast to industrialized countries, there is limited literature from LMICs including India. Infectious conditions, high child and infant mortality and maternal deaths occupy the centre space in public health research. As a result, childhood disability was never accorded a public health priority, which in turn is reflected in scant literature on needs of caregivers of children with disabilities in India. A review of literature identified ten Indian studies on needs of caregivers of children with locomotor, intellectual and hearing impairment [46–55]. Majority of the data were from qualitative studies. Caregiver either referred to both parents or only the mother.

Two qualitative studies documented the needs of caregivers of children with hearing impairment. Gupta et al. used focus group discussions to understand the needs of ten caregivers. Results of this study showed that there was a general lack of awareness about symptoms and diagnosis of the condition. Parents expressed a need for creating awareness about disability. There was a demand for prenatal counselling and education regarding causes, diagnosis and early detection of hearing loss [53]. In a study reported by Thakre in Maharashtra, India, parents expressed information needs about the causes of hearing loss and availability of rehabilitation services. Parents also wished to seek guidance about frequency of testing and interpretation of test results. Parents expressed the necessity of frequent consultations with health professionals to better understand the prognosis [54].

Studies on needs of caregivers of children with intellectual disability include in addition to idiopathic intellectual disability, conditions like Down syndrome and autism. Caregivers' reported challenges during pre-diagnosis and diagnosis phase, in accessing medical care, identifying appropriate schooling and social welfare services, and facing societal attitudes towards children with intellectual disability. Lack of information about disability and rehabilitative services were the major concerns of families. Information on availability and accessibility of healthcare resources, specialized schools, innovative teaching strategies, daycare centres, aware-ness building on the condition, psychosocial support for dealing with the stress were some of the needs mentioned [47, 50].

Peshwaria et al. reported the results of a study conducted in Bhopal city in 1995. The study included 218 caregivers of children with intellectual disability. The study used a structured needs assessment schedule (National Institute of Mentally Handicapped-Family Needs Assessment Schedule, NIMH-FAMNS) to elicit infor-mation on disability, child management, services, vocational planning, sexuality, marriage, institutional care, emotional support, finance, family relationships, future planning, government benefits and legislation. Findings showed that parents reported an urgent need for information about government schemes, benefits and legislation devised for mentally challenged children. Other priority needs expressed by the parents were related to information about the disability, vocational opportunities and planning for the future of the child [55].

Autism was a frequently researched area with studies investigating caregivers needs across the life span. Since autism is manifested as a spectrum of symptoms, confirmatory diagnosis of the condition was perceived as a complex and time-consuming process by caregivers. Key challenges documented included the need to visit to multiple care providers for diagnosis and travelling long distances for seeking autism-specific interventions. Lack of skilling of caregivers for nurturing ADL and guidance for managing behavioural issues were recognized as key issues [46, 47, 52]. Edwardraj et al. in a study set in Tamil Nadu state reported needs related to availability and accessibility of health care and the need for specialized schools for children with intellectual disability [46].

For children with cerebral palsy, transportation and functional limitations arising from spasticity were identified as key issues. Sen and Goldbart identified caregiver concerns about daily activities, such as bathing, feeding and commuting [46]. Assis-tance and guidance for these activities and mealtimes were identified as one of the needs in case of children with multiple disabilities. Since caregivers were not appro-priately counselled and empowered, children were fed inadequately and at inap-propriate positions leading to multiple complications [46, 51]. More than 90% of children with locomotor disability had no access to assistive devices resulting in limited participation in social activities. Additionally, studies reported caregivers demand for respite facility. Sending children to schools was looked at as means of respite suggesting paucity of organized respite care [50, 56]. Visiting multiple health-care providers was logistically and financially challenging for families, indicating the need for home visits by community health workers or other health professionals [56, 57].

Majority of the studies documented feelings of stigma, discrimination and social isolation among families, identifying the need for widespread community awareness programmes [47, 50, 52, 58]. Studies reported physical and verbal abuse of children. A Tanzanian study noted that children with intellectual disabilities especially those who were non-verbal were at risk of physical and sexual abuse and experienced mistreatment such as being beaten, pushed and burnt [59]. In studies conducted in Malawi and Uganda, more than 90% of children with intellectual disability had experienced violence at some or the other time. However, negligible proportion of them had accessed child protection schemes due to lack of disability inclusive planning, lack of decentralized social protection services and financial barriers to seeking care [60].

Developing a Model to Address Caregivers Needs in India

A study conducted in Pune city of Maharashtra, India used both quantitative and qualitative methods to identify the needs of caregivers of children with cerebral palsy, autism and intellectual disabilities, in order to understand the public health and social sector factors responsible for the expressed needs. A total of 114 mothers of cerebral palsy, autism and intellectual disability was interviewed. The study observed that the process of diagnosis was shaped by factors such as instruction from peers to wait and watch after observing delayed milestones, the use of alternative therapies and belief that they might assist in attaining a cure, the feeling of receiving incomplete information from doctors and mistrust of medical practitioners. The information needs identified caregiver needs for holistic information about the condition including prognosis, knowledge on subsequent course of action, and counselling and support to deal with the shock of the diagnosis. Caregivers reported that the referral from medical services to rehabilitation care was weak. Parents frequently suspected that doctors were unaware of rehabilitation services. The other major need defined by parents was the lack of a team of medical and rehabilitation professionals who could guide parents on the progress of the child, especially in terms of the need to continue rehabilitation therapies. Lack of support around the diagnosis was identified as a major issue.

As caregivers transitioned from the shock of diagnosis to caring for the child, they identified the need for guidance on managing the activities of daily living of the child. Caregivers expressed frustration at their inability to communicate with the child and the challenge of teaching the child the activities of daily living. At school going age, parents were confronted with the choice of mainstream school or special school. Primary caregivers also experienced lack of sensitivity among teachers. Recreational facilities for children were another challenge. Parenting a child with a disability in absence of any social welfare support resulted in significant financial burden on caregivers. The major expenditure was on medicines, rehabilitation therapists, schooling and transportation. Some caregivers mentioned that they had health insurance, but the required care for the child was not reimbursed by insurance providers.

Discussion

The purpose of this review was to document the needs of caregivers of children with disability. The literature showed several studies on family needs pertaining to different disabling conditions, such as hearing and visual impairment, loco-motor disabilities including cerebral palsy and Duchene muscular dystrophy, intellectual disabilities like autism, idiopathic intellectual disability, Down syndrome and multiple disabilities. Intellectual disability was found to be the most commonly studied conditions while visual impairment was least explored for assessing caregiver needs.

Caregiver needs reported in the literature identified differences in expressed needs between caregivers in industrialized countries and those from LMICs. These differences were based on availability, accessibility and affordability of services. Studies from developing countries primarily reported information needs while families from high-income countries, where information and support services are already in place, caregivers had more advanced needs in terms of respite care centres, specialized schools and parent support groups. Information needs of caregivers from LMICs included the need for information about the disabling condition, need to know about treatment and referral services, educational and occupational opportunities, child-care and community services, assistive devices, laws, regulations, disability benefits, and need to access professional support networks. As compared to developed countries, literature on caregivers' needs from India was found to be scant, suggesting the need to conduct more comprehensive studies on childhood disability. This would create an evidence base for designing context-specific care models for children with disabilities and their families.

Services for children with disabilities, caused by birth defects and developmental disabilities, are yet to be established in several LMICs, including India. While growth monitoring is done, early child development programmes are not a component of child health services. Without such services, diagnosis of a child with a developmental delay or a disability is dependent on parents reporting their concerns to medical practitioners. Referral pathways for linking medical services to early intervention and rehabilitation services are in a fledgling state, challenging paediatricians, who are unaware of where to refer caregivers. Doctors in general, have not been provided appropriate training in necessary rehabilitation interventions. With therapists in private practice, the decision to use therapy services remains at the parents' discretion, and on the ability to pay. The review identifies the need to put in place services and referral pathways, so that an integrated package of services can address the concerns raised by caregivers of children with disabilities.

This review brings forth some priority research areas which can guide policy-makers. Designing and testing health education material to help women understand the causes and prevention opportunities for birth defects and developmental disabilities is critical. Research to develop resources to address inadequate knowledge about developmental disabilities, medical and rehabilitative services and availability and utility of assistive devices is required. Developing a validated psychosocial counselling protocol may help alleviating the confusion faced by the caregivers at the time of diagnosis. Scattered services and having to run from pillar to post for information highlights the need for creating a common platform for dissemination of such information during the course of treatment, and developing and testing referral systems. The existing Indian literature shows dearth of evidence of social sector interventions. This includes day care and respite care facilities, professional parental training on ADL of the child, parent support groups and livelihood opportunities for adults. There should be research on delivery and scalability of such interventions and ways of mainstreaming them through existing public health and welfare infrastructures.

References

1. Siebes R, Marjolijn K, Willem G, Mattijs A, Jongmans M (2012) Needs of families with children who have a physical disability: a literature review. Crit Rev Phys Rehabil Med 24:85–108
2. Adler K, Salantera S, Leino H, Gradel B (2015) An integrated literature review of the knowledge needs of parents with children with special health care needs and of instruments to assess these needs. Infants Young Child 28:46–71
3. NHS England. The NHS and caring for a disabled child—England. Information for families. Available from https://contact.org.uk/media/477893/nhs_englandfinallastupdatedsept2013v5.5_low_res_web.pdf. Accessed 31 May 2018
4. Vasilopoulao E, Nisbet J (2015) The quality of life of parents of children with autism spectrum disorder: a systematic review. Res Autism Spectr Disord 23:36–49
5. Pousada M, Guillamón N, Hernández-Encuentra E, Muñoz E, Redolar D, Boixadós M et al (2013) Impact of caring for a child with cerebral palsy on the quality of life of parents: a systematic review of the literature. J Dev Phys Disabil 25(5):545–577
6. Shield L, Pratt J, Hunter J (2006) Family centered care: a review of qualitative studies. J Clin Nurs 15:1317–1323
7. Lomax EMR (1996) Small and special: the development of hospitals for children in Victorian Britain. Med Hist Suppl 16:1–217
8. Bowlby J (1944) Forty-four juvenile thieves: their characteristics and home life (I). Int J Psycho-Anal 25:19–53
9. Bowlby J (1973) Attachment and loss, volume II: separation anxiety and anger. Basic Books, University of Minnesota, New York
10. Burlingham D, Freud A (1942) Young children in war time. Allen & Unwin Ltd., London
11. Goodband S, Jennings K (1992) Parent care: a US experience in Indianapolis. In: Cleary J (ed) Caring for children in hospital: parents and nurses in partnership. Square Press, London
12. Cleary J (1992) Caring for children in hospital: parents and nurses in partnership. Scutari Press, London
13. AAP agenda for children: medical home. American Academy of Paediatrics. Available at https://www.aap.org/en-us/about-the-aap/aap-facts/AAP-Agenda-for-Children-Strategic-Plan/Pages/AAP-Agenda-for-Children-Strategic-Plan-Medical-Home.aspx. Accessed 10 Aug 2019
14. Reichow B, Servili C, Yasamy M, Barbui C, Saxena S (2013) Non-specialist psychosocial interventions for children and adolescents with intellectual disability or lower-functioning autism spectrum disorders: a systematic review. PLoS Med 10:132–140
15. Kurani D, Nerurka A, Miranda L, Jawadwala F, Prabhulkar D (2009) Impact of parent's involvement and engagement in a learning readiness programme for children with severe and profound intellectual disability. J Intellect Disabil 13:269–289

16. Zuurmond M, Banion D, Gladstone M, Carsamar S, Kerac M, Baltussen M, Tann C, Nyante G, Polack S (2018) Evaluating the impact of a community-based parent training programme for children with cerebral palsy in Ghana. PLoS ONE 13:e0202096

17. Derguy C, Michel G, Ballara K, Bouvard M (2015) Assessing needs in parents of children with autism spectrum disorder: a crucial preliminary step to target relevant issues for support programs. J Intellect Dev Disabil 40(2):156–166

18. Ludlow A, Skelly C, Rohleder P (2012) Challenges faced by parents of children diagnosed with autism spectrum disorder. J Health Psychol 17:702–711

19. Lindsay S, Klassen A, Fellin M, Esses V (2012) Working with immigrant families raising a child with a disability: challenges and recommendations for healthcare and community service providers. Disabil Rehabil 34:2007–2017

20. Sloper P, Turner S (1992) Service needs of families of children with severe physical disability. Child Care Health Dev 18:259–282

21. Bailey D, Simeonsson R (1988) Assessing needs of families with handicapped infants. J Spec Educ 22:117–127

22. Mansoor D, Halabi M, Khamis A, Kowash M (2018) Oral health challenges facing Dubai children with autism spectrum disorder at home and in accessing oral health care. Eur J Paediatr Dent 19:127–133

23. Almasri N, O'Neil M, Palisano R (2013) Predictors of needs for families of children with cerebral palsy. Disabil Rehabil 36:210–219

24. Paschal A, Wilroy J, Hawley S (2016) Unmet needs for dental care in children with special health care needs. Prev Med Rep 3:62–67

25. Fitzpatrick E, Angus D, Smith A, Graham I (2008) Parents' needs following identification of childhood hearing loss. Am J Audiol 17:38–49

26. Freedmon R, Boyer N (2000) The power to choose: supports for families caring for individuals with developmental disabilities. Health Soc Work 25:59–68

27. Vilaseca R, Gracia M, Beltran F, Pinatella D (2015) Needs and supports of people with intellectual disability and their families in Catalonia. J Appl Res Intellect Disabil 6:21–29

28. Houtrow A, Jones J, Ghandour R, Strikland B, Newacheck P (2012) Participation of children with special health care needs in schools and the community. Acad Pediatr 12:326–334

29. Piskur B, Beurskens J, Jogmans M, Ketelaar M, Smeets R (2005) What do parents need to enhance participation of their school-aged child with a physical disability? A cross sectional study in the Netherlands. Child Care Health Dev 41:1–9

30. Bennett K, Hay D (2007) The role of family in the development of social skills in children with physical disabilities. Int J Disabil Dev Educ 54:381–397

31. Saetermoe C, Gómez J, Bámaca M, Gallardo C (2004) A qualitative enquiry of caregivers of adolescents with severe disabilities in Guatemala City. Disabil Rehabil 26:1032–1047

32. Gray C (2005) Inclusion, impact and need: young children with a visual impairment. Child Care Pract 11:179–190

33. Parsons S, Lewis A, Ellins J (2009) The views and experiences of parents of children with autistic spectrum disorder about educational provision: comparison with parents of children with other disabilities from an online survey. Eur J Spec Needs Educ 24:37–58

34. Papageorgiou V, Kalyva E (2010) Self-reported needs and expectations of parents of children with autism spectrum disorders who participate in support groups. Res Autism Spectr Disord 14:653–660

35. Resch J, Mireles G, Benz M, Grenwelge C, Peterson R, Zhang D (2010) Giving parents a voice: a qualitative study of the challenges experienced by parents of children with disabilities. Rehabil Psychol 55:139–150

36. Buran C, Sawin K, Grayson P, Criss S (2009) Family needs assessment in cerebral palsy clinic. J Spec Pediatr Nurs 14:86–93

37. Palisano R, Rosenbaum P, Walter S, Russell D, Wood E, Galuppi B (1997) Development and reliability of a system to classify gross motor function in children with cerebral palsy. Dev Med Child Neurol 39:214–223

38. Huang Y, Kellett U, John W (2012) Being concerned: caregiving for Taiwanese mothers of a child with cerebral palsy. J Clin Nurs 21:189–197
39. Strunk A, Pickler R, McCain L, Ameringer S, Myers B (2004) Managing the health care needs of adolescents with autism spectrum disorder: the parents' experience. Fam Syst Health 32:328–337
40. Farmer J, Marien W, Mary J, Clark R, Sherman A, Selva T (2004) Primary care supports for children with chronic health conditions: identifying and predicting unmet family needs. J Pediatr Psychol 29:355–367
41. Hendriks A, De Moor J, Franken W (2000) Service needs of parents with motor or multiply disabled children in Dutch therapeutic toddler classes. Clin Rehabil 14:506–517
42. Dabebneh K, Fayez M, Bataineh O (2012) Needs of parents caring for children with physical disabilities. a case study in Jordan. Int J Spec Educ 27:120–133
43. Hodgetts S, Zwaigenbaum L, Nicholas D (2015) Profile and predictors of service needs for families of children with autism spectrum disorders. Autism 19:673–683
44. Siklos S, Kerns K (2006) Assessing need for social support in parents of children with autism and Down syndrome. J Autism Dev Disord 36:921–933
45. Linda S, Horenstein B, Baverly L (1995) Migrant Hispanic families of young children: an analysis of parent needs and family support. Educ Cult 12:7–15
46. Sen R, Goldbart J (2005) Partnership in action: introducing family-based intervention for children with disability in urban slums of Kolkata, India. Int J Disabil Dev Educ 52:275–311
47. Edwardraj S, Mumtaj K, Prasad J, Kuruvilla Jacob K (2010) Perceptions about intellectual disability: a qualitative study from Vellore, South India. J Intellect Disabil Res 54:736–748
48. Sahay A, Prakash J, Khaique A, Kumar P (2013) Parents of intellectual disabled children: a study of their needs and expectations. Int J Humanit Soc Sci Invent 71:1–8
49. Kaur R, Arora H (2010) Attitudes of family members towards mentally handicapped children and family burden. Delhi Psychiatry J 13:70–74
50. Divan G, Vajaratkar V, Desai M, Strik L, Patel V (2012) Challenges, coping strategies, and unmet needs of families with a child with autism spectrum disorder in Goa, India. Autism Res 5:190–200
51. Yousafzai A, Pagedar S, Wirz S, Filteau S (2003) Beliefs about feeding practices and nutrition for children with disabilities among families in Dharavi, Mumbai. Int J Rehabil Res 26:33–41
52. Desai M, Divan G, Wertz F, Patel V (2012) The discovery of autism: Indian parents' experiences of caring for their child with an autism spectrum disorder. Transcult Psychiatry 49:613–637
53. Gupta N, Sharma A, Singh P (2014) Generating an evidence base for information, education and communication needs of the community regarding deafness: a qualitative study. Indian J Community Med 35(3):420–423
54. Thakre S (2012) Qualitative analysis of parents' experience of hearing loss of their school going children of a rural area of Nagpur. J Res Med Sci 17:764–771
55. Peshwaria R, Menon D, Ganguly R, Roy S, Pillay R, Gupta A (1995) Understanding Indian families: having persons with mental retardation. National Institute for the Mentally Handicapped. Available at https://niepid.nic.in/Understanding%20Indian%20Families.pdf. Accessed Nov 2018
56. Chiluba B, Moya G (2017) Caring for a cerebral palsy child: a caregiver's perspective at the University Teaching Hospital, Zambia. BMC Res Notes 10:724–730
57. Qayyum A, Lasi S, Rafique G (2013) Perceptions of primary caregivers of children with disabilities in two communities from Sindh and Balochistan, Pakistan. Disabil CBR Inclus Dev 24:131–142
58. Ansari NJR, Dhongade RK, Lad PS, Borade A, Suvarna YG et al (2016) Study of parental perceptions on health and social needs of children with neuro-developmental disability and its impact on the family. J Clin Diagn Res 10:15–20

59. Ambikile J, Outwater A (2012) Challenges of caring for children with mental disorders: experiences and views of caregivers attending the outpatient clinic at Muhimbili National Hospital, Dar es Salam-Tanzania. Child Adolesc Psychiatry Ment Health 6:1–11
60. Banks L, Kelly S, Kyegombe N, Kuper H, Devries K (2017) If he could speak, he would be able to point out who does those things to him: experience of violence and access to child protection among children with disabilities in Uganda and Malawi. PLoS ONE 12:1–17

Chapter 16
Quality of Life and Psychosocial Needs of Caregivers of Children with Birth Defects: A Case Study of Haemophilia

Supriya Phadnis

Abstract Disabling birth defects significantly impact the life of caregivers and siblings. The article uses haemophilia, a lifelong bleeding disorder, as a case study to illustrate the impact of a congenital disorder, with specific context to India, where, assured services for treatment are available in limitation. The article presents the results of a qualitative enquiry of mothers of children and young adults with haemophilia, in order to illustrate the impact of a chronic, heritable condition on caregivers. The article discusses that despite the different clinical types, the challenges encountered by caregivers are similar. As the psychosocial support improves the quality of life, this article contends that psychosocial interventions must be an integral component of a birth defects programme.

Keywords Haemophilia · India · Caregivers · Quality of life · Chronic disease

The impact of birth defects ranges from mild to severe. Children with severe birth defects have disabilities or complex medical needs. Several of these children transit to adulthood with dependence on caregivers for daily living. One of the major public health challenges of birth defects, especially in low- and middle-income countries (LMICs) like India, is in organizing services for medical care and rehabilitation for children with disabling birth defects. Addressing psychosocial health of family caregivers is another important component, as several studies have documented that parenting a child with a disability or a chronic medical condition impacts the quality of life (QoL). The chronicity of the condition impacts siblings and disrupts family functioning. While industrialized nations have established psychosocial support services to assist parents, counselling and support services are unavailable through general public health services in India and most LMICs. If available, they may be delivered through existing health staff for disease-specific conditions, such as HIV or TB. Birth defects services are also not in place, so that the essential service framework through which psychosocial support services could be offered is underdeveloped.

S. Phadnis (✉)
Healthcare Management, Goa Institute of Management, Sanquelim Campus, Poreim, Sattari, Goa 403505, India
e-mail: supriya.phadnis@gim.ac.in

© Springer Nature Singapore Pte Ltd. 2021
A. Kar (ed.), *Birth Defects in India*,
https://doi.org/10.1007/978-981-16-1554-2_16

351

The chronic nature of congenital disorders, the frequent medical and therapy needs, family economic status and the lack of an organized government service providing subsidized care are some of the factors exerting physical, emotional and economic demands on families. Counselling and support for families with children with medical conditions and disabilities is an unmet need in LMICs. The purpose of this chapter is to discuss the challenges faced by families while raising children with birth defects by presenting the results of a qualitative enquiry into the challenges encountered by mothers of children with haemophilia. The purpose is to illustrate the need for including counselling and support as essential components of a birth defects service, within the maternal and child health services in India. The term caregiver is used specifically to indicate family caregivers, that is, parents or close relatives.

Haemophilia A is an inherited single-gene disorders, affecting 1 in 5000 male births [1]. The condition is characterized by frequent bleeding episodes in the weight carrying joints of the body and in soft tissues. Essentially, each bleeding episode, if untreated or poorly treated, is excruciatingly painful. The patient has to curtail activities and immobilize the affected part until the bleeding resolves. Published data highlights the fact that a patient with severe haemophilia A can bleed about 11 times in a year [2]. The repeated disruption in routine life due to bleeding episodes interrupts schooling and employment. School dropout or unemployment is high among patients. Dependence on caregivers, even in adulthood, affects the emotional well-being of the patient and family members [3].

The primary treatment for haemorrhagic episodes is through infusion of clotting factor concentrate (CFC) VIII for haemophilia A [4]. CFC is an orphan drug [5] and, therefore, expensive. When CFC is available and subsidized through government health services, haemophilia can be appropriately managed, and as seen in industrialized nations, patients have normal life expectancy [6]. In LMICs, the lack of access to CFC is one of the crucial factors that severely compromises the QoL of patients. For caregivers, one of the major psychosocial stressors is the cost of providing treatment. In India, CFC is only within the reach of limited number of families. As such, most patients use first aid, that is, immobilization and cold compression for managing bleeding episodes. As all haemorrhages cannot be controlled in this way, families need to purchase CFC. A study estimated that if all haemorrhagic episodes were to be appropriately treated, the out-of-pocket expenditure would be catastrophic to 70% of families [7]. Such catastrophic economic consequences can be circumvented if there is a government haemophilia service [8]. Although such a service has been initiated in India, and CFC is being provided through district hospitals, the supply is erratic. Furthermore, patients have to travel to district hospitals, which makes access to treatment extremely challenging.

Another psychosocial stressor is the inherited nature of haemophilia. As a sex-linked disorder, haemophilia has a typical criss-cross inheritance, with the disorder being transmitted from a carrier mother to a son. In many instances however, the mutation may be de novo; that is, it originates in the son. Haemophilia has gender implications, as maternal guilt and self-blame are overarching emotions. Fathers are more likely to blame the mother for transmitting the condition to the son, or the mother's guilt may precipitate the feeling that fathers are likely to blame them for the condition of their child [9]. There is a high risk of marital discord. Carriers of this condition, that is mothers and sisters of the patient are also likely to have a higher risk of perceiving stigma. An important component of psychosocial/genetic counselling is to address the feeling of guilt and blame in affected families.

The impact of haemophilia on the psychosocial health of patients and parents has been documented in several studies [10–14]. These reviews and studies identify three major domains of psychosocial stress. The first psychosocial challenge is faced at the time of diagnosis and treatment, as the acceptance of the diagnosis remains as a challenge for parents. Parental denial that the child has a lifelong condition, that will require medical care and that can be potentially life-threatening and life-limiting, is the first response on learning the diagnosis. The feeling of guilt, anxiety about bleeding episodes and worry about the child's future affect both parents. The psychosocial burden is more pronounced among mothers who are the primary caregivers of patients. As haemophilia is transmitted from the mother to the son, maternal guilt is exacerbated. Parental responses after diagnosis vary and may range from child neglect to parental overprotection in an effort to reduce the risk of trauma-induced bleeding [13, 14].

Even though provision of prophylactic treatment in developed nations has reduced the risk of life-threatening bleeding episodes, parental stress about the safety of the child remains a lifelong psychosocial stressor. The experience of repeated bleeding episodes is responsible for disruption of routine family life and social activities. Sibling neglect due to overwhelming parental attention to the patient is noted [12].

Transition from childhood to adolescence is a third hurdle for patients and parents. Rebellion of adolescents being a natural part of this transition causes health risks, as patients practise unsafe behaviours that enhance the risk of bleeding. Parental interventions and instructions during childhood, aimed at preventing bleeding, such as not participating in specific types of sporting activities and adopting lifestyles that reduce the risk of injury of any form, are not easily accepted by the adolescent. Rather, behaviours considered risky for people with haemophilia are adopted in the need to feel accepted by peers. Caring becomes difficult, aggravating parental anxiety and worry [15, 16]. Disclosure of haemophilia to peers is often an issue for patients at this stage, as they wish to avoid the differential treatment given by their friends and society. Not informing parents about a bleeding episode due to the feeling of being a burden to the family is reported. The increased anxiety about and fear of other complications such as transfusion transmitted infections and development of inhibitors are highly elevated in parents during their child's adolescence [12].

Table 16.1 summarizes the main psychosocial stressors affecting parents of patients with haemophilia as reviewed from studies conducted in developed countries. The major stressors result as the consequences of diagnosis of haemophilia, the challenges of understanding and managing treatment of the disorder including home management and medical emergencies, family challenges including issues regarding upbringing of the child and inter-family relationships, and social and psychosocial challenges faced by parents. These psychosocial stressors are likely to be aggravated in India, due to lack of accessibility, availability and affordability of haemophilia treatment. Parental guilt arising from inability to provide treatment and inability to

Table 16.1 Psychosocial stressors of parents of children with haemophilia

		Psychosocial stressors	References
1	Diagnosis and treatment	Acceptance of initial diagnosis and early management of disease Challenges involved in early prophylaxis and treatment at home	[11, 12, 17]
2	Family challenges	Raising a child with the disorder Fostering independence Managing marital and family life	[11, 12, 17]
3	Social challenges	Social isolation Overprotection Schooling	[11, 12, 18, 19]
4	Psychological challenges of parents	Managing anxiety, depression, anger and frustration	[11, 15, 18]
5	Financial concerns	Organizing finances for treatment, assuring the future of the child	[2, 7, 20]

prevent the sequelae of poorly treated bleeding, such as progressive disability, is a likely trigger of psychosocial stress. These experiences are over and above the challenge of acceptance of a chronic disorder in the child, which requires significant psychosocial adjustment by parents [11, 12].

Psychosocial Stressors of Haemophilia: A Case Study

The psychosocial stressors of haemophilia were investigated through a study that conducted in-depth interviews of Indian mothers of patients with haemophilia. The respondents were between 25 and 55 years of age, while the age of the son ranged from 2 to 25 years. Only one of the participants was illiterate, but all others had over 15 years of education. The illiterate woman was widowed and supported her son, a daughter and herself by supplying packed lunches (dabba). All other respondents were homemakers. In terms of education of children, two were school going, three were college going, while one patient was in the last stage of his doctoral research in the USA. All these children and young adults had severe haemophilia.

The qualitative enquiry revealed the unique psychosocial stressors, influenced by the sociocultural and healthcare context in India. Negotiating medical care was challenging for most mothers. The key issues identified were, poor knowledge about haemophilia among clinicians, which resulted in delayed diagnosis of haemophilia, insensitivity of doctors, cost of treatment, unavailability of treatment product and the use of alternative therapies to reduce expenditure.

The experience of diagnosis of haemophilia was dependent on the knowledge about haemophilia among medical care providers. While haemophilia was immediately diagnosed, 25 and 16 years prior to the date of the interview by knowledgeable haematologists, doctors at a well-known hospital in a major city were unable to understand the cause of severe haematoma's after venepuncture, two years prior to the date of the interview. The type of practitioner approached also determined the experience of diagnosis of haemophilia. As practitioner consultancy fee was lower in the traditional medical sector, participants reported approaching several such practitioners in the hope of receiving a diagnosis and appropriate treatment. Despite being relatively cheaper, it did involve spending considering amounts of money and resulted in a delayed diagnosis. For example, one of the participants reported spending "*around 15 to 20,000 rupees on tests*" before haemophilia was diagnosed when the patient had a medical emergency that necessitated hospital admission. The patient was eleven years of age at that time.

The major underlying concern among all respondents was the cost of haemophilia treatment, the debt caused by expenditure during medical emergencies, concern about the availability of clotting factor concentrate and ability to find a healthcare provider who would be able to infuse clotting factor concentrate. The family of one of the participants had migrated to the city in order to ensure the access to CFC, while another participant was reluctant to leave the city for social engagements due to concerns about access to CFC and a doctor with the skills to infuse the patient. One of the respondents had accepted the fact that she would be unable to provide treatment. "*This is a disease where there is always shortage of money, so there is no use of financial planning*".

Even as there was concern about the ability to provide sufficient treatment to the patient, there was an underlying fear of development of inhibitors (factor VIII alloantibodies that reduce the effectiveness of treatment), due to the frequent use of CFC. One of the respondents stated, "*I always worry about development of antibodies, what will happen if he develops inhibitors? I just have no answer*". At least three participants used traditional Ayurvedic medicines in a search for a cheaper treatment option. One participant mentioned that she learnt "*Reiki*" so that her son would get some strength to cope with the pain. In rural areas, traditional practices were used. "*Whenever his knee would swell, villagers would heat the 'khurapa' (a sharp object used in farming) and burn his knee to reduce swelling, He would scream but we were helpless ...*".

One of the overarching themes that emerged from the interviews was the effect of medical communication at the time of diagnosis. Issues identified were inability of the doctor to communicate the difficult diagnosis, even though participants perceived that "*the doctor appeared to understand the diagnosis*". One participant was asked by the doctor "*not to worry, it is not a serious disease*". Another doctor subsequently told the parents "*who told you that it is not a problem? It is the biggest problem in your life ... I don't want to hide it from you*".

The impact of insensitive medical communication was still felt by participants. One participant recounted her experience. "*I do not wish to take the name of the doctor. But the expression of this doctor was that, what is the use of investing your effort in such a child? ... Now tell me whether it was worth investing our efforts in my son or not. He is more efficient than anybody*". (The patient was completing doctoral studies in the USA.)

The lack of access to information was reflected in all interviews. The medical provider was the main source of information. In lieu of medical communication, information was collected from other educated family members or the Internet. The illiterate participant did not feel the need to seek information on haemophilia.

Lack of treatment and poor information on haemophilia management was one of the factors that emphasized the debilitating consequences of haemophilia for mothers. One of the consequences was the mother's perception of the child, reflected by the descriptors of the patient used by the mother. The child was described with phrases like *"less than a normal child"*, *"inferior to other children"* and *"my son will always have an inferiority complex when he will start going to school"*. One of the participants would console herself by comparing her child with children with intellectual disabilities *"... At least he is able to learn, he can do some other things that mentally challenged children cannot do ..."*. One of the participants wished that the child had died as she could not bear the child's suffering. *"I always feel that. You may say I am peculiar, I am frustrated. But when he is in pain, I just cannot [bear to] see him. I know that he cannot enjoy his life. When we go out, he is always surrounded by normal children"*. In contrast to these mothers, the illiterate mother perceived haemophilia to be a routine ailment which her son had to survive with. Perception of stigma was also perceived by the child, as participants reported that the children or young adults refused to attend activities of the extended family or refused to disclose the diagnosis of haemophilia to friends.

With no information on managing life with haemophilia, several respondents mentioned the need to meet an adult patient in order to be assured about the longevity of their child. All except the illiterate participant mentioned a feeling of relief after meeting older, employed patients. *"... I felt good after meeting this person, as I thought that my son can definitely survive this long"*. Meeting patients who were married and employed helped reassure participants. *"... If he can become an engineer, he can get married, so can my son. If his parents can tackle this disease and make their son independent, so can we ..."*.

Other themes that emerged related to the upbringing of the child. Notable was the overemphasis on education, as this was believed to assure good employment in future, so that the patient would be able to bear the cost of treatment of haemophilia. In order to ensure a good education, mothers went to various lengths to make certain that the child would be appropriately schooled. As disability-friendly measures were lacking in schools, mothers reported carrying the child up and down stairs to the classroom. As bleeding episodes were usually treated with first aid, resulting in loss of school days, mothers would be primarily responsible for homework until such time that maternal assistance with studies was limited by her own education. Patient's career choices were primarily influenced by parents, with an emphasis on choice of sedentary careers, overriding the views of the young adult. One mother reported that her son had disobeyed parental decisions of a sedentary career in accounting and had decided to pursue engineering. The illiterate mother was undecided as to what opportunities education would provide.

Maternal concerns on the occurrence of bleeding during the school hours were expressed by nearly all of the participants. With the exception of the illiterate participant, participants did not experience any difficulty with the school system, as teachers were cooperative and accommodative. The illiterate participant recounted an incidence where the son experienced corporal punishment, so that he had thigh bleeds which kept him at home for one month. According to her, there was a high turnover of teachers, who were unconcerned about the child.

All participants accepted that they had overprotected their child, justifying that this was necessary as "… *he is not normal and needs this kind of protection*". The illiterate mother did not have time for overprotection. As she stated "*I don't have time to even look after him when he has a bleed. If I start overprotecting him then we won't have anything to eat*". The impact of parental overprotection resulted in rebellious behaviour that was difficult for mothers to handle.

Marriage was mentioned by all participants as a method for ensuring support, when the parents would not be present to care for the patient. The issue of marriage also raised concerns about disclosure of haemophilia. Participant's opinions ranged from selecting a girl from a poor family so that there would be no need to disclose the disorder, or selecting a physically handicapped girl as an equivalent match for the son. One participant however believed in full disclosure so that "*the girl would not feel as though she had been trapped in marriage*". One participant wished for a "*love marriage*" (as against a marriage arranged by parents), so that the girl would accept the disorder. In contrast, the illiterate participant looked forward to her son's marriage, as she would obtain another helping hand in her small business.

With the exception of one, all participants had a daughter. (Being a sex-linked recessive disorder, female siblings of the patient are at a risk of being a carrier of haemophilia, with the possibility of transmitting the disorder to the next generation.) All participants accepted the fact that because of the medical needs of the son, they had ignored the daughter. "*I have this guilty feeling that I never gave her (daughter) the proper time that she deserved. Sometimes when he would have painful bleeding, I would just be with him for a long time. I never even thought that my daughter would be hungry* (and that) *I should give her something to eat*". One of the participants mentioned that they had forced their daughter to select a career in medicine so that she would be able to treat her brother in future. This participant ascribed her daughter's introvert nature to the fact that the girl would be asked to sit quietly with the boy and not to play in fear of an injury. "*Since we would restrict (patient's) activities, this also had an effect on his sister; she became quiet and reserved*".

All respondents mentioned that spousal support played an important role in coping with the situation. Some respondents, however, narrated difficulties in marital relationship. "*He (husband) never directly blames me, but sometimes if something happens to our son, he blames me for not taking care of him, and says that it is my only responsibility. I realize that he is frustrated because I have given him an affected son*". With the exception of the illiterate participant, all other participants raised the issue of acceptance by in-laws. Not being blamed by their husband and in-laws was a big emotional relief for the participants. This was not always the case however, as in some cases the in-laws would make veiled accusations against the woman. "*It was a big relief for me because my in-laws and other relatives suspected that my parents did not disclose the reality before marriage. But now since the (carrier testing) reports are negative, they cannot blame us, but still they blame me for giving them such a grandson".*

All participants accepted the fact that due to such a chronic disease, mothers do not have time to socialize, at least till the child is grown up and can take care of himself. One respondent stated that *"when a mother has a child affected with such a disorder, she does not have any social life". "I cannot even speak to my friends over the phone because it will divert my attention* (from the child) *and I feel it is necessary to minimize the social interaction".* All participants reported that family plans had to be cancelled if the son had a bleed.

There were instances where diagnosis of the son's haemophilia resulted in an understanding that there were other maternal relatives who had haemophilia. In one case, the diagnosis led to the understanding that the maternal granduncle and uncle both had haemophilia. One family understood that the participant's three brothers had probably expired in childhood due to haemophilia-related causes. Another participant expressed relief that her brother (aged 40 years) now had access to treatment, only after her son was diagnosed with haemophilia and doctors had investigated and explained the family history of the disorder.

There was an overarching sense of guilt among mothers, and various explanations were provided to assure why she could not be blamed for the child's haemophilia. One of the respondents mentioned that the family history of haemophilia was realized after diagnosis of her son. This participant said that her husband never blamed her. *"… if you had known about haemophilia at least one of your brothers would have been alive …".* Another participant stated that *"women should not be blamed. Husbands should be counselled about not blaming their wives for this disease as she feels equally bad about her son's condition".* Another participant emphasized her ignorance about her carrier status *"… my brothers are normal, (I have) also asked my great-grandmother who confirmed that there was no one with such symptoms in our family … my mother thinks that I ate something wrong because of which my son got this disease".* Together with guilt and blame, all participants stated the need for support from the husband.

Participants had varied responses about disclosure of the daughter's carrier status at the time of her marriage. A participant mentioned that though her daughter was a doctor, the daughter had refused carrier testing. *"… Her in-laws and her husband are in the medical field and since it is a love-marriage, they all knew about our problem and had no problems … hence we were relieved".* Participants had not done the carrier testing of their daughters but intended to do it before marriage. All of them expressed their worry about disclosure while arranging the marriage, but were clear about the need to inform the prospective groom and his family about their carrier status. *"It is better that we inform them about this problem to avoid later complications in my daughter's life".* The illiterate participant was not worried about her daughter's marriage as she had been informed that there was *"an injection"* that had to be taken when her daughter would conceive. *"I am going to give her that injection which will tell us whether her son will get this haemophilia or not. So once her marriage is fixed, I will tell her in-laws about this injection. They will understand, as I am going to spend all the money that will be needed for this injection*". In all likelihood, this respondent was describing prenatal diagnosis and genetic testing.

This qualitative description identifies psychosocial stressors that are congruent with earlier findings, as well as additional factors, arising from the sociocultural and healthcare contexts. Figure 16.1 shows the stressors that were identified in this case study. The low awareness about this rare disorder among general healthcare providers and the limited numbers of doctors with the specialty training in patient management compounded the stress of limited availability and high costs of CFC. The need for professional support to provide information to parents, provide skills in home management of haemorrhagic episodes, and especially, the warning signs of bleeding, were evident. One of the overarching needs was to address maternal guilt and blame, and resolve the issues relating to disclosure of the hereditary nature of haemophilia. Negotiating the school system and sensitizing these institutions about haemophilia and accessibility issues were other important factors.

Evidences show that parents of children with different types of chronic medical conditions or disabilities have similar concerns and issues. Like haemophilia, birth defects as a whole are costly medical conditions, exerting significant financial burden on the healthcare system as well as on families [21]. Conditions like congenital heart defects, orofacial clefts and spina bifida require major surgical interventions that may result into medical costs, while conditions like Down syndrome require rehabilitation services throughout the lifespan [22, 23]. In addition to these direct costs, families encounter the indirect costs of birth defects, resulting from considerable investment of time required to care for the child, often at the cost of job loss of the primary caregiver [24–26]. Research indicates that especially in developing countries where

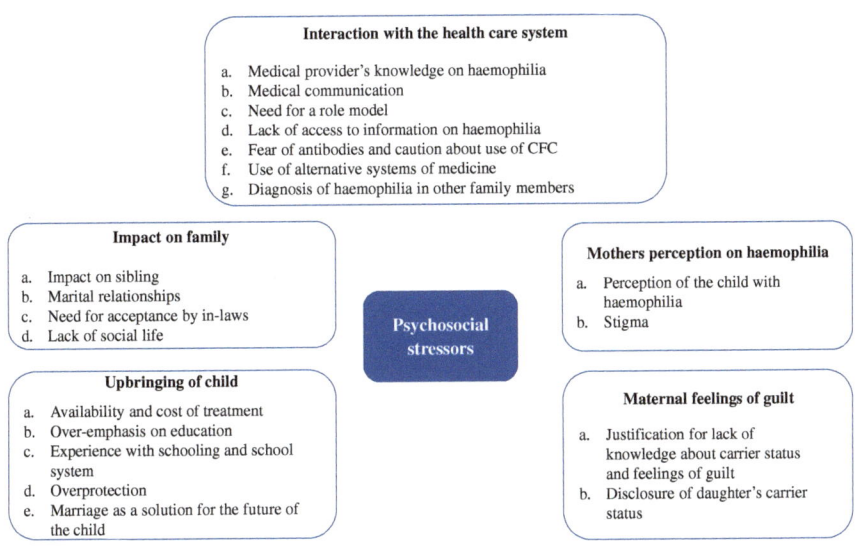

Fig. 16.1 Psychosocial stressors of caregivers of children with haemophilia

the financial burden arises from the inability to provide public services, family income is insufficient to pay for medical care in the medium and long term [27–29].

Even in the industrialized countries, birth defects are a significant economic burden. One of the surveys carried out in the USA showed that out of 78,771 families caring for children with disabilities, about 52% families had difficulty in paying medical bills as compared to 32% of families a child with a disability [23]. Another study identified that about 40% of families of children having special health needs experienced significant financial burden due to repetitive need of medical care [30]. A study on families with a child with spina bifida incurred medical expenditures that were 13 times greater than children without spina bifida [31]. These findings indicate that even in settings that provide services for children with special healthcare needs, families still remain vulnerable to catastrophic out-of-pocket expenditures [32].

In addition to financial stressors, the chronic nature of these conditions exerts multiple strains on caregivers such as increased time demands and hindrance of social participation [33, 34]. Under these demanding situations, caregivers risk mental health issues [35]. Studies have described parental stress resulting from the impact of childhood disability on family life [36–38]. The results of these studies pointed out that between parents, mothers were more affected with depression and anxiety. In one of the studies, elevated levels of depressive symptoms were seen among mothers of children with epilepsy [39] and children with developmental disabilities [40], and higher psychological distress was reported among parents of children with pervasive developmental disorders [41–44].

The provision of psychosocial support has been shown to reduce feelings of guilt and anxiety of parents [15, 16]. Studies have identified the importance of educating parents about the disorder and its management. Providing health education measures has been reported to improve the quality of life of caregivers [3, 16, 45, 46]. Psychosocial support to affected families thus remains an integral component of a comprehensive care programme for birth defects. It is recommended that under the umbrella of comprehensive care, providing treatment should be a primary concern that must be accompanied by prevention and most importantly psychosocial support to affected families. It is also of utmost importance that such comprehensive care programme should be strictly governed by guidelines so as to provide optimum care to patients and families.

References

1. Haldane JBS (1935) The rate of spontaneous mutation of a human gene. J Genet 31:317–326
2. Dharmarajan S, Gund P, Phadnis S, Lohade S, Lalwani A, Kar A (2012) Treatment decisions and usage of clotting factor concentrate by a cohort of Indian haemophilia patients. Haemophilia e18
3. Phadnis S, Kar A (2017) The impact of a haemophilia education intervention on the knowledge and health related quality of life of parents of Indian children with haemophilia. Haemophilia 23(1):82–88

4. Fischer K, van den Berg HM (2014) Prophylaxis. In: Christine A, Lee CA, Berntorp EE, Hoots KW (eds) Textbook of haemophilia, 3rd edn. Wiley-Blackwell, pp 33–37

5. EURORDIS Rare Diseases Europe. About orphan drugs. https://www.eurordis.org/about-orp han-drugs. Accessed 20 Aug 2020

6. Iorio A, Marchesini E, Marcucci M, Stobart K, Chan AKC (2011) Clotting factor concentrates given to prevent bleeding and bleeding-related complications in people with hemophilia A or B. Cochrane Database Syst Rev Issue 9, Art. No. CD003429. https://doi.org/10.1002/146 51858.CD003429.pub4

7. Dharmarajan S, Phadnis S, Gund P et al (2014) Out-of-pocket and catastrophic expenditure on treatment of haemophilia by Indian families. Haemophilia 20(3):382–387

8. Singh P, Mukherjee K (2017) Cost-benefit analysis and assessment of quality of care in patients with hemophilia undergoing treatment at National Rural Health Mission in Maharashtra, India. Value Health Reg Issues 12:101–106

9. James CA, Hadley DW et al (2006) How does the mode of inheritance of a genetic condition influence families? A study of guilt, blame, stigma, and understanding of inheritance and reproductive risks in families with X-linked and autosomal recessive diseases. Genet Med 8(4):234–242

10. Bullinger M, von Mackensen S, Haemo-QoL Group (2003) Quality of life in children and families with bleeding disorders. J Pediatr Hematol Oncol 25:64–67

11. Beeton K, Neal D et al (2007) Parents of children with haemophilia: a transforming experience. Haemophilia 13:570–579

12. Psychosocial aspects of life with haemophilia (2010) Literature review, prepared for Novo Nordisk as part of the HERO initiative. https://www.novonordisk.com/images/about_us/cha nging_possibilities/HEROLitReview.pdf. Accessed 20 Aug 2020

13. Ross J (2004) Perspectives of hemophilia carriers, Revised edn. World Federation of Hemophilia, Montreal. https://www.wfh.org/2/docs/Publications/Genetic_counsel/TOH-8_E nglish_Carriers.pdf. Accessed 20 Aug 2020

14. Cassis FRMY et al (2012) Psychosocial aspects of haemophilia: a systematic review of methodologies and findings. Haemophilia 18:e101–e114

15. Kang S et al (2010) Development, implementation and evaluation of a new self-help programme for mothers of haemophilic children in Korea: a pilot study. Haemophilia 16:130–135

16. Bottos AM, Zanon E et al (2007) Psychological aspects and coping styles of parents with haemophilic child undergoing a programme of counseling and psychological support. Haemophilia 13:305–310

17. Kind P (1999) The EuroQol instrument: an index of health-related quality of life. In: Spilker B (ed) Quality of life in 15 different cultural groups worldwide. Health Psychol 18:495–505

18. Wiedebusch S, Pollmann H et al (2008) Quality of life, psychosocial strains and coping in parents of children with haemophilia. Haemophilia 14:1014–1022

19. Saviolo-Negrin N, Cristante F et al (1999) Psychological aspects and coping of parents with a haemophilic child: a quantitative approach. Haemophilia 5:63–68

20. Kar A, Mirkazemi R, Singh P et al (2007) Disability in Indian patients with haemophilia. Haemophilia 13(4):398–404

21. Boulet SL, Molinari NA, Grosse SD (2008) Health care expenditures for infants and young children with Down syndrome in a privately insured population. J Paediatr 153(2):241–246

22. Waitzman N, Scheffler RM, Romano PS (1996) The cost of birth defects: estimates of the value of prevention. University Press of America, Lanham

23. Dawson AL, Cassell CH et al (2014) Hospitalizations and associated costs in a population-based study of children with Down syndrome born in Florida. Birth Defects Res A 100(11):826–836

24. Olson DH (1996) Family stress and coping: a multi-system perspective. In: Cusinato M (ed) Research and needs across the world. Milano LED, pp 73–105

25. Beckman PJ (1991) Comparison of mothers and fathers perceptions of the effect of young children with and without disabilities. Am J Ment Retard 95:585–595

26. Dyson LL (1991) Families of young children with handicaps: parental stress and family functioning. Am J Ment Retard 95:623–629

27. Carcao MD (2012) The diagnosis and management of congenital hemophilia. Semin Thromb Hemost 38:727–734
28. Goudie A, Havercamp S et al (2010) Caring for children with disabilities in Ohio: the impact on families. In: Council ODD (ed). Columbus
29. Shaw D, Riley GA (2008) The impact on parents of developments in the care of children with bleeding disorders. Haemophilia 14:65–67
30. Kuhlthau K, Hill KS, Yucel R, Perrin JM (2005) Financial burden for families of children with special health care needs. Matern Child Health J 9(2):207–218
31. Ouyang L, Grosse SD et al (2007) Health care expenditures of children and adults with spina bifida in a privately insured U.S. population. Birth Defects Res A Clin Mol Teratol 79:552–558
32. Keeling D, Tait C, Makris M (2008) Guideline on the selection and use of therapeutic products to treat haemophilia and other hereditary bleeding disorders. A United Kingdom Haemophilia Center Doctors' Organisation (UKHCDO) guideline approved by the British Committee for Standards in Haematology. Haemophilia 14:671–684
33. Montes G, Halterman JS (2007) Child care problems and employment among families with preschool-aged children with autism in the United States. Pediatrics 122(1):e202–e208. Epub 2008/07/04. https://doi.org/10.1542/peds.2007-3037
34. Nes RB, Hauge LJ et al (2014) The impact of child behaviour problems on maternal employment: a longitudinal cohort study. J Fam Econ Issues 35:351–361
35. Murphy NA, Christian B, Caplin DA et al (2007) The health of caregivers for children with disabilities: caregivers perspectives. Child Care Health Dev 33(2):180–187
36. Dutreil S, Rice J, Merritt D et al (2011) Parents Empowering Parents (PEP) Program: understanding its impact on the bleeding disorders community. Haemophilia 17:e895–e900
37. Beresford BA (1994) Resources and strategies: how parents cope with the care of a disabled child. J Child Psychol Psychiatry 83:171–200
38. Dyson LL (1993) Response to the presence of a child with disabilities: parental stress and family functioning overtime. Am J Ment Retard 2:207–218
39. Ferro MA, Speechley KN (2009) Depressive symptoms among mothers of children with epilepsy: a review of prevalence, associated factors, and impact on children. Epilepsia 50(11):2344–2354
40. Singer GH (2006) Meta-analysis of comparative studies of depression in mothers of children with and without developmental disabilities. Am J Ment Retard AJMR 111(3):155–169
41. Yamada A, Suzuki M et al (2007) Emotional distress and its correlates among parents of children with pervasive developmental disorders. Psychiatry Clin Neurosci 61(6):651–657
42. Lach LM, Kohen DE, Garner RE et al (2009) The health and psychosocial functioning of caregivers of children with neurodevelopmental disorders. Disabil Rehabil 31(9):741–752
43. Schneider M, Steele R, Cadell S, Hemsworth D (2011) Differences on psychosocial outcomes between male and female caregivers of children with life-limiting illnesses. J Pediatr Nurs 26(3):186–199
44. Farmer JE, Marien WE, Clark MJ, Sherman A, Selva TJ (2004) Primary care supports for children with chronic health conditions: identifying and predicting unmet family needs. J Pediatr Psychol 29(5):355–367
45. Wijesinghe CJ, Cunningham N, Fonseka P, Hewage CG, Østbye T (2015) Factors associated with caregiver burden among caregivers of children with cerebral palsy in Sri Lanka. Asia Pac J Public Health 27(1):85–95
46. Cullen LA, Barlow JH (2004) A training and support programme for caregivers of children with disabilities: an exploratory study. Patient Educ Couns 55(2):203–209